Quantum Probability and Related Topics

QP–PQ: Quantum Probability and White Noise Analysis*

Managing Editor: W. Freudenberg
Advisory Board Members: L. Accardi, T. Hida, R. Hudson and
K. R. Parthasarathy

QP–PQ: Quantum Probability and White Noise Analysis

*For the complete list of the published titles in this series, please visit:
www.worldscientific.com/series/qp-pq

QP–PQ
Quantum Probability and White Noise Analysis
Volume XXIX

Quantum Probability and Related Topics

Proceedings of the 32nd Conference

Levico Terme, Italy   29 May – 4 June 2011

Editors

Luigi Accardi

University of Rome II "Tor Vergata", Italy

Franco Fagnola

Politecnico di Milano, Italy

World Scientific

NEW JERSEY · LONDON · SINGAPORE · BEIJING · SHANGHAI · HONG KONG · TAIPEI · CHENNAI

Published by

World Scientific Publishing Co. Pte. Ltd.

5 Toh Tuck Link, Singapore 596224

USA office: 27 Warren Street, Suite 401-402, Hackensack, NJ 07601

UK office: 57 Shelton Street, Covent Garden, London WC2H 9HE

Library of Congress Cataloging-in-Publication Data
International Conference on Quantum Probability and Related Topics (32nd : 2011 : Levico Terme, Italy)
  Quantum probability and related topics : proceedings of the 32nd conference, Levico Terme, Italy, 29 May–4 June 2011 / edited by Luigi Accardi (University of Rome II, Tor Vergata, Italy) & Franco Fagnola (Politecnico di Milano, Italy).
     pages cm. -- (QP-PQ, quantum probability and white noise analysis ; vol. 29)
     Includes bibliographical references and index.
     ISBN 978-981-4447-53-9 (hardcover : alk. paper)
     1. Probabilities--Congresses. 2. Quantum theory--Congresses. I. Accardi, L. (Luigi), 1947–
editor of compilation. II. Fagnola, Franco, editor of compilation. III. Title.
     QC174.17.P68I59 2011
     530.1201'5192--dc23

                                                                    2012040685

British Library Cataloguing-in-Publication Data
A catalogue record for this book is available from the British Library.

Printed in Singapore.

PREFACE

The present volume contains contributions of the 32nd International Conference on Quantum Probability and Related Topics held in Levico Terme (Italy) in June 2011 as the annual meeting of the Association for Quantum Probability and Infinite Dimensional Analysis (AQPIDA). The conference was sponsored by CIRM Bruno Kessler Foundation, Centro V. Volterra and MIUR PRIN 2009 project "Quantum Probability and Applications to Information Theory".

The goal of the conference was to communicate new results in quantum probability and infinite dimensional analysis and related fields. A wide range of topics was discussed – white noise and higher powers of white noise, quantum Markov semigroups and applications to quantum statistical mechanics, quantum stochastic calculus, stochastic differential equations, quantum Lévy processes, product systems and interacting Fock spaces, algebraic probability, realizations of classical processes, quantum probability in bio-systems, complexity, entanglement and measurement – showing the highly interdisciplinary activity of the field and deep connections with many areas of mathematics.

Selected contributions highlighting the latest developments, collected after the conference, refereed by one or more experts, are now presented in this volume that will be of great value for researchers, applied mathematicians and graduate students.

We gratefully acknowledge the financial support of all sponsors. Special thanks go to Dr. Alessandro Toigo who patiently typesetted this volume and the CIRM secretary Augusto Micheletti who took care of the logistics.

Luigi Accardi   Roma and Milano, September 2012
Franco Fagnola

CONTENTS

CENTRAL EXTENSION OF VIRASORO TYPE SUBALGEBRAS OF THE ZAMOLODCHIKOV-w_∞ LIE ALGEBRA

LUIGI ACCARDI

Centro Vito Volterra, Università di Roma Tor Vergata
via Columbia, 2– 00133 Roma, Italy
E-mail: accardi@Volterra.mat.uniroma2.it

ANDREAS BOUKAS

Centro Vito Volterra, Università di Roma Tor Vergata
via Columbia, 2– 00133 Roma, Italy
E-mail: andreasboukas@yahoo.com

It is known that the centerless Zamolodchikov–w_∞ ∗–Lie algebra of conformal field theory does not admit nontrivial central extensions, but the Witt ∗–Lie algebra, which is a sub–algebra of w_∞, admits a nontrivial central extension: the Virasoro algebra. Therefore the following question naturally arises: *are there other natural sub–algebras of w_∞ which admit nontrivial central extensions other than the Virasoro one?* We show that for certain infinite dimensional closed subalgebras of w_∞, which are natural generalizations of the Witt algebra the answer is negative.

Keywords: Witt algebra; Virasoro algebra; Central extensions

1. Introduction

The centerless Virasoro (or Witt)-Zamolodchikov-w_∞ ∗–Lie algebra (cf.[3]-[6]) is the infinite dimensional ∗–Lie algebra, with generators

$$\{\hat{B}_k^n \; : \; n \in \mathbb{N}, n \geq 2, k \in \mathbb{Z}\} \tag{1}$$

commutation relations

$$[\hat{B}_k^n, \hat{B}_K^N] = (k\,(N-1) - K\,(n-1))\,\hat{B}_{k+K}^{n+N-2} \tag{2}$$

and involution

$$\left(\hat{B}_k^n\right)^* = \hat{B}_{-k}^n \qquad (3)$$

where $\mathbb{N} = \{1, 2, \ldots\}$.

The central extensions of the w_∞ algebra have been widely studied in the physical literature. In particular, Bakas proved in Ref. 3 that the w_∞ *– Lie algebra does not admit non-trivial central extensions. That was done by showing that, after a suitable contraction which yields the w_∞ commutation relations, the central terms appearing in the algebra W_∞, which is defined as a $N \to \infty$ limit of the Zamolodchikov type Lie algebras W_N, vanish. A direct proof of the triviality of all central extensions of w_∞ based on the cocycle definition of a central extension and avoiding the ambiguities that arise from passing to the (non-unique) limit of W_N, was given in Ref. 1.

The *–Lie sub–algebra of the w_∞ algebra, generated by the family $\{\hat{B}_k^2 : k \in \mathbb{Z}\}$ is the Witt algebra which admits the Virasoro non trivial central extension

$$[\hat{B}_m^2, \hat{B}_n^2] = (m - n)\,\hat{B}_{m+n}^2 + \delta_{m+n,0}\,m\,(m^2 - 1)\,E \qquad (4)$$

where traditionally $E = c/12$ where $c \in \mathbb{C}$ is the "central charge".

In this paper we examine whether certain infinite dimensional closed sub-algebras of w_∞, which are natural generalizations of the Witt algebra, can also be non-trivially centrally extended.

2. Closed subalgebras of w_∞

In this section we investigate the structure of the Lie sub–algebras of w_∞. More precisely, we investigate which subsets of the generators of w_∞ are such that the Lie algebra (resp. *–Lie algebra) generated by them, is a proper sub–algebra of w_∞. To this goal notice that, if $\hat{S}$ is any subset of the generators of w_∞, then there exists a unique partition $\{\hat{S}_+, \hat{S}_0, \hat{S}_-\}$ of $\hat{S}$ defined by

$$\hat{S}_+ := \{\hat{B}_k^n \in \hat{S} \; : \; k > 0\}$$
$$\hat{S}_0 := \{\hat{B}_k^n \in \hat{S} \; : \; k = 0\}$$
$$\hat{S}_- := \{\hat{B}_k^n \in \hat{S} \; : \; k < 0\}$$

From (3) we know that a generator $\hat{B}_k^n$ is self–adjoint if and only if $k = 0$. Therefore $\hat{S}_0$ is a self–adjoint set. Moreover the set $\hat{S}$ generates a $*$–sub–algebra if and only if $(\hat{S}_+)^* = \hat{S}_-$. Denote $\hat{\mathcal{L}}(\hat{S})$ the Lie sub–algebra of w_∞ generated by $\hat{S}$. From (2), we see that the sets $\hat{S}_+$, $\hat{S}_-$ generate Lie sub–algebras $\hat{\mathcal{L}}(\hat{S}_+)$, $\hat{\mathcal{L}}(\hat{S}_-)$ of $\hat{\mathcal{L}}(\hat{S})$, while $\hat{S}_0$ generates a Lie $*$–sub–algebra $\hat{\mathcal{L}}(\hat{S}_0)$. Denote $\mathbb{N}_{\geq 2} := \{n \in \mathbb{N} : n \geq 2\}$. The map $\hat{B}_k^n \mapsto (n, k) \in \mathbb{N}_{\geq 2} \times \mathbb{Z}$ defines a one–to–one correspondence between the set of generators (1) and the set $\mathbb{N}_{\geq 2} \times \mathbb{Z}$. Therefore the sub–set of generators $\hat{S}$ will be in one–to–one correspondence with a subset $\mathcal{S} \subseteq \mathbb{N}_{\geq 2} \times \mathbb{Z}$. The images of the subsets $\hat{S}_\varepsilon$ where $\varepsilon \in \{+, 0, -\}$ under this correspondence will be denoted by $\mathcal{S}_\varepsilon$.

We want to study the following problem: *which subset of* $\mathbb{N}_{\geq 2} \times \mathbb{Z}$ *corresponds to those generators (1) which belong to* $\hat{\mathcal{L}}(\hat{S})$ *(resp.* $\hat{\mathcal{L}}(\hat{S}_\varepsilon)$, $\varepsilon \in \{+, 0, -\}$)? This sub–set will be denoted by $\mathcal{L}(\mathcal{S})$ (resp. $\mathcal{L}(\mathcal{S}_\varepsilon)$, $\varepsilon \in \{, 0, -\}$). The answer to this question, for a generic $\hat{S}$, is very difficult therefore we begin to analyze a simpler problem, namely: *Can we construct interesting families of subsets* $\hat{S} \subseteq \mathbb{N}_{\geq 2} \times \mathbb{Z}$ *with the property that the linear span of such a subset is a proper Lie* $*$–*sub–algebra of* w_∞? Notice that, if $\hat{B}_k^n, \hat{B}_K^N \in \hat{S}$, then from (2) one sees that, if $k(N - 1) - K(n - 1) \neq 0$ then the generator $\hat{B}_{k+K}^{n+N-2} \in \hat{\mathcal{L}}(\hat{S})$.

Moreover, the set $\mathbb{N}_{\geq 2} \times \mathbb{Z}$ is an associative semi–group under the composition law

$$(n, k) \dot{+} (N, K) := (n + N - 2, k + K) \tag{5}$$

In fact, it is the product of the semi–group $\mathbb{N}_{\geq 2}$ with composition law

$$n \dot{+} N := n + N - 2 \tag{6}$$

and the (semi–) group $\mathbb{Z}$ with the usual addition. Thus the set $\mathcal{L}(\mathcal{S})$ will be contained in the sub–semi–group of $\mathbb{N}_{\geq 2} \times \mathbb{Z}$ generated by $\mathcal{S}$. Conversely, if $\mathcal{S}_0$ is any sub–semi–group of $\mathbb{N}_{\geq 2} \times \mathbb{Z}$, then the linear span of $\hat{S}_0 := \{\hat{B}_k^n : (n, k) \in \mathcal{S}_0\}$ is a Lie–sub–algebra of w_∞ and it is a Lie $*$–sub–algebra if and only if $\mathcal{S}_0$ is a self–adjoint subset under the involution

$$(n, k) \in \mathbb{N}_{\geq 2} \times \mathbb{Z} \mapsto (n, -k) \in \mathbb{N}_{\geq 2} \times \mathbb{Z}$$

For this reason, it is interesting to study the sub–semi–groups of $\mathbb{N}_{\geq 2} \times \mathbb{Z}$ under the composition law (Ref. 5). An interesting class of these semigroups are those of the form

$$\mathcal{S} = \mathcal{S}_1 \times \mathcal{S}_2 \qquad (7)$$

where $\mathcal{S}_1$ is a sub–semi–group of $\mathbb{N}_{\geq 2}$ with composition law (6) and $\mathcal{S}_2$ a sub–semi–group of $\mathbb{Z}$. The composition law (6) has an identity, given by the number 2, which is in $\mathbb{N}_{\geq 2}$. Hence $\{2\} \times \mathbb{Z}$ is a self–adjoint sub–semi–group of $\mathbb{N}_{\geq 2} \times \mathbb{Z}$. Therefore the linear span of the set $\{\hat{B}_k^2 : k \in \mathbb{Z}\}$ is a Lie $*$–sub–algebra of w_∞ which is precisely the Witt (or centerless Virasoro) algebra.

Notice that $\{2\}$ is the only finite sub–semi–group of $\mathbb{N}_{\geq 2}$. In fact if S is such a semigroup and $n \in S$, then $\forall \nu \in \mathbb{N} \cup \{0\}$

$$n \dot{+} \ldots \dot{+} n \quad (\nu - \text{times}) = \nu n - 2(\nu - 1) = \nu(n - 2) + 2 \in S \quad (8)$$

and, for varying ν, this is a finite set if and only if $n = 2$. Notice also that the sub–semi–group of $\mathbb{N}_{\geq 2}$ generated by the single element $n \in \mathbb{N}_{\geq 2}$ is the set of elements of the form (8) for $\nu \in \mathbb{N} \cup \{0\}$. Denoting by S_n this semi–group, one has that $S_n \times \mathbb{Z}$ is a self–adjoint sub–semi–group of $\mathbb{N}_{\geq 2} \times \mathbb{Z}$. Therefore $\forall n \in \mathbb{N}_{\geq 2}$ the linear span of the set

$$\{\hat{B}_k^{\nu(n-2)+2} : \nu \in \mathbb{N} \cup \{0\}, k \in \mathbb{Z}\}$$

is a closed Lie $*$–sub–algebra of w_∞. Letting $N = n - 2 \geq 0$ and (for fixed N)

$$W_k^n := \hat{B}_k^{n\,N+2}$$

we arrive at the following definition.

Definition 2.1. For any natural integer $N \geq 0$ we denote w_N the $*$–Lie subalgebra of w_∞ defined by

$$w_N := span\{W_k^n : n \in \mathbb{N} \cup \{0\}, k \in \mathbb{Z}\}$$

with Lie brackets (inherited from w_∞)

$$[W_k^n, W_m^l] = ((k\,l - m\,n)\,N + (k - m))\,W_{k+m}^{n+l} \qquad (9)$$

For $N = 0$, w_0 is the Witt algebra.

The question of the existence of non–trivial central extensions of w_N is the subject of this paper.

Notice that w_N is a direct generalization of the Witt algebra w_0. Furthermore, notice that the Witt algebra is the vector space generated by the generators of the form $\{\hat{B}_k^{\varphi(k)} : k \in \mathbb{Z}\}$ where φ is the constant function $\varphi(k) = 2$, $\forall k \in \mathbb{Z}$.

One may wonder if there exist other functions $\varphi : \mathbb{Z} \to \mathbb{N}_{\geq 2}$ with this property. The following Lemma shows that this is not the case.

Lemma 2.1. *Let $\varphi : \mathbb{Z} \to \mathbb{N}_{\geq 2}$ be a function such that the linear span of $\{\hat{B}_k^{\varphi(k)} : k \in \mathbb{Z}\}$ is a $*$–Lie algebra. Then φ is the constant function $\varphi(k) = 2$, $\forall k \in \mathbb{Z}$.*

Proof. The condition $(\hat{B}_k^{\varphi(k)})^* = \hat{B}_{-k}^{\varphi(k)}$ for all $k \in \mathbb{N}$ implies that $\varphi(k) = \varphi(-k)$. This, together with the condition

$$[\hat{B}_k^{\varphi(k)}, \hat{B}_{-k}^{\varphi(-k)}] = 2k(\varphi(k) - 1)B_0^{\varphi(k) \dot{+} \varphi(-k)} ; \quad \forall k \in \mathbb{Z}$$

gives that, $\forall k \in \mathbb{Z}$

$$\varphi(0) = \varphi(k) \dot{+} \varphi(k) = 2\varphi(k) - 2 \Leftrightarrow 2\varphi(k) = \varphi(0) + 2 \Leftrightarrow \varphi(k) = \frac{1}{2}\varphi(0) + 1$$

But then the condition

$$[\hat{B}_0^{\varphi(0)}, \hat{B}_k^{\varphi(k)}] = -k(\varphi(0) - 1)\hat{B}_k^{\varphi(k) \dot{+} \varphi(0)}$$

gives that

$$\varphi(k) \dot{+} \varphi(0) = \varphi(k) \Leftrightarrow \varphi(0) = 2$$

Therefore $\forall k \in \mathbb{Z}$, $\varphi(k) = \frac{1}{2}\varphi(0) + 1 = 2$. $\qquad\square$

A class of examples not of product type, i.e. defined by semi–groups not of the form (7), might be built as follows. Suppose that $[\hat{B}_k^n, \hat{B}_{k'}^{n'}] = 0$, $[\hat{B}_k^n, \hat{B}_{k''}^{n''}] \neq 0$, and $[\hat{B}_{k'}^{n'}, \hat{B}_{k''}^{n''}] \neq 0$. Then the $*$–algebra generated by $\{\hat{B}_k^n, \hat{B}_{k'}^{n'}, \hat{B}_{k''}^{n''}\}$ should not be of product type.

3. Abelian sub–algebras of w_∞

Lemma 3.1. *Any subset of the set*

$$\mathcal{A}_0 := \{\hat{B}_0^n : n \in \mathbb{N}_{\geq 2}\} \tag{10}$$

consists of commuting self–adjoint generators. The set (10) is a maximal set with this property and generates a maximal Abelian $*$–sub–algebra of w_∞.

Proof. The commutativity of the set (10) is clear from (2). The same identity shows that if $X \in W_\infty$, then $\forall n \in \mathbb{N}_{\geq 2}$, $[\hat{B}_0^n, X]$ is a linear combination of the (linearly independent) generators of the form $\hat{B}_k^n$ with $k \neq 0$. Therefore either $X \in \mathcal{A}_0$ or X cannot commute with $\mathcal{A}_0$. This proves maximality. That $\mathcal{A}_0$ is a $*$–sub–algebra follows from the fact that the generators are self–adjoint. $\qquad\square$

Lemma 3.2. *If a subset $\hat{S}$ of generators of the form (1) contains an element of the form $\hat{B}_0^n$, then $\hat{S}$ can be a commutative subset if and only if*

$$\hat{B}_k^m \in \hat{S} \Rightarrow k = 0 \tag{11}$$

Proof. From Lemma 3.1 we know that (11) is a sufficient condition for commutativity of $\hat{S}$. Let us prove that, under the conditions of the Lemma, it is also necessary. Suppose that $\hat{B}_k^m \in \hat{S}$ and that $k \neq 0$. Then (2) implies that $0 = [\hat{B}_0^n, \hat{B}_k^m] = k(m-1)\hat{B}_k^{n+m-2}$. Since by assumption $m, n \geq 2$ and $\hat{B}_k^{n+m-2} \neq 0$, it follows that $k = 0$, against the assumption. $\qquad\square$

Lemma 3.3. *Two generators $\hat{B}_k^n$, $\hat{B}_K^N$ with k, $K \neq 0$, commute if and only if $\mathrm{sgn}\,(k) = \mathrm{sgn}\,(K) =: \pm$ and there exist $p, q \in \mathbb{N} \cup \{0\}$ mutually prime, such that, for some $k', K' \geq 1$: $(n, k) = (1 + qk', \pm pk')$ and $(n, K) = (1 + qK', \pm pK')$.*

Proof. We have that

$$0 = [\hat{B}_k^n, \hat{B}_K^N] = (k(N-1) - K(n-1))\hat{B}_{k+K}^{n+N-2}$$

Since $\hat{B}_{k+K}^{n+N+2} \neq 0$, this is equivalent to $k(N-1) - K(n-1) = 0$. Since $N, n \geq 2$, this is possible if and only if k and K have the same sign. In this case the condition is equivalent to

$$\frac{k}{n-1} = \frac{K}{N-1} =: \pm\frac{p}{q}$$

where p and q are mutually prime natural integers and the $\pm$ sign is the common sign of k and K. This means that $k = \pm pk'$, $n-1 = qk'$ and $K = \pm pK'$, $N-1 = qK'$ where the sign $\pm$ is the same in both cases and k', $K' \geq 1$. This is equivalent to the statement in the Lemma. $\qquad\square$

Definition 3.1. A half–line in $\mathbb{N}_{\geq 2} \times \mathbb{Z}$ is a subset either of the form

$$H_{\varepsilon,p,q} := \{(1 + qk, \varepsilon pk) : k \in \mathbb{N} \cup \{0\}\}$$

where $\varepsilon \in \{\pm 1\}$ and $q, p \in \mathbb{N} \cup \{0\}$ are mutually prime, or of the form

$$H_{1,0,q} := \{(1 + qk, 0) : k \in \mathbb{N} \cup \{0\}\}$$

Theorem 3.1. *Each of the three sets of indices $H_{1,0,1} = \{(1 + k, 0) : k \in \mathbb{N} \cup \{0\}\}$, $H_{+,1,1} = \{(1 + k, k) : k \in \mathbb{N} \cup \{0\}\}$ and $H_{-,1,1} = \{(1 + k, -k) : k \in \mathbb{N} \cup \{0\}\}$ defines a maximal family of mutually commuting generators.*

Proof. We know from Lemma 3.1 that $H_{1,0,1}$ is a mutually commuting family. The same is true for $H_{+,1,1}$ and $H_{-,1,1}$ because of Lemma 3.3. Now let $\hat{S}$ be a mutually commuting family of generators (1). If $\hat{S}$ contains a generator of the form $\hat{B}_0^n$, for some $n \in \mathbb{N}_{\geq 2}$, from Lemma 3.2 we know that $\hat{S} \subseteq H_{1,0,1}$. If this is not the case, then from Lemma 3.3 we know that $\hat{S}$ is contained in some half–line $H_{\varepsilon,p,q}$ in $\mathbb{N}_{\geq 2} \times \mathbb{Z}$ with $p \neq 0$. But all half–lines of this type, with $\varepsilon = +1$ (resp. $\varepsilon = -1$), are contained in $H_{+,1,1}$ (resp. $H_{-,1,1}$) and this implies the statement. $\qquad\square$

Notice that, of the three families listed in Theorem 3.1, only $H_{1,0,1}$ generates a $*$–sub–algebra.

4. Basic facts on central extensions of Lie algebras

If L and $\widetilde{L}$ are two complex Lie algebras, we say that $\widetilde{L}$ is a one-dimensional *central extension* of L with *central element* E if there is a Lie algebra exact sequence $0 \mapsto \mathbb{C}E \mapsto \widetilde{L} \mapsto L \mapsto 0$ where $\mathbb{C}E$ is the one-dimensional trivial Lie algebra and the image of $\mathbb{C}E$ is contained in the center $Cent(L)$ of $\widetilde{L}$ i.e.,

$$[l_1, E]_{\widetilde{L}} = 0 \qquad , \qquad \forall l_1 \in L$$

where $[\cdot, \cdot]_{\widetilde{L}}$ are the Lie brackets in $\widetilde{L}$. For $*$–Lie algebras we also require that the central element E is self–adjoint, i.e

$$(E)^* = E \tag{12}$$

A 2-*cocycle* on L is a bilinear form $\phi : L \times L \mapsto \mathbb{C}$ on L satisfying, for all $l_1, l_2 \in L$, the skew-symmetry condition

$$\phi(l_1, l_2) = -\phi(l_2, l_1)$$

(in particular $\phi(l, l) = 0$ for all $l \in L$) and the 2-*cocycle* identity:

$$\phi([l_1, l_2]_L, l_3) + \phi([l_2, l_3]_L, l_1) + \phi([l_3, l_1]_L, l_2) = 0 \tag{13}$$

One-dimensional central extensions of L are classified by 2-cocycles in the sense that $\widetilde{L}$ is a central extension of L if and only if, as vector space, it is the direct sum

$$\widetilde{L} = M \oplus \mathbb{C} E$$

where M is a Lie algebra isomorphic to L, and there exists a 2-cocycle on L such that, for all $l_1, l_2 \in L$, the Lie brackets in $\widetilde{L}$ are given by

$$[l_1, l_2]_{\widetilde{L}} = [l_1, l_2]_L + \phi(l_1, l_2) E \tag{14}$$

where, in the right hand sides of (14), L is identified to $L \oplus \{0\} \subseteq L \oplus \mathbb{C} E$, and $\phi : L \times L \mapsto \mathbb{C}$ is a 2-*cocycle* on L,

$$[l_1, l_2]_{\widetilde{L}} = [l_1, l_2]_L + \phi(l_1, l_2) E$$

where $[\cdot, \cdot]_L$ are the Lie brackets in L. A central extension is *trivial* if the corresponding 2-cocycle ϕ is uniquely determined by a linear function $f : L \mapsto \mathbb{C}$ through the identity

$$\phi(l_1, l_2) = f([l_1, l_2]_L) \qquad , \qquad \forall l_1, l_2 \in L \tag{15}$$

Such a 2-cocycle is called a 2-*coboundary*, or a *trivial* 2-cocycle. Two extensions are called *equivalent* if each of them is a trivial extension of the other. This is the case if and only if the difference of the corresponding 2-cocycles is a trivial cocycle. A central extension $\widetilde{L}$ of L is called *universal* whenever there exists a homomorphism from $\widetilde{L}$ to any other central extension of L. A Lie algebra L possesses a universal central extension if and only if L is *perfect* (i.e. $L = [L, L]$). In this case, the universal central extension of L is unique up to isomorphism.

Notice that the 2-cocycle identity (13) implies that, if $l_c \in Cent(L)$ is an element of the center of L, then

$$\phi([l_1, l_2]_L, l_c) = 0 \qquad ; \quad \forall l_1, l_2 \in L$$

i.e. l_c is ϕ–orthogonal to the derived set $[L, L]$ of L. Similarly (15) implies that a necessary condition for the 2-cocycle ϕ to be trivial is that the center of L is ϕ–orthogonal to the whole algebra L. Because of (14) this is equivalent to say that the center of L is mapped into the center of $\widetilde{L}$. Therefore a sufficient condition for a 2-cocycle ϕ on L to be non trivial is that there exist $l_c \in Cent(L)$ and $x \in L \setminus [L, L]$ such that

$$\phi(x, l_c) \neq 0$$

This practical rule is useful for Lie algebras L with a *large* derivative $[L, L]$.

5. Central extensions of w_N

Throughout this section we assume that $\widetilde{w_N}$ is a central extension of w_N, where $N > 0$ is fixed. For $N = 0$, the Witt algebra w_0 admits the well-known non-trivial Virasoro central extension

$$[W_k^0, W_m^0] = (k - m) W_{k+m}^0 + \delta_{k+m,0} \, m \, (m^2 - 1) \, E$$

We denote by $c(n, k; l, m)$ the value assumed by the corresponding 2–cocycle on the pair of generators (W_k^n, W_m^l), i.e.:

$$c(n, k; l, m) := \phi(W_k^n, W_m^l) \in \mathbb{C} \qquad (16)$$

$$[W_k^n, W_m^l] = ((k \, l - m \, n) \, N + (k - m)) \, W_{k+m}^{n+l} + c(n, k; l, m) \, E$$

The skew-symmetry of ϕ and the adjointness condition (12) imply respectively that:

$$c(n, k; l, m) = -c(l, m; n, k) \tag{17}$$

$$c(n, k; l, m) = -\overline{c(n, -k; l, -m)} \tag{18}$$

If at least one of n, l is negative we set

$$c(n, k; l, m) = 0 \tag{19}$$

Lemma 5.1. *The derived set of the w_N $*$–Lie algebra is itself.*

Proof. From (9) we see that the derived set of the w_N $*$–Lie algebra is

$$Der(W_N) := \{W_{k+m}^{n+l} : (k\,l - m\,n)\,N + (k - m) \neq 0, n, l \in \mathbb{N} \cup \{0\}, k, m \in \mathbb{Z}\}$$

Choosing $(n, k) = (0, 0)$ we see that $Der(W_N)$ contains the generators of the form W_m^l with $l \in \mathbb{N} \cup \{0\}$ and $m \in \mathbb{Z} \setminus \{0\}$. Choosing $n = 0$ and $(k, m) = (1, -1)$ we see that $Der(W_N)$ also contains the generators of the form W_0^l such that $l\,N + 2 \neq 0$ which is always true for all $l \in \mathbb{N} \cup \{0\}$. $\square$

Combining the remark after equation (15), with Lemma 5.1 one deduces that, in any central extension of W_N, the central element is mapped to the central element of the extension so that, for any $l \in \mathbb{N} \cup \{0\}$ and $m \in \mathbb{Z}$

$$c(0, 0; l, m) = 0 \tag{20}$$

Lemma 5.2. *On the w_N generators W_k^n, for the family $\{c(n, k; l, m)\}$ defined by (16), the 2–cocycle identity (13) is equivalent to*

$$((k_1\,n_2 - k_2\,n_1)\,N + (k_1 - k_2))\,c(n_1 + n_2, k_1 + k_2; n_3, k_3) \tag{21}$$
$$+((k_2\,n_3 - k_3\,n_2)\,N + (k_2 - k_3))\,c(n_2 + n_3, k_2 + k_3; n_1, k_1)$$
$$+((k_3\,n_1 - k_1\,n_3)\,N + (k_3 - k_1))\,c(n_3 + n_1, k_3 + k_1; n_2, k_2) = 0$$

Conversely any family $\{c(n, k; l, m)\}$ satisfying (21) defines, through (16), a 2–cocycle on w_N.

Proof. For all n_i, k_i, where $i = 1, 2, 3$, making use of (17) we have

$$0 = \phi([W_{k_1}^{n_1}, W_{k_2}^{n_2}], W_{k_3}^{n_3}) + \phi([W_{k_2}^{n_2}, W_{k_3}^{n_3}], W_{k_1}^{n_1}) + \phi([W_{k_3}^{n_3}, W_{k_1}^{n_1}], W_{k_2}^{n_2})$$

$$= ((k_1 \, n_2 - k_2 \, n_1) \, N + (k_1 - k_3)) \, \phi(W_{k_1+k_2}^{n_1+n_2}, W_{k_3}^{n_3})$$

$$+ ((k_2 \, n_3 - k_3 \, n_2) \, N + (k_2 - k_3)) \, \phi(W_{k_2+k_3}^{n_2+n_3}, W_{k_1}^{n_1})$$

$$+ ((k_3 \, n_1 - k_1 \, n_3) \, N + (k_3 - k_1)) \, \phi(W_{k_3+k_1}^{n_3+n_1}, W_{k_2}^{n_2})$$

$$= ((k_1 \, n_2 - k_2 \, n_1) \, N + (k_1 - k_3)) \, c(n_1 + n_2, k_1 + k_2, n_3, k_3)$$

$$+ ((k_2 \, n_3 - k_3 \, n_2) \, N + (k_2 - k_3)) \, c(n_2 + n_3, k_2 + k_3, n_1, k_1)$$

$$+ ((k_3 \, n_1 - k_1 \, n_3) \, N + (k_3 - k_1)) \, c(n_3 + n_1, k_3 + k_1, n_2, k_2)$$

The converse is clear due to the linear independence of the generators. $\square$

We notice that the sum of the first and third (resp. second and fourth) arguments in the three 2-cocycle values $c(n_2 + n_3, k_2 + k_3; n_1, k_1)$, $c(n_1 + n_2, k_1 + k_2; n_3, k_3)$ and $c(n_3 + n_1, k_3 + k_1; n_2, k_2)$ appearing in (21) is equal to $n_1 + n_2 + n_3$ (resp. $k_1 + k_2 + k_3$). We are thus led to the following definition.

Definition 5.1. Given natural integers $n_1, n_2, n_3 \geq 0$ and $k_1, k_2, k_3 \in \mathbb{Z}$, define $S \in \mathbb{N} \cup \{0\}$ and $M \in \mathbb{Z}$ by:

$$S := n_1 + n_2 + n_3 \quad ; \quad M := k_1 + k_2 + k_3$$

and

$$\psi_{S,M}(n_i, k_i) := c(S - n_i, M - k_i; n_i, k_i) \ ; \ i \in \{1, 2, 3\} \tag{22}$$

Corollary 5.1. *The skew-symmetry condition (17) becomes*

$$\psi_{S,M}(n_i, k_i) = -\psi_{S,M}(S - n_i, M - k_i)$$

and (21) is equivalent to

$$
\begin{aligned}
&((k_1\, n_2 - k_2\, n_1)\, N + (k_1 - k_2))\, c(S - n_3, M - k_3; n_3, k_3) &(23)\\
&+((k_2\, n_3 - k_3\, n_2)\, N + (k_2 - k_3))\, c(S - n_1, M - k_1; n_1, k_1)\\
&+((k_3\, n_1 - k_1\, n_3)\, N + (k_3 - k_1))\, c(S - n_2, M - k_2; n_2, k_2) = 0
\end{aligned}
$$

or in ψ-form

$$
\begin{aligned}
&((k_1\, n_2 - k_2\, n_1)\, N + (k_1 - k_2))\, \psi_{S,M}(n_3, k_3) &(24)\\
&+((k_2\, n_3 - k_3\, n_2)\, N + (k_2 - k_3))\, \psi_{S,M}(n_1, k_1)\\
&+((k_3\, n_1 - k_1\, n_3)\, N + (k_3 - k_1))\, \psi_{S,M}(n_2, k_2) = 0
\end{aligned}
$$

Proof. The proof follows directly from Definition 5.1. $\qquad\square$

Proposition 5.1. *For any $\lambda \in \mathbb{R}$ the family $\{c(n, k; l, m)\}$, defined by*

$$c(n, k; l, m) := \delta_{k+m,0}\, \lambda\, k \qquad (25)$$

defines, through (16), a 2–cocycle on w_N.

Proof. Condition (17) is verified by inspection and (18) follows from the fact that λ is real. We want to prove that (24) this is satisfied by the family $\{c(n, k; l, m)\}$, defined by (25). Direct substitution shows that, if the family $\{c(n, k; l, m)\}$ is defined by (25), then $\psi_{S,M}$, defined by (22), satisfies (24). Moreover, $\psi_{S,M}(n_i, k_i) = \delta_{M,0}\, \lambda\, k_i$ implies that $c(S - n_i, M - k_i; n_i, k_i) = \delta_{M,0}\, \lambda\, k_i$. For $i = 1$ we get $c(S - n_1, M - k_1; n_1, k_1) = \delta_{M,0}\, \lambda\, k_1$ which for $n_3 = 0$ becomes $c(n_2, k_2 + k_3; n_1, k_1) = \delta_{M,0}\, \lambda\, k_1$. Letting $k_2 + k_3 := K$ we have that

$$c(n_2, K; n_1, k_1) = \delta_{k_1+K,0}\, \lambda\, k_1$$

i.e. $c(n, k; l, m) = \delta_{k+m,0}\, \lambda\, k$. $\qquad\square$

Proposition 5.2. *The central extension*

$$[W_k^n, W_m^l] = ((k\,l - m\,n)\,N + (k - m))\,W_{k+m}^{n+l} + \delta_{k+m,0}\,\lambda\,k\,E$$

of w_N is trivial.

Proof. We look for a linear complex-valued function f defined on w_N such that

$$f\left([W_k^n, W_m^l]\right) = \delta_{k+m,0}\,k\,\lambda \tag{26}$$

By the w_N commutation relations (9) and the linearity of f, equation (Ref. 26) is equivalent to

$$((k\,l - m\,n)\,N + (k - m))\,f\left(W_{k+m}^{n+l}\right) = \delta_{k+m,0}\,k\,\lambda \tag{27}$$

For $k + m \neq 0$ this is equivalent to

$$f\left(W_x^{n+l}\right) = 0 \quad ; \quad \forall x \in \mathbb{Z} \setminus \{0\} \tag{28}$$

For $k + m = 0 \Leftrightarrow m = -k$ (27) is equivalent to

$$((k\,l + k\,n)\,N + 2k)\,f\left(W_0^{n+l}\right) = k\lambda \Leftrightarrow ((l + n)\,N + 2)\,f\left(W_0^{n+l}\right) = \lambda$$

$$\Leftrightarrow f\left(W_0^{n+l}\right) = \frac{\lambda}{(l + n)\,N + 2}$$

and this, together with (28) uniquely defines a linear functional f with the required property. Therefore the central extension of w_N is trivial. $\quad\square$

Lemma 5.3. *Let $z \in \mathbb{C}$. If $z = 2\,\bar{z}$ then $z = 0$.*

Proof. If $z = x + i\,y$, $x, y \in \mathbb{R}$, then $z = 2\,\bar{z}$ implies that $x = 2x$ and $y = -2\,y$. Therefore $x = y = 0$. $\quad\square$

Lemma 5.4. *In the notation of Definition 5.1, let $S \in \mathbb{N} \cup \{0\}$, $M = 0$ and $N > 0$. Then:*

(i) $\psi_{S,0}(0, 1) = c(S, -1; 0, 1) = 0$

(ii) For all $k \in \mathbb{Z}$, $\psi_{S,0}(0, -k) = c(S, k; 0, -k) = 0$

(iii) For all $n \geq 0$ and $k \in \mathbb{Z}$, $c(S - n, k; n, -k) = 0$

Notice that (iii) $\Rightarrow$ (ii) $\Rightarrow$ (i).

Proof. (i) For $n_2 = S - n_1$, $n_3 = 0$, $k_1 = 0$, $k_2 = -1$ and $k_3 = 1$, (21) yields

$$(n_1 N + 1) c(S, -1; 0, 1) = \tag{29}$$
$$((S - n_1) N + 2) c(S - n_1, 0; n_1, 0) + (n_1 N + 1) c(S - n_1, -1; n_1, 1)$$

For $n_3 = n_1$, $n_2 = S - 2 n_1$, $k_1 = 1$, $k_2 = -1$ and $k_3 = 0$, (21) yields

$$((S - n_1) N + 2) c(S - n_1, 0; n_1, 0) = \tag{30}$$
$$(1 + n_1 N) c(S - n_1, -1; n_1, 1) - (n_1 N + 1) c(S - 2 n_1, -1; 2 n_1, 1)$$

Substituting (30) in (29) we obtain

$$(n_1 N + 1) c(S, -1; 0, 1) = (n_1 N + 1) c(S - n_1, -1; n_1, 1)$$

$$-(n_1 N + 1) c(S - 2 n_1, -1; 2 n_1, 1) + (n_1 N + 1) c(S - n_1, -1; n_1, 1)$$

which for $n_1 = S$, since by (19) $c(-S, -1; 2 S, 1) = 0$, after dividing out $(S N + 1)$, yields with the use of (17) and (18)

$$c(S, -1; 0, 1) = 2 c(0, -1; S, 1) = -2 c(S, 1; 0, -1) = 2 \overline{c(S, -1; 0, 1)}$$

which, by Lemma 5.3, implies that $c(S, -1; 0, 1) = 0$.

(ii) For $n_1 = S$, $n_2 = 0$, $n_3 = 0$, $k_1 = k$, $k_2 = 1$, $k_3 = -(k + 1)$, letting $a_k := c(S, k; 0, -k)$, since by (i) $a_{-1} = 0$, (21) yields

$$(k - S N - 1) a_{k+1} = (k + 2) a_k$$

which implies that $a_k = 0$ for all k.

(iii) For $k_1 = k \neq 0$, $k_2 = -k$, $k_3 = 0$, $n_1 = S - n$, $n_2 = 0$ and $n_3 = n$, after dividing out $k \neq 0$ and using $c(S, k; 0, -k) = 0$, (21) yields

$$c(S - n, k; n, -k) = -\frac{(S - n) N + 2}{n N + 1} c(S - n, 0; n, 0)$$

for all $k \neq 0$. Similarly, for $k_1 = -k \neq 0$, $k_2 = k$, $k_3 = 0$, $n_1 = 0$, $n_2 = n$ and $n_3 = S - n$, (21) yields

$$c(S - n, k; n, -k) = -\frac{n\,N + 2}{(S - n)\,N + 1}\, c(S - n, 0; n, 0)$$

for all $k \neq 0$. Thus

$$\frac{(S - n)\,N + 2}{n\,N + 1}\, c(S - n, 0; n, 0) = \frac{n\,N + 2}{(S - n)\,N + 1}\, c(S - n, 0; n, 0) \qquad (31)$$

If $S = 2n$ then $c(S - n, 0; n, 0) = c(n, 0; n, 0) = 0$ by (17). If $S \neq 2n$ then $c(S - n, 0; n, 0) = 0$ by (31). □

Proposition 5.3. *Let $S \in \mathbb{N} \cup \{0\}$ and $M \in \mathbb{Z}$. In the notation of Definition 5.1, all non-trivial 2–cocycles $\psi_{S,M}(n, k)$ on w_N are given by*

$$\psi_{S,M}(n, k) = \delta_{S,0}\, \delta_{M,0}\, k\, (k^2 - 1)$$

Proof. Case (i): $S = 0$. Then $n_1 + n_2 + n_3 = 0$ and so $n_1 = n_2 = n_3 = 0$ which means that we are reduced to the standard Witt-Virasoro case $W_k^0 = \hat{B}_k^2$. Therefore, the only non-trivial cocycle is

$$\psi_{S,M}(n, k) = \psi_{0,M}(n, k) = \delta_{M,0}\, k\, (k^2 - 1)$$

Case (ii): $S \neq 0$ and $M \neq 0$. For $n_3 = k_3 = 0$, using $c(n_2, k_2; n_1, k_1) = -c(n_1, k_1; n_2, k_2)$, $n_1 + n_2 = S$ and $k_1 + k_2 = M$, (21) yields

$$(k_1\,(n_2\,N + 1) - k_2\,(n_1\,N + 1))\, c(S, M; 0, 0) - (k_2 + k_1)\, c(n_1, k_1; n_2, k_2) = 0$$

which, letting $n_2 = n$, $k_2 = k$, $n_1 = S - n$ and $k_1 = M - k$, implies that

$$\psi_{S,M}(n, k) = c(S - n, M - k; n, k)$$

$$= ((M - k)\,(n\,N + 1) - k\,((S - n)\,N + 1))\, c(S, M; 0, 0) = 0$$

by (20).

Case (iii): $S \neq 0$ and $M = 0$. For $k_3 = 0$, $n_1 = 0$, $k_1 \neq 0$, using Lemma 5.4 (ii) and (iii), (24) yields

$$\psi_{S,0}(n_2, k_2) = \frac{n_2\, N + 1 - \frac{k_2}{k_1}}{n_3\, N + 1}\, \psi_{S,0}(n_3, 0) + \frac{k_2}{k_1}\, \psi_{S,0}(0, k_1) = 0 \qquad (32)$$

and the result follows by the arbitrariness of n_2 and k_2.   □

The next corollary shows that there are no non-trivial central extensions of w_N other than the Virasoro one.

Corollary 5.2. *The non-trivial central extensions of w_N are given by*

$$[W_k^n, W_m^l] = ((k\, l - m\, n)\, N + (k - m))\, W_{k+m}^{n+l} + \delta_{n,0}\, \delta_{l,0}\, \delta_{k+m,0}\, m\, (m^2 - 1)\, E$$

Thus only the Virasoro sector of w_N can be extended in a non-trivial way.

Proof. By Proposition 5.3, in the notation of Definition 5.1,

$$\psi_{S,M}(n_1, k_1) = c(S - n_1, M - k_1; n_1, k_1) = \delta_{S,0}\, \delta_{M,0}\, k_1\, (k_1^2 - 1)$$

i.e.,

$$c(n_2 + n_3, k_2 + k_3; n_1, k_1) = \delta_{n_1 + n_2 + n_3, 0}\, \delta_{k_1 + k_2 + k_3, 0}\, k_1\, (k_1^2 - 1)$$

which, letting $n_3 = k_3 = 0$, $n_1 = n$, $k_1 = k$, $n_2 = l$ and $k_2 = m$ implies that

$$c(n, k; l, m) = \delta_{n+l, 0}\, \delta_{k+m, 0}\, m\, (m^2 - 1) = \delta_{n,0}\, \delta_{l,0}\, \delta_{k+m,0}\, m\, (m^2 - 1) \quad □$$

References

1. L. Accardi and A. Boukas, *Infinite Dimensional Anal. Quantum Probab. Related Topics*, **12**, No. 2 (2009) 193–212.
2. L. Accardi and A. Boukas, *Stochastics*, **81** (2009), no. 3-4, 201–218.
3. I. Bakas, *Commun. Math. Phys.* **134** (1990) 487-508.
4. V. A. Fateev and A. B. Zamolodchikov, *Nuclear Phys. B* **280** (4) (1987) 644–660.
5. J. Fuchs and C. Schweigert, *Cambridge Monographs on Mathematical Physics*, Cambridge University Press (1997).
6. S. V. Ketov, *Conformal field theory*, World Scientific (1995).
7. C.N. Pope, *Lectures on W algebras and W gravity*, Lectures given at the Trieste Summer School in High-Energy Physics, August 1991.
8. A.B. Zamolodchikov, *Teoret. Mat. Fiz.* **65** (3)(1985) 347–359.

ENTANGLEMENT PROTECTION AND GENERATION UNDER CONTINUOUS MONITORING

ALBERTO BARCHIELLI and MATTEO GREGORATTI

Politecnico di Milano, Department of Mathematics,
Piazza Leonardo da Vinci, I-20133 Milano, Italy.
Also: Istituto Nazionale di Fisica Nucleare, Sezione di Milano.

Entanglement between two quantum systems is a resource in quantum information, but dissipation usually destroys it. In this article we consider two qubits
without direct interaction. We show that, even in cases where the entanglement
is destroyed by the open system dynamics, the entanglement can be preserved
or created by the mere monitoring of the environment, just by filtering the
state of the qubits. While the systems we study are very simple, we can show
examples with entanglement protection or entanglement birth, death, rebirth
due to monitoring.

Keywords: Entanglement; Concurrence; Dissipative dynamics; Continuous observation; A priori state; A posteriori state.

1. Introduction

Entanglement is an intrinsically quantum type of correlation among quantum systems, which is of fundamental importance in quantum information.[1]
The behaviour of entanglement under dissipative dynamics has been studied
extensively,[2–4] either to find means to protect entanglement against decoherence, either to understand how to use a dissipative dynamics to create
entanglement. Usually dissipation tends to destroy entanglement, at least
when the two quantum systems do not interact directly. Sometimes this dis-
entaglement can be completed even in a finite time[4–6] and this phenomenon
has been called *entanglement sudden death* (ESD). However, dissipation can
create entanglement too; this happens when the two parties interact with a
common bath,[3,4,7–10] even if they do not interact directly, and we can have
entanglement birth, death, rebirth. Entanglement can be preserved or generated also by controlling the composite system by means of measurement
based feedback.[11–13]

Preservation of entanglement can be obtained also by pure monitoring of

the system,[14–16] that is by an indirect measurement that acquires information thanks to the observation of its environment, but that does not perturb
the system. Quantum trajectory theory allows for describing a continuous
monitoring[17,18] and in such a theory we have to distinguish between the *a
posteriori state*, the conditional state given the observed output, and the
a priori state, the mean state, satisfying a master equation. It is possible
that the a posteriori states are entangled, while the a priori state is not. By
using the *concurrence*[19] as a measure of entanglement it has been shown
that the pure monitoring can slow down the decay of the entanglement.[14]

The aim of our paper is indeed to study the effect of monitoring on the a
posteriori entanglement when the a priori dynamics washes out any initial
entanglement. More precisely, we consider the case of the open dynamics
of two qubits in the Markovian regime and we model their global evolution by a Hudson-Parthasarathy equation. This approach allows to clearly
characterize the Markovian evolutions representing two qubits, which do
interact or do not interact, directly or through a common bath. Section 2
is devoted to the HP evolutions and to such a characterization; we recall
also how to introduce measurements continuous in time and how to get the
corresponding *stochastic Schrödinger equation* (SSE) and *stochastic master
equation*, which are the starting points to study the dynamical behaviour
of the monitored system and of its entanglement. In Section 3 we consider
the case of no direct or indirect interaction between two qubits. When only
local detection operators are involved, we show that, by pure monitoring,
the decay of entanglement can be slowed down and, in special cases, even
stopped independently of the qubit initial state (entanglement protection).
In cases with non-local detection operators, we show that, now depending on the qubit initial state, entanglement can even be created by pure
monitoring (entanglement generation). In Section 4 we study a case of indirect interaction between the two qubits through a common bath. We show
that, even if the a priori dynamics completely destroys any entanglement, a
proper monitoring scheme can maximally entangle any initial qubit state.

1.1. *Two qubits*

We consider two qubits; for each qubit we denote by $|1\rangle$ the *up* state and
by $|0\rangle$ the *down* state. By σ_x, σ_y, σ_z we denote the Pauli matrices. In
$\mathscr{H} = \mathbb{C}^2 \otimes \mathbb{C}^2$ the canonical basis (or *computational basis*)[1] is

$$|u_1\rangle = |11\rangle, \quad |u_2\rangle = |10\rangle, \quad |u_3\rangle = |01\rangle, \quad |u_4\rangle = |00\rangle, \tag{1}$$

and the *Bell basis*[20] is

$$|\beta_0\rangle = \frac{1}{\sqrt{2}}\left(|00\rangle + |11\rangle\right), \qquad |\beta_i\rangle = \sigma_i \otimes \mathbb{1}|\beta_0\rangle, \quad i = 1, 2, 3. \qquad (2)$$

The set of *statistical operators* is $\mathscr{S}(\mathscr{H})$ and the one of linear operators is $\mathscr{L}(\mathscr{H})$. A *local operator* is a linear operator that acts non-trivially only on one of the factors of $\mathbb{C}^2 \otimes \mathbb{C}^2$, i.e. it has the form $A \otimes \mathbb{1}$ or $\mathbb{1} \otimes A$ with $A \in \mathscr{L}(\mathbb{C}^2)$. The two qubits are independent if their state is a product state $\rho = \rho_1 \otimes \rho_2$. The *separable states*[21] are the statistical operators that admit a convex decomposition into product states, so that the correlation between the two qubits has a classical explanation; the other statistical operators are said to be *entangled*. The *maximally entangled states* are the pure states which, by partial trace on one of the two factors, reduce to maximally chaotic states, that is $\mathbb{1}/2$. The projection on one of the Bell vectors (2) is a maximally entangled state.

1.2. *Concurrence*

A very useful measure of entanglement is the *concurrence*, introduced by Wootters.[19] Let us consider a generic vector $\varphi \in \mathscr{H}$ and expand it on the canonical basis (1)

$$\varphi = \varphi_{11}|11\rangle + \varphi_{10}|10\rangle + \varphi_{01}|01\rangle + \varphi_{00}|00\rangle. \qquad (3)$$

Let T be the complex conjugation of the coefficients in the canonical basis:

$$\mathsf{T}\varphi = \overline{\varphi_{11}}|11\rangle + \overline{\varphi_{10}}|10\rangle + \overline{\varphi_{01}}|01\rangle + \overline{\varphi_{00}}|00\rangle. \qquad (4)$$

Let us define

$$\chi_\varphi := \langle \mathsf{T}\varphi | \sigma_y \otimes \sigma_y \varphi \rangle = 2\left(\varphi_{10}\varphi_{01} - \varphi_{11}\varphi_{00}\right), \qquad C_\varphi := |\chi_\varphi|. \qquad (5)$$

When $\|\varphi\| = 1$, C_φ is the *concurrence* of the pure state φ. In general, if φ is not normalized and $\psi = \frac{\varphi}{\|\varphi\|}$, then

$$C_\psi = \frac{C_\varphi}{\|\varphi\|^2}. \qquad (6)$$

Note that $C_{\beta_j} = 1$ and $C_{u_j} = 0$.

If ρ is a generic statistical operator, the concurrence is defined by

$$C_\rho := \inf \sum_i p_i C_{\psi_i}, \qquad (7)$$

where the infimum is taken over all decompositions of ρ in pure states, $\rho = \sum_i p_i |\psi_i\rangle\langle\psi_i|$, see for instance [4] p. 231. We have $0 \leq C_\rho \leq 1$, $\forall \rho \in \mathscr{S}(\mathscr{H})$, with $C_\rho = 0$ if and only if ρ is separable and $C_\rho = 1$ if and only if ρ is maximally entangled.

A subclass of states, for which it is easy to compute the concurrence, is the one of the "X" states:[4,6] in the canonical basis, an X state has non-vanishing matrix elements only in the two main diagonals. The projection on a Bell vector is an X state. For any X state ρ, by setting $\rho_{ij} = \langle u_i | \rho u_j \rangle$, we have $\rho_{jj} \geq 0$, $\rho_{ij} = \overline{\rho_{ji}}$, $\sum_{j=1}^{4} \rho_{jj} = 1$, $\rho_{11}\rho_{44} \geq |\rho_{14}|^2$, $\rho_{22}\rho_{33} \geq |\rho_{23}|^2$; moreover, the concurrence is given by[6]

$$C_\rho = 2 \max \{0, C_1, C_2\}, \tag{8a}$$

$$C_1 = |\rho_{23}| - \sqrt{\rho_{11}\rho_{44}}, \qquad C_2 = |\rho_{14}| - \sqrt{\rho_{22}\rho_{33}}. \tag{8b}$$

Finally, let A and B be linear operators on $\mathbb{C}^2$. In studying the dynamics of the concurrence, the following formulae will be very useful:

$$\chi_{(A\otimes B)\varphi} = (\det_{\mathbb{C}^2} A)(\det_{\mathbb{C}^2} B)\chi_\varphi, \tag{9a}$$

$$\langle T\varphi | (\sigma_y A) \otimes \sigma_y \varphi \rangle = \langle TA \otimes \mathbb{1}\varphi | \sigma_y \otimes \sigma_y \varphi \rangle = \frac{1}{2} (\mathrm{Tr}_{\mathbb{C}^2} A) \chi_\varphi. \tag{9b}$$

2. Global evolution and continuous measurements

The way to understand whether two qubits interact or do not interact, directly or indirectly, is to look at the unitary dynamics of the two qubits plus their environment. In the Markov regime this can be done by starting from a quantum stochastic differential equation à la Hudson and Parthasarathy (HP equation)[22] and this is also a clear way to introduce continuous mesurements.[17,23]

As before the system space is $\mathscr{H}$, while we take as environment space the symmetric Fock space $\mathscr{K} = \Gamma[L^2(\mathbb{R}; \mathfrak{Z})]$; $\mathfrak{Z}$ is a complex Hilbert space, which will be only finite dimensional in the present paper. Let $U_t = \mathrm{e}^{-\mathrm{i}t H_T}$, $H_T = H_T^*$, denote the unitary (Hamiltonian) global evolution in $\mathscr{K} \otimes \mathscr{H}$. We suppose that the free environment evolution is $\Theta_t = \mathrm{e}^{-\mathrm{i}t E_0}$, the second quantization of the left shift, with its free Hamiltonian E_0. Then, the global evolution in interaction picture with respect to Θ_t is

$$V(t) = \Theta_t^* U_t = \mathrm{e}^{\mathrm{i}E_0 t} \mathrm{e}^{-\mathrm{i}t H_T}, \qquad t \geq 0,$$

which, in the Markov regime, can be defined directly by a HP-equation.

2.1. HP evolutions

We fix a basis $\{|z\rangle\}_{z\in Z}$ in the Hilbert space $\mathfrak{Z}$. Let $a_z(t)$ and $a_z^\dagger(t)$ be the fundamental Bose field operators in $\Gamma[L^2(\mathbb{R}; \mathfrak{Z})]$ and $A_z(t) = \int_0^t a_z(s)\mathrm{d}s$,

$A_z^\dagger(t) = \int_0^t a_z^\dagger(s)ds$, $\Lambda_{zw}(t) = \int_0^t a_z^\dagger(s)a_w(s)ds$ be the fundamental integrators of quantum stochastic calculus.

Let us consider the HP-equation[22] for unitary operators on $\mathcal{K} \otimes \mathcal{H}$

$$dV(t) = \left[\sum_{z,w \in Z} (S_{zw} - \delta_{zw})\, d\Lambda_{zw}(t) - \sum_{z,w \in Z} L_z^* S_{zw}\, dA_w(t) \right.$$
$$\left. + \sum_{z \in Z} L_z\, dA_z^\dagger(t) - iH\,dt - \frac{1}{2}\sum_{z \in Z} L_z^* L_z\, dt \right] V(t); \quad (10)$$

the initial condition is $V(0) = \mathbb{1}$. By taking

(1) H, L_z, $S_{zw} \in \mathscr{L}(\mathcal{H})$ (bounded operators), $\forall z, w \in Z$,
(2) $H = H^*$,
(3) $S \in \mathscr{U}(\mathfrak{Z} \otimes \mathcal{H})$ (unitary operators), where $S = \sum_{zw} |z\rangle\langle w| \otimes S_{zw}$,

the solution of (10) is indeed unique and unitary. Every operator is identified with its natural extension to $\mathcal{K} \otimes \mathcal{H}$.

By using the time ordered exponentials introduced by Holevo,[24] the solution $V(t)$ can be represented as

$$V(t) = \overleftarrow{\exp}\left\{ -i\int_0^t \left[\sum_{zw} K_{zw} a_z^\dagger(s)a_w(s) - \sum_{zw} L_z^* \left(\frac{K}{\mathbb{1} - S^*}\right)_{zw} a_w(s) \right.\right.$$
$$\left.\left. + \sum_{zw} \left(\frac{K}{S - \mathbb{1}}\right)_{zw} L_z a_w^\dagger(s) + H + \sum_{zw} L_z^* \left(\frac{K - \sin K}{4\left(\sin(K/2)\right)^2}\right)_{zw} L_w \right] ds \right\}, \quad (11)$$

where $S = e^{-iK}$, with a selfadjoint operator K on $\mathfrak{Z} \otimes \mathcal{H}$.

Moreover, we have that U_t, defined by $U_t := \Theta_t V(t)$ for $t \geq 0$, and by $U_t := U_{-t}^*$ for $t \leq 0$, is a unitary strongly continuous group. So, we can interprete U_t as the evolution operator of a closed system, Θ_t as the free evolution of the fields and $V(t)$ as the total evolution in the interaction picture with respect to Θ_t.

The interaction between $\mathcal{H}$ and $\mathcal{K}$ is regulated by the system operators H, L_z and S_{zw}; the corresponding global Hamiltonian H_T is a very singular unbounded operator which could even encode the whole interaction just in the shape of its domain.[25] Anyway, thanks to representation (11), the global Hamiltonian H_T has the heuristic expression

$$H_T = E_0 + \sum_{zw} K_{zw}\, a_z^\dagger(0)\, a_w(0) - \sum_{zw} L_z^* \left(\frac{K}{\mathbb{1} - S^*}\right)_{zw} a_w(0)$$
$$+ \sum_{zw} \left(\frac{K}{S - \mathbb{1}}\right)_{zw} L_z\, a_w^\dagger(0) + H + \sum_{zw} L_z^* \left(\frac{K - \sin K}{4\left(\sin(K/2)\right)^2}\right)_{zw} L_w, \quad (12)$$

which allows to read more explicitly the interaction between the systems. In the special case $L = 0$ we have

$$H_T = E_0 + \sum_{zw} K_{zw} a_z^\dagger(0) \, a_w(0) + H, \tag{13}$$

while for $K = 0$, i.e. $S = \mathbb{1}$, we get

$$H_T = E_0 - \mathrm{i} \sum_z L_z^* \, a_z(0) + \mathrm{i} \sum_z L_z \, a_z^\dagger(0) + H. \tag{14}$$

As initial state let us take $|e(v)\rangle\langle e(v)| \otimes \rho_0$, where $\rho_0 \in \mathscr{S}(\mathscr{H})$ is the initial system state and $e(v)$ is the coherent vector in $\mathscr{K} = \Gamma[L^2(\mathbb{R}; 3)]$ with argument v in $L^2(\mathbb{R}; 3)$. At the end it will be possible to take v only locally square integrable.

Then, thanks to the properties of the HP-equation, the dynamics of the reduced system state

$$\eta(t) := \mathrm{Tr}_{\mathscr{K}} \left\{ U(t) \left(|e(v)\rangle\langle e(v)| \otimes \rho_0 \right) U(t)^* \right\}$$
$$= \mathrm{Tr}_{\mathscr{K}} \left\{ V(t) \left(|e(v)\rangle\langle e(v)| \otimes \rho_0 \right) V(t)^* \right\} \tag{15}$$

is given[17,22] by the master equation $\dot{\eta}(t) = \mathcal{L}(t)[\eta(t)]$ with Liouville operator

$$\mathcal{L}(t)[\tau] = -\mathrm{i}[H(t), \tau] + \sum_z \left(\tilde{L}_z(t) \tau \tilde{L}_z(t)^* - \frac{1}{2} \left\{ \tilde{L}_z(t)^* \tilde{L}_z(t), \tau \right\} \right), \tag{16a}$$

$$\tilde{L}_z(t) := L_z + \sum_w (S_{zw} - \delta_{zw}) \, v_w(t), \tag{16b}$$

$$H(t) := H + \frac{\mathrm{i}}{2} \sum_{zw} \left[\overline{v_z(t)} \, (S_{wz}^* + \delta_{zw}) \, L_w + \overline{v_z(t)} S_{zw} v_w(t) - \mathrm{h.c.} \right]. \tag{16c}$$

Of course the reduced evolution depends on the global dynamics (10) and on the environment initial state. But this correspondence is not injective at all, so that it is not enough to know the Liouvillian $\mathcal{L}$ to know the system/environment interaction.

2.2. From the HP-equation to the SSE

The fields that have already interacted with $\mathscr{H}$ can be manipulated in various ways and then monitored continuously in time. In this way we avoid to further perturb the dynamics of $\mathscr{H}$, but, at the same time, as we indirectly acquire information on its state, the dynamics of $\mathscr{H}$ turns out to be conditioned by the observed output. In the typical case of quantum optics the system is a photoemissive source and the output fields are mixed up by means of beam splitters and optical fibers and detected by photon counters (direct, homodyne, heterodyne detection).[17] In general, we identify a measurement in continuous time by a family of commuting selfadjoint field operators which can be chosen as follows.

The manipulation of the fundamental fields is represented by a unitary, possibly time dependent, matrix $u_{iz}(t)$,

$$\sum_{i \in Z} \overline{u_{iz}(t)}\, u_{iw}(t) = \delta_{zw}, \qquad \sum_{z \in Z} u_{iz}(t)\overline{u_{jz}(t)} = \delta_{ij},$$

and produces the new field operators

$$B_i(t) := \sum_{z \in Z} \int_0^t u_{iz}(s)\,\mathrm{d}A_z(s),$$

$$\hat{\Lambda}_{ij}(t) := \sum_{z,w \in Z} \int_0^t \overline{u_{iz}(s)}\, u_{jw}(s)\,\mathrm{d}\Lambda_{zw}(s), \qquad i,j \in Z.$$

Then, we set $\dim \mathfrak{Z} = d + d'$ and choose as observables the commuting selfadjoint operators (interaction picture)

$$B_i(s)+B_i^\dagger(s), \quad \hat{\Lambda}_{kk}(s), \quad i=1,\ldots,d, \ k=d+1,\ldots,d+d', \quad s \geq 0. \tag{17}$$

The global evolution (10), the environment initial (coherent) state $|e(v)\rangle\langle e(v)|$ and the observed fields (17) together determine both the distribution of the output processes and the a posteriori dynamics of the system $\mathscr{H}$, that is the evolution of $\mathscr{H}$ as a function of the observed outputs; both of them depending on the system initial state ρ_0. As we observe a maximal family of compatible fields, the a posteriori evolution preserves the purity of the system states and thus the problem of dynamics and observation can be reduced to a classical linear SSE:[18,23]

$$\mathrm{d}\varphi(t) = \mathrm{K}(t)\varphi(t_-)\mathrm{d}t + \sum_{j=1}^d R_j(t)\varphi(t_-)\mathrm{d}W_j(t)$$

$$+ \sum_{k=1}^{d'} \left[\left(\frac{J_k(t)}{\sqrt{\lambda_k}} - \mathbb{1} \right)\varphi(t_-)\mathrm{d}N_k(t) + \frac{\lambda_k}{2}\varphi(t_-)\mathrm{d}t \right], \tag{18}$$

$$\mathrm{K}(t) := -\mathrm{i}H_0(t) - \frac{1}{2}\sum_{j \in Z} R_j(t)^* R_j(t), \tag{19}$$

$$H_0(t) := H + \frac{\mathrm{i}}{2} \sum_{z,w \in Z} \left(\overline{v_z(t)} S_{wz}^* L_w - L_w^* S_{wz} v_z \right), \tag{20a}$$

$$R_j(t) := \sum_{z \in Z} u_{jz}(t) \left(L_z + \sum_{w \in Z} S_{zw} v_w(t) \right), \qquad J_k(t) := R_{d+k}(t); \quad (20b)$$

the initial condition is $\varphi(0) = \psi_0 \in \mathscr{H}$, $\|\psi_0\| = 1$. Equation (18) is a stochastic differential equation for a $\mathscr{H}$-valued stochastic process $\varphi(t)$ in a filtered probability space, say $(\Omega, \mathscr{F}, (\mathscr{F}_t), \mathbb{Q})$, where W_j, N_k are independent Wiener and Poisson processes, each N_k with rate λ_k. The solution $\varphi(t)$ is taken continuous from the right and $\varphi(t_-)$ in the right hand side means that the value of the solution is taken before of the possible jump at time t. The solution $\varphi(t)$ is a function of the initial condition ψ_0 and of the trajectories of the processes W_j and N_k up to time t.

Equation (18) can be translated in the language of stochastic processes $\sigma(t)$ taking values among positive operators on $\mathscr{H}$. Indeed, if $A(t, s)$ is the fundamental solution of Eq. (18), or the *propagator* from time s to t, taken $\rho_0 \in \mathcal{S}(\mathscr{H})$, the stochastic process $\sigma(t) := A(t, 0)\rho_0 A(t, 0)^*$ satisfies the linear stochastic master equation

$$d\sigma(t) = \mathcal{L}(t)[\sigma(t_-)]dt + \sum_{j=1}^{d} (R_j(t)\sigma(t_-) + \sigma(t_-)R_j(t)^*)dW_j(t)$$

$$+ \sum_{k=1}^{d'} \left[\left(\frac{J_k(t)\sigma(t_-)J_k(t)^*}{\lambda_k} - \sigma(t_-) \right) (dN_k(t) - \lambda_k \, dt) \right], \quad (21)$$

where $\mathcal{L}(t)$ is the Liouville operator defined in Eqs. (16).

Starting from Eq. (18) or Eq. (21) one can get both the distribution of the outputs and the a posteriori dynamics of $\mathscr{H}$.

Of course, the joint distribution of the compatible field observables $B_i^\dagger(t) + B_i(t)$ and $\hat{\Lambda}_{kk}(t)$ is given by the Born rule based on their joint projection valued measure and the initial system/field state. Anyway, it can be obtained directly from Eq. (18) or Eq. (21), as it is the joint distribution of the processes W_j, N_k under the *physical probability* on $(\Omega, \mathscr{F}_T)$:

$$\mathbb{P}_T(d\omega) = p_T(\omega)\mathbb{Q}(d\omega), \qquad p_t = \mathrm{Tr}\{\sigma(t)\}. \quad (22)$$

Moreover, by defining $\rho(t) := \frac{\sigma(t)}{p_t}$ when $p_t > 0$, and by taking an arbitrary state for $\rho(t)$ when $p_t = 0$, we obtain the so called a *posteriori state*, the conditional state to be attributed to the system, having observed the realization of *all* the processes W_j and N_k up to time t. Correspondingly, let us call $\sigma(t)$ the *non-normalized a posteriori state*

In particular, regarding the distribution of the outputs, by Girsanov theorem we can say that under the physical probability $\mathbb{P}_T$

$$\widehat{W}_j(t) := W_j(t) - \int_0^t m_j(s)\mathrm{d}s, \qquad m_j(t) := 2\,\mathrm{Re}\,\mathrm{Tr}\,\{R_j(t)\rho(t_-)\}, \quad (23)$$

$j = 1, \ldots, d$, is a d-dimensional standard Wiener process, while $N_k(t)$ is a counting process of stochastic intensity $\mu_k(t) = \mathrm{Tr}\,\{J_k(t)^* J_k(t)\rho(t_-)\}$.

As we observe the fields without introducing any new disturbance on $\mathscr{H}$, we have that its a priori state, that is the mean of its a posteriori states, coincides with its reduced state (15) in absence of measurement:

$$\eta(t) = \mathbb{E}_{\mathbb{P}_T}[\rho(t)] = \mathbb{E}_{\mathbb{Q}}[\sigma(t)], \qquad t \in [0, T]. \quad (24)$$

Thus, the continuous measurement gives an unravelling (with a physical interpretation) to the open dynamics (16). Of course, if we change the observed fields for a given global evolution and a given environmental initial state, we get a different unravelling of the same open evolution.

2.3. *Interacting and non-interacting subsystems*

Let us finally consider a bipartite system $\mathscr{H} = \mathscr{H}_1 \otimes \mathscr{H}_2$ with its environment $\mathscr{K} = \Gamma[L^2(\mathbb{R}; 3)]$ and their HP-evolution (10). We are interested in the case of no direct interaction between the two subsystems $\mathscr{H}_1$ and $\mathscr{H}_2$, but, because of the common environment $\mathscr{K}$, the two subsystems could have or not have an indirect interaction.

If the global Hamiltonian H_T were bounded, we could say that $\mathscr{H}_1$ and $\mathscr{H}_2$ do not interact directly if the global Hamiltonian is

$$H_T = H_0 + H_1 + H_2 + H_{01} + H_{02}$$

where $H_0 = H_0^* \in \mathscr{L}(\mathscr{K})$ is the free Hamiltonian of the environment, $H_1 = H_1^* \in \mathscr{L}(\mathscr{H}_1)$ is the free Hamiltonian of $\mathscr{H}_1$, $H_2 = H_2^* \in \mathscr{L}(\mathscr{H}_2)$ is the free Hamiltonian of $\mathscr{H}_2$, while $H_{01} \in \mathscr{L}(\mathscr{K} \otimes \mathscr{H}_1)$ and $H_{02} \in \mathscr{L}(\mathscr{K} \otimes \mathscr{H}_2)$ give the interaction, respectively, of $\mathscr{H}_1$ with $\mathscr{K}$ and of $\mathscr{H}_2$ with $\mathscr{K}$.

Analogously, dealing with HP-evolutions, we say that $\mathscr{H}_1$ and $\mathscr{H}_2$ do not interact directly if, in the heuristic representation (12) of the global Hamiltonian H_T, each one of the operators K_{zw}, $\sum_w \left(\frac{K}{S-1}\right)_{zw} L_w$ and $H + \sum_{zw} L_z^* \left(\frac{K - \sin K}{4\left(\sin(K/2)\right)^2}\right)_{zw} L_w$ is the sum of local operators. This property is independent of the basis $\{|z\rangle\}_{z \in Z}$ chosen in 3.

In the case $L = 0$, this means $H = H_1 + H_2$, with $H_\ell = H_\ell^* \in \mathscr{L}(\mathscr{H}_\ell)$, and $K = K_1 + K_2$, with $K_\ell = K_\ell^* \in \mathscr{L}(3 \otimes \mathscr{H}_\ell)$.

In the case $K = 0$, this means $H = H_1 + H_2$, with $H_\ell = H_\ell^* \in \mathscr{L}(\mathscr{H}_\ell)$, and each $L_z = L_z^{(1)} + L_z^{(2)}$, with $L_z^{(\ell)} \in \mathscr{L}(\mathscr{H}_\ell)$.

An important subcase is when the subsystems $\mathscr{H}_1$ and $\mathscr{H}_2$ do not have any kind of interaction, either direct or indirect. In other words, this means that each subsystem $\mathscr{H}_\ell$ has its own environment $\mathscr{K}_\ell$ and that there is no interaction between $\mathscr{H}_1$ and $\mathscr{K}_1$ on one side and $\mathscr{H}_2$ and $\mathscr{K}_2$ on the other. Thus, we say that $\mathscr{H}_1$ and $\mathscr{H}_2$ do not interact, either directly or indirectly, if there exists a decomposition $\mathsf{3} = \mathsf{3}_1 \oplus \mathsf{3}_2$, that is a decomposition $\mathscr{K} = \Gamma[L^2(\mathbb{R}; \mathsf{3})] = \Gamma[L^2(\mathbb{R}; \mathsf{3}_1)] \otimes \Gamma[L^2(\mathbb{R}; \mathsf{3}_2)] = \mathscr{K}_1 \otimes \mathscr{K}_2$ such that, chosen a basis $\{|z\rangle\}_{z \in Z_1}$ in $\mathsf{3}_1$ and a basis $\{|z\rangle\}_{z \in Z_2}$ in $\mathsf{3}_2$ and considering the heuristic representation (12) of the global Hamiltonian in the basis $\{|z\rangle\}_{z \in Z_1 \cup Z_2}$ in $\mathsf{3}$, each addendum is an operator on $\mathscr{H}_1 \otimes \mathscr{K}_1$ or on $\mathscr{H}_2 \otimes \mathscr{K}_2$. This means that K_{zw} belongs $\mathscr{L}(\mathscr{H}_\ell)$ when both $z, w \in Z_\ell$, while it is null otherwise, that $\sum_w \left(\frac{K}{S-\mathbb{1}}\right)_{zw} L_w$ belongs to $\mathscr{L}(\mathscr{H}_\ell)$ when $z \in Z_\ell$, and that $H + \sum_{zw} L_z^* \left(\frac{K - \sin K}{4\left(\sin(K/2)\right)^2}\right)_{zw} L_w$ is the sum of local operators. This property is independent of the bases chosen in $\mathsf{3}_1$ and $\mathsf{3}_2$.

In the case $K = 0$, this means $H = H_1 + H_2$, with $H_\ell = H_\ell^* \in \mathscr{L}(\mathscr{H}_\ell)$, and $L_z \in \mathscr{L}(\mathscr{H}_1)$ for $z \in Z_1$, $L_z \in \mathscr{L}(\mathscr{H}_2)$ for $z \in Z_2$.

Let us remark that the Liouvillian (16) is not enough to understand whether the subsystems $\mathscr{H}_1$ and $\mathscr{H}_2$ do or do not interact.

3. No direct or indirect interaction

Let us start by the last case presented in the previous section, when the two qubits do not interact either directly or indirectly through a common bath, and let us study the role of a complete continuous measurement. We consider only the case $S = \mathbb{1}$, so that we need to take $Z = Z_1 \cup Z_2$, $Z_1 \cap Z_2 = \emptyset$,

$$L_z = \begin{cases} \hat{L}_z \otimes \mathbb{1} & \text{for } z \in Z_1, \\ \mathbb{1} \otimes \hat{L}_z & \text{for } z \in Z_2, \end{cases} \qquad H = H_1 \otimes \mathbb{1} + \mathbb{1} \otimes H_2. \qquad (25)$$

As it will be useful in the following, from now on we give evidence to the tensor product structure of the various operators we need. Then, from Eqs. (16) we get the Liouville operator $\mathcal{L}(t) = \mathcal{L}_1(t) \otimes \mathbb{1} + \mathbb{1} \otimes \mathcal{L}_2(t)$ with

$$\mathcal{L}_i(t)[\tau] := -\mathrm{i}[H_i(t), \tau] + \sum_{z \in Z_i} \left(\hat{L}_z \tau \hat{L}_z^* - \frac{1}{2}\left\{\hat{L}_z^* \hat{L}_z, \tau\right\}\right),$$

$$H_i(t) := H_i + \mathrm{i} \sum_{z \in Z_i} \left(\overline{v_z(t)}\, \hat{L}_z - v_z(t)\hat{L}_z^*\right).$$

Recall that v is the argument in the environment initial coherent state. Moreover, in the case of a complete observation, we obtain the SSE (18) with $R_j(t)$ and $J_k(t)$ given by Eq. (20b), $\mathsf{K}(t) = \mathsf{K}_1(t) \otimes \mathbb{1} + \mathbb{1} \otimes \mathsf{K}_2(t)$,

$$\mathsf{K}_i(t) := -\mathrm{i}H_i - \frac{1}{2} \sum_{z \in Z_i} \left(\hat{L}_z^* \hat{L}_z + 2v_z(t) \hat{L}_z^* + |v_z(t)|^2 \right).$$

Let us start by considering a pure initial state $\rho_0 = |\psi_0\rangle\langle\psi_0|$, so that $\sigma(t) = |\varphi(t)\rangle\langle\varphi(t)|$ and $p_t = \|\varphi(t)\|^2$, cf. Eqs. (18), (21), (22). Now the random a posteriori states are given by $\rho(t) = |\psi(t)\rangle\langle\psi(t)|$ with $\psi(t) = \varphi(t)/\|\varphi(t)\|$ and the a priori states by $\eta(t) = \mathbb{E}_{\mathbb{P}_T}[\rho(t)] = \mathbb{E}_{\mathbb{Q}}[\sigma(t)]$, see Sect. 2.

3.1. *The a posteriori concurrence*

By the definition of concurrence in the case of pure states (5), (6), we can introduce the random *a posteriori concurrence*

$$C_{\rho(t)} \equiv C_{\psi(t)} = \frac{|\chi_{\varphi(t)}|}{\|\varphi(t)\|^2} \tag{26}$$

and the *mean a posteriori concurrence*

$$\mathbb{E}_{\mathbb{P}_T}\left[C_{\psi(t)}\right] = \mathbb{E}_{\mathbb{Q}}\left[|\chi_{\varphi(t)}|\right], \quad 0 \le t \le T. \tag{27}$$

By the definition of concurrence for mixed states (7) and of a priori states (24), we get that the *a priori concurrence* is bounded by the mean a posteriori concurrence:

$$C_{\eta(t)} \le \mathbb{E}_{\mathbb{P}_T}\left[C_{\psi(t)}\right]. \tag{28}$$

By the linear SSE and Itô's formula we get the stochastic differential of $\chi_{\varphi(t)}$, which we shall need in the following,

$$\mathrm{d}\chi_{\varphi(t)} = \epsilon(t)\mathrm{d}t + \sum_{j=1}^{d} \ell_j(t)\chi_{\varphi(t)} \, \mathrm{d}W_j(t)$$

$$+ \sum_{k=1}^{d'} \left[q_k(t)\mathrm{d}N_k(t) + \lambda_k \chi_{\varphi(t)} \, \mathrm{d}t \right], \tag{29}$$

where

$$\epsilon(t) := \mathrm{Tr}_{\mathbb{C}^2}\left\{\mathsf{K}_1(t) + \mathsf{K}_2(t)\right\} \chi_{\varphi(t)} + \sum_{j=1}^{d} \langle \mathrm{T}R_j(t)\varphi(t)|\sigma_y \otimes \sigma_y R_j(t)\varphi(t)\rangle,$$

$$\ell_j(t) := \sum_{z \in Z} u_{jz}(t) \left(\mathrm{Tr}_{\mathbb{C}^2} \, \hat{L}_z + 2v_z(t) \right), \tag{30}$$

$$q_k(t) := \frac{1}{\lambda_k} \langle \mathsf{T} J_k(t) \varphi(t) | \sigma_y \otimes \sigma_y J_k(t) \varphi(t) \rangle - \chi_{\varphi(t)}.$$

By writing

$$\hat{L}_z = \sum_{i=1}^{3} h_{zi} \sigma_i + r_z, \tag{31}$$

we get

$$\ell_j(t) = 2 \sum_z u_{jz}(t) \big(r_z + v_z(t) \big), \tag{32}$$

$$\mathrm{Tr}_{\mathbb{C}^2} \{K_1(t) + K_2(t)\} = -i \, \mathrm{Tr}_{\mathbb{C}^2} \{H_1 + H_2\}$$
$$- \sum_{z \in Z} \left\{ \sum_{i=1}^{3} |h_{zi}|^2 + |r_z|^2 + |v_z(t)|^2 + 2v_z(t) \, \overline{r_z} \right\}. \tag{33}$$

Let us stress that the operators $R_j(t)$ and $J_k(t)$ are not in general local operators, but sums of local operators. By this fact we cannot write in a more explicit form the coefficients $\epsilon(t)$ and $q_k(t)$.

3.2. *Only local detection operators*

As already said, in this section we are considering only local operators in the dynamics: every qubit has its own environment and there is neither direct nor indirect interaction between the two qubits. Now we consider the case in which also the detection operators are local, that is

$$R_j(t) = R_j^0(t) \otimes \mathbb{1} \qquad \text{or} \qquad R_j(t) = \mathbb{1} \otimes R_j^0(t). \tag{34}$$

This means that we observe separately the two environments. With this further assumption, the stochastic differential (29) becomes the closed equation

$$d\chi_{\varphi(t)} = \chi_{\varphi(t)} \left(\kappa(t)dt + \sum_j \ell_j(t)dW_j(t) + \sum_k \left(\frac{d_k(t)}{\lambda_k} - 1 \right) dN_k(t) \right), \tag{35}$$

$$\kappa(t) = \mathrm{Tr}_{\mathbb{C}^2} \{K_1(t) + K_2(t)\} + \sum_{k=1}^{d'} \lambda_k + \sum_{j=1}^{d} \det_{\mathbb{C}^2} R_j^0(t),$$

$$\ell_j(t) = \mathrm{Tr}_{\mathbb{C}^2} \, R_j^0(t), \qquad d_k(t) = \det_{\mathbb{C}^2} R_{d+k}^0(t). \tag{36}$$

Equation (35) can be explicitly solved and, by stochastic calculus, we get

$$\mathbb{E}_{\mathbb{P}_T}\left[C_{\psi(t)}\right] = \mathbb{E}_{\mathbb{Q}}\left[C_{\varphi(t)}\right] = C_{\psi_0}\exp\left\{-\int_0^t c(s)\mathrm{d}s\right\}, \tag{37}$$

$$c(t) := \sum_{k=1}^{d'}\left(\lambda_k - |d_k(t)|\right) - \frac{1}{2}\sum_{j=1}^{d}\left(\mathrm{Im}\,\ell_j(t)\right)^2 - \mathrm{Re}\,\kappa(t).$$

The first important result is that $c(t)$ does not depend on the initial state of the qubits, but only on the operators involved in the reduced dynamics and in the observation. This result is a slight generalization of the analogous one in Ref. [14]. By using (31), we get, by straightforward calculations,

$$R_j^0(t) = \sum_{i=1}^{3}\tilde{h}_{ji}(t)\sigma_i + \frac{\ell_j(t)}{2}, \qquad \tilde{h}_{ji}(t) := \sum_{z\in Z}u_{jz}(t)h_{zi}, \tag{38}$$

$$d_k(t) = \frac{\ell_{d+k}(t)^2}{4} - \sum_{i=1}^{3}\tilde{h}_{(d+k)i}(t)^2, \qquad c(t) = \sum_{j\in Z}c_j(t), \tag{39}$$

$$c_j(t) = 2\sum_{i=1}^{3}\left(\mathrm{Re}\,\tilde{h}_{ji}(t)\right)^2 \geq 0, \qquad j \leq d, \tag{40}$$

$$c_j(t) = \frac{1}{4}|\ell_j(t)|^2 - |d_{j-d}(t)| + \sum_{i=1}^{3}\left|\tilde{h}_{ji}(t)\right|^2 \geq 0, \quad j > d. \tag{41}$$

By the fact that c does not depend on the initial state of the qubits we can extend the result to the case of an initial mixed state ρ_0 and we get

$$C_{\eta(t)} \leq \mathbb{E}_{\mathbb{P}_T}\left[C_{\rho(t)}\right] = C_{\rho_0}\exp\left\{-\int_0^t c(s)\mathrm{d}s\right\}; \tag{42}$$

we always assume complete observation. Note that the mean a posteriori concurrence is non-increasing. Moreover,

$$\int_0^{+\infty}c(s)\mathrm{d}s = +\infty \quad \Rightarrow \quad \lim_{t\to+\infty}\mathbb{E}_{\mathbb{P}_t}\left[C_{\rho(t)}\right] = 0, \tag{43}$$

and, if $c(t) = c > 0$, the mean a posteriori concurrence decreases exponentially.

For what concerns the a priori states $\eta(t)$, when the master equation involves only local operators, one can have the phenomenon of entanglement sudden death (ESD).[4,14] Note that no revival is possible for the concurrence of $\eta(t)$ due to the bound given by the mean a posteriori concurrence (42).

Also the a posteriori concurrence, without the mean, can be studied. From the SDEs (18) for $\varphi(t)$ and (35) for $\chi_{\varphi(t)}$, we can compute the stochastic differential of the concurrence $C_{\psi(t)} = |\chi_{\varphi(t)}| / \|\varphi(t)\|^2$; in terms of the new Wiener process (23), the final result is the closed SDE

$$dC_{\psi(t)} = C_{\psi(t)} \left\{ \sum_{j=1}^{d} \left[n_j(t) d\widehat{W}_j(t) - c_j(t) \right] \right.$$

$$\left. + \sum_{k=1}^{d'} \left[\left(\frac{|d_k(t)|}{\mu_k(t)} - 1 \right) (dN_k(t) - \mu_k(t)\,dt) - c_{d+k}(t)\,dt \right] \right\}, \quad (44)$$

where the $c_j(t)$ are given by Eq. (40), the $c_{d+k}(t)$ by Eq. (41), the $d_k(t)$ by Eq. (39) and

$$n_j(t) := \operatorname{Re} \ell_j(t) - m_j(t) = -2 \sum_{i=1}^{3} \left(\operatorname{Re} \tilde{h}_{ji}(t) \right) \langle \psi(t)|\mathbf{s}_i \psi(t) \rangle, \quad (45)$$

$$\mu_k(t) = \left\| \left(\sum_{i=1}^{3} \tilde{h}_{(k+d)\,i}(t)\mathbf{s}_i + \frac{\ell_{k+d}(t)}{2} \right) \psi(t) \right\|^2,$$

$$\tilde{h}_{ji}(t)\mathbf{s}_i = \begin{cases} \tilde{h}_{ji}(t)\,\sigma_i \otimes \mathbf{1} & \text{if } R_j(t) = R_j^0(t) \otimes \mathbf{1}, \\ \tilde{h}_{ji}(t)\,\mathbf{1} \otimes \sigma_i & \text{if } R_j(t) = \mathbf{1} \otimes R_j^0(t). \end{cases}$$

The solution of the SDE (44) is given by the stochastic exponential

$$C_{\psi(t)} = C_{\psi_0} \exp\left\{ \sum_{j=1}^{d} \left[\int_0^t n_j(s)\,d\widehat{W}_j(s) - \int_0^t \left(c_j(s) + \frac{n_j(s)^2}{2} \right) ds \right] \right.$$

$$\left. - \sum_{k=1}^{d'} \int_0^t \left(c_{d+k}(s) + |d_k(s)| - \mu_k(s) \right) ds \right\} \prod_{0 < s \le t} \prod_{k=1}^{d'} \left| \frac{d_k(s)}{\mu_k(s)} \right|^{\Delta N_k(s)}. \quad (46)$$

3.2.1. *Diffusive case*

Here we consider the purely diffusive case $(d' = 0)$. Now, in Eq. (37) the decay intensity of the mean a posteriori concurrence is $c(t) = \sum_{j=1}^{d} c_j(t)$ with $c_j(t)$ given by Eq. (40), while the random a posteriori concurrence reduces to

$$C_{\psi(t)} = C_{\psi_0} \exp\left\{ \sum_{j=1}^{d} \left[\int_0^t n_j(s)\,d\widehat{W}_j(s) - \int_0^t \left(c_j(s) + \frac{n_j(s)^2}{2} \right) ds \right] \right\}, \quad (47)$$

with $n_j(t)$ given by Eq. (45). Let us stress that neither c_j nor n_j depend on the trace of the operators $R_j^0(t)$.

Note that, while the a priori states $\eta(t)$ can suddenly loose any entanglement (ESD), this is a.s. impossible for the a posteriori state (with complete observation).

In the particular case of all the R_j^0's selfadjoint there is decay of the a posteriori concurrence, but, thanks to the freedom in the choice of the matrix u, by a change of phase we can pass from this case to the case of all the R_j^0's anti-selfadjoint, for which there is no decay for every initial qubit state ($n_j = c_j = 0$). Therefore, without changing the master equation, i.e. without changing the dynamical behaviour of the concurrence of the a priori state, one gets the complete entanglement protection by the choice of a phase in the detection operators. The case of all the R_j^0 anti-selfadjoint gives $\|\varphi(t)\| = $ constant and the SSE describes two independent random unitary evolutions.

3.2.2. *Jump case*

Let us consider the purely jump case, i.e. $d = 0$ and $J_k(t) = J_k^0(t) \otimes \mathbb{1}$ or $J_k(t) = \mathbb{1} \otimes J_k^0(t)$, for which we have $\mathbb{E}_{\mathbb{P}_T}\left[C_{\psi(t)}\right] = C_{\psi_0} e^{-\int_0^t c(s)\, \mathrm{d}s}$,

$$C_{\psi(t)} = C_{\psi_0} \exp\left\{-\sum_{k=1}^{d'} \int_0^t \left(c_k(s) + |d_k(s)| - \mu_k(s)\right) \mathrm{d}s\right\}$$

$$\times \prod_{0 < s \leq t} \prod_{k=1}^{d'} \left|\frac{d_k(s)}{\mu_k(s)}\right|^{\Delta N_k(s)},$$

$$d_k(t) = \det{}_{\mathbb{C}^2} J_k^0(t) = \frac{\ell_k(t)^2}{4} - \sum_{i=1}^{3} \tilde{h}_{ki}(t)^2, \qquad \mu_k(t) = \|J_k(t)\psi(t)\|^2,$$

$$c(t) = \sum_{k=1}^{d'} c_k(t), \qquad c_k(t) = \frac{1}{4}|\ell_k(t)|^2 - |d_k(t)| + \sum_{i=1}^{3}\left|\tilde{h}_{ki}(t)\right|^2 \geq 0.$$

One can check that $c_k(t) = 0$ if and only if $\mathrm{Im}\left(\overline{\tilde{h}_{kj}(t)}\,\tilde{h}_{ki}(t)\right) = 0$, $\mathrm{Re}\left(\overline{\ell_k(t)}\,\tilde{h}_{ki}(t)\right) = 0$, $i, j = 1, 2, 3$. Again, in some cases, one can protect the entanglement by tuning the detection operators without changing the mean dynamics, for instance by changing the unitary matrix $u(t)$. Let us give some examples.

A jump operator such as $J_k^0 = h\sigma_i + \ell/2$ contributes[14] with

$$c_k = |h|^2 + \frac{|\ell|^2}{4} - \sqrt{\left(|h|^2 + \frac{|\ell|^2}{4}\right)^2 - \frac{1}{2}\left(\mathrm{Re}\,\overline{h}\,\ell\right)^2};$$

note that this contribution is zero when $\mathrm{Re}\,\overline{h}\,\ell = 0$, while its maximum contribution is $c_k = |h|^2 + |\ell|^2/4 - \sqrt{|h|^4 + |\ell|^4/16}$, reached when $\mathrm{Re}\,\overline{h}\,\ell = \pm|h\,\ell|$.

The jump operator $J_k^0 = \alpha\sigma_\pm + \beta$ contributes[14] with $c_k = |\alpha|^2/2$.

Let us consider the term

$$\gamma_-\left(\sigma_- \bullet \sigma_+ - \frac{1}{2}\{\sigma_+\sigma_-, \bullet\}\right) + \gamma_+\left(\sigma_+ \bullet \sigma_- - \frac{1}{2}\{\sigma_-\sigma_+, \bullet\}\right) \qquad (48)$$

in the Liouville operator with $\gamma_+ \geq 0$, $\delta > 0$, $\gamma_- = \delta + \gamma_+$. Three different choices of detection operators, but which give rise to the same dissipative term (48) in the master equation, are:

(1) $J_- = \sqrt{\gamma_-}\,\sigma_-$ and $J_+ = \sqrt{\gamma_+}\,\sigma_+$, which contribute to c with $\gamma_+ + \delta/2$;

(2) $J_1 = \sqrt{\gamma_+}\,\sigma_1$, $J_2 = \sqrt{\gamma_+}\,\sigma_2$, $J_3 = \sqrt{\delta}\,\sigma_-$, which contribute to c with $\delta/2$;

(3) $J_1 = \frac{1}{\sqrt{2}}\left(\sqrt{\gamma_+}\,\sigma_+ + \sqrt{\gamma_-}\,\sigma_-\right)$, $J_2 = \frac{1}{\sqrt{2}}\left(\sqrt{\gamma_+}\,\sigma_+ - \sqrt{\gamma_-}\,\sigma_-\right)$, which contribute to c with $\frac{1}{2}\left(\sqrt{\gamma_-} - \sqrt{\gamma_+}\right)^2$.

Note that $\frac{1}{2}\left(\sqrt{\gamma_-} - \sqrt{\gamma_+}\right)^2 \leq \frac{\delta}{2} \leq \gamma_+ + \frac{\delta}{2}$. Given the dissipative term (48) in the Liouville operator, the choice (3) is the best one to slow down the disentanglement [14, Eq. (19)].

For what concerns the random a posteriori concurrence, if $C_{\psi(0)} > 0$, $C_{\psi(t)}$ can vanish only if $d_k(s) \equiv \det_{\mathbb{C}^2} J_k^0(s) = 0$ for some k and some s, as in the case of $\sigma_\pm$. In the jump case we can have ESD for some trajectories, eventually for all trajectories. The exponential decay of the mean concurrence is due to the randomness of the time of death.

3.3. An example with general detection operators

Let us now consider a concrete model of non-interacting qubits plus an environment. We want to show, in a very simple model, how the mere choice of the detection operators changes the behaviour of the a posteriori concurrence and how much this behaviour is different from the one of the a priori concurrence. The Liouville operator is fixed, but different choices of detection operators are studied.

As starting point (25) we take $Z = \{1, 2\}$, $v(t) = 0$,

$$L_1 = \hat{L}_1 \otimes \mathbb{1}, \quad L_2 = \mathbb{1} \otimes \hat{L}_2, \quad \hat{L}_1 = \hat{L}_2 = \sqrt{\frac{\gamma}{2}} \sigma_x, \quad \gamma > 0,$$

$$H_1 = H_2 = \frac{\omega_0}{2} \sigma_z, \quad \omega_0 \in \mathbb{R}.$$

The Liouville operator turns out to be

$$\mathcal{L} = \mathcal{L}_0 \otimes \mathbb{1} + \mathbb{1} \otimes \mathcal{L}_0, \quad \mathcal{L}_0[\tau] = -\mathrm{i} \frac{\omega_0}{2} [\sigma_z, \tau] - \frac{\gamma}{4} [\sigma_x, [\sigma_x, \tau]];$$

we can also write

$$\mathcal{L}[\eta] = -\mathrm{i} \frac{\omega_0}{2} [\sigma_z \otimes \mathbb{1} + \mathbb{1} \otimes \sigma_z, \eta] - \gamma \eta$$
$$+ \frac{\gamma}{2} (\sigma_x \otimes \mathbb{1} \, \eta \, \sigma_x \otimes \mathbb{1} + \mathbb{1} \otimes \sigma_x \, \eta \, \mathbb{1} \otimes \sigma_x). \quad (49)$$

The master equation with Liouville operator (49) with $\omega_0 \neq 0$ has a unique equilibrium state given by $\eta_{\mathrm{eq}} = \mathbb{1}/4$. When $\omega_0 = 0$, we have more equilibria, the statistical operators that are diagonal in the canonical basis. In any case the equilibrium states are separable.

3.3.1. *Concurrence of the a priori state*

If one writes down the master equation with Liouville operator (49), one sees that it decomposes in subsystems of equations that can be solved analytically. However, to simplify the analysis of the dynamics and the computation of the concurrence, it is worthwhile to consider the subclass of the "X" states given in Section 1.2. By checking the master equation with generator (49) in the canonical basis, one can see that the class of X states is preserved.

Case $\omega_0 \neq 0$. By the fact that there is a unique equilibrium state proportional to the identity, we get

$$\lim_{t \to +\infty} \rho_{23}(t) = \lim_{t \to +\infty} \rho_{14}(t) = 0, \quad \lim_{t \to +\infty} \rho_{jj}(t) = \frac{1}{4}.$$

Then, if the initial X state has positive concurrence, it exists a finite time $t_D > 0$ for which $C_{\rho(t_D)} = 0$ and we have entanglement sudden death.

Case $\omega_0 = 0$. In this case there is not a unique equilibrium state. As we shall see, the a priori concurrence is always limited by the exponential decay (50) (local detection operators, diffusive case); one can also check that this limit is saturated when the initial state is a Bell state. But we can have also ESD; for instance, take as initial state $\rho_0 = |\psi_0\rangle\langle\psi_0|$, $\psi_0 = \frac{1}{\sqrt{2}}(|10\rangle + i|01\rangle) = \frac{1+i}{2}(|\beta_1\rangle + |\beta_2\rangle)$, which is again an X state. By solving the master equation and computing the concurrence by formulae (8) we find ESD at the time $t_D = -\frac{1}{\gamma}\ln(\sqrt{2}-1)$; moreover, for $t \in [0, t_D]$ the a priori concurrence is given by $C_{\eta(t)} = \frac{1}{2}(1 + e^{-\gamma t})^2 - 1$.

3.3.2. *Local detection operators*

We start by considering local detection operators. The unitary matrix u which fixes the observed fields in Sect. 2.2 is taken to be $u_{jz} = \delta_{jz}\,e^{i\phi_j}$, $\phi_j \in [0, 2\pi]$. Then, the detection operators (20b), (34) reduce to

$$R_1 = \sqrt{\frac{\gamma}{2}}\,e^{i\phi_1}\sigma_x \otimes \mathbb{1}, \qquad R_2 = \sqrt{\frac{\gamma}{2}}\,e^{i\phi_2}\mathbb{1} \otimes \sigma_x.$$

Diffusive case. Let us start by an observation of homodyne/heterodyne type: $d = 2$, $d' = 0$. Then, by Eqs. (38)–(42), (45), (47) we get the a posteriori concurrence

$$C_{\psi(t)} = C_{\psi_0}e^{-ct}\exp\left\{\sum_{j=1}^{2}\left[\int_0^t n_j(s)\,d\widehat{W}_j(s) - \frac{1}{2}\int_0^t n_j(s)^2 ds\right]\right\}$$

and the mean a posteriori concurrence $\mathbb{E}_{\mathbb{P}_T}\left[C_{\rho(t)}\right] = C_{\rho_0}e^{-ct}$, where

$$0 \leq c = \gamma\left[(\cos\phi_1)^2 + (\cos\phi_2)^2\right] \leq 2\gamma,$$

$$n_1(t) = \sqrt{2\gamma}\cos\phi_1\,\langle\psi(t)|\sigma_x \otimes \mathbb{1}\,\psi(t)\rangle,$$

$$n_2(t) = \sqrt{2\gamma}\cos\phi_2\,\langle\psi(t)|\mathbb{1} \otimes \sigma_x\,\psi(t)\rangle.$$

The important feature of this model is that it shows the dependence on the measuring phases: the decay constant c can take any value in the closed interval $[0, 2\gamma]$. Note that c does not depend on ω_0. Finally, by the bound (42) for the a priori concurrence, we get

$$C_{\eta(t)} \leq C_{\rho_0}e^{-2\gamma t}. \tag{50}$$

Jump case. Now let us consider a counting observation, with the same detection operators: $d' = 2$, $d = 0$, $J_k = R_k$. From (46) we can check that the a posteriori concurrence turns out to be non-random and constant: $C_{\psi(t)} = C_{\psi_0}$. This is due to the fact that the jump operators are proportional to local unitaries. Thus, any initial entanglement can be perfectly protected just by a proper monitoring of the environment. Let us stress that the a priori concurrence always vanishes for long times and sometimes even in a finite time.

3.3.3. *Non local detection operators*

We give now an example of detection with non-local operators for the same non-interacting qubits. Now we measure in a non-local way the environments of the qubits, but we do not change the interaction with the environments and, thus, their a priori dynamics. We consider only the diffusive case $(d = 2$, $d' = 0)$ and we take the unitary matrix u of Section 2.2 to be

$$u = \frac{1}{\sqrt{2}} \begin{pmatrix} e^{i(\theta+\phi)} & e^{i(\theta-\phi)} \\ ie^{i(\theta+\phi)} & -ie^{i(\theta-\phi)} \end{pmatrix} ;$$

then, we get

$$R_1 = \frac{e^{i\theta}\sqrt{\gamma}}{2} \left(e^{i\phi}\sigma_x \otimes \mathbb{1} + e^{-i\phi}\mathbb{1} \otimes \sigma_x \right) ,$$

$$R_2 = \frac{e^{i\theta}\sqrt{\gamma}}{2} \left(ie^{i\phi}\sigma_x \otimes \mathbb{1} - ie^{-i\phi}\mathbb{1} \otimes \sigma_x \right) .$$

By particularizing the general formulae of Sect. 3.1 we obtain that the stochastic differential of $\chi_{\varphi(t)}$ does not contain the white noise term and we have

$$\dot{\chi}_{\varphi(t)} = -\gamma\chi_{\varphi(t)} + \gamma e^{2i\theta}\mathcal{D}(t), \qquad \mathcal{D}(t) := \langle \mathrm{T}\varphi(t)|\sigma_z \otimes \sigma_z\varphi(t)\rangle. \tag{51}$$

Again by stochastic differentiation, we get

$$\dot{\mathcal{D}}(t) = \gamma e^{2i\theta}\chi_{\varphi(t)} - \gamma\mathcal{D}(t) - i\omega_0\mathcal{E}(t), \tag{52}$$
$$\mathcal{E}(t) := \langle \mathrm{T}\varphi(t)| \left(\sigma_z \otimes \mathbb{1} + \mathbb{1} \otimes \sigma_z \right) \varphi(t)\rangle.$$

By differentiation of $\mathcal{E}$ we get more complicated expressions, including terms with stochastic differentials. Anyway, from Eqs. (51), (52) we obtain

$$\chi_{\varphi(t)} \pm \mathcal{D}(t) = e^{-\gamma_\pm t}\left(\chi_{\psi_0} \pm \mathcal{D}(0)\right) \mp i\omega_0 \int_0^t e^{-\gamma_\pm(t-s)}\mathcal{E}(s)\mathrm{d}s, \tag{53}$$

$$\gamma_\pm := \gamma \left(1 \pm e^{2i\theta} \right) .$$

In this model one can have a variety of behaviours for the mean concurrence, such as revivals and creation of concurrence in the long run. Let us see this in the simplest case.

The case $\omega_0 = 0$. In this case we have

$$\chi_{\varphi(t)} = \frac{1}{2}\,\mathrm{e}^{-\gamma_+ t}\big(\chi_{\psi_0} + \mathcal{D}(0)\big) + \frac{1}{2}\,\mathrm{e}^{-\gamma_- t}\big(\chi_{\psi_0} - \mathcal{D}(0)\big). \tag{54}$$

Being non-random, by Eqs. (26), (27), we get $\mathbb{E}_{\mathbb{P}_T}\big[C_{\psi(t)}\big] = \big|\chi_{\varphi(t)}\big|$, for all $T \geq t$.

If $\mathrm{e}^{2\mathrm{i}\theta} \neq \pm 1$, we get $\mathrm{Re}\,\gamma_\pm > 0$. Then, the mean a posteriori concurrence decays exponentially at long times, but, depending on the initial state of the qubits, it can have also revivals. For instance, by taking ψ_0 such that $\chi_{\psi_0} = 0$ and $\mathcal{D}(0) \neq 0$, we have

$$\big|\chi_{\varphi(t)}\big| = \frac{1}{2}\,|\mathcal{D}(0)|\,\big|\mathrm{e}^{-\gamma_+ t} - \mathrm{e}^{-\gamma_- t}\big|.$$

If $\mathrm{e}^{2\mathrm{i}\theta} = 1$, we get $\gamma_+ = 2\gamma$, $\gamma_- = 0$ and

$$\big|\chi_{\varphi(t)}\big| = \frac{1}{2}\,\big|\mathrm{e}^{-2\gamma t}\big(\chi_{\psi_0} + \mathcal{D}(0)\big) + \big(\chi_{\psi_0} - \mathcal{D}(0)\big)\big|.$$

So, depending on the initial state of the qubits, some concurrence can survive (entanglement protection) or can be created in the long run (entanglement generation). The case $\mathrm{e}^{2\mathrm{i}\theta} = -1$ is similar.

4. An example with indirect interaction

In this last section we consider the case of indirect interaction between two qubits and, by means of an explicit model, we show that a very extreme scenario can occur: the interaction with the environment completely destroys any entanglement between the qubits, if no measurement is performed, while the same interaction generates maximally entangled states, independently of the initial state of the qubits, if the environment is simply continuously monitored after the interaction. Indeed, while in the long run the a priori state of the qubits becomes maximally chaotic, and thus separable, their a posteriori state becomes maximally entangled for every output of the continuous measurement.

We consider a couple of qubits $\mathscr{H} = \mathscr{H}_1 \otimes \mathscr{H}_2$ interacting with a sort of continuous flow $\mathscr{K} = \Gamma[L^2(\mathbb{R}; 3)]$ of quadruples of qubits $3 = 3_1 \otimes 3_2 \otimes 3_1' \otimes 3_2'$.

Let us denote by $\{|i\rangle\}_{i=0,1}$ the canonical basis in $\mathscr{H}_1 = \mathscr{H}_2 = 3_1 = 3_2 = 3'_1 = 3'_2 = \mathbb{C}^2$ and then let us introduce the flip operator F_ℓ in $3_\ell \otimes \mathscr{H}_\ell$:

$$F_\ell = \sum_{ij} |ij\rangle\langle ji| = F_\ell^* = F_\ell^{-1} = e^{-i\frac{\pi}{2}(F_\ell - 1)}.$$

Let us choose in 3 the basis generated by the Bell bases in $3_1 \otimes 3_2$ and in $3'_1 \otimes 3'_2$, that is $\{|\beta_x \otimes \beta_{x'}\rangle\}_{x,x'=0,\dots,3}$.

We consider the HP evolution (10) generated by the interaction

$$H = 0, \qquad L = 0,$$

$$S = F_1 F_2 = F_2 F_1 = e^{-i\frac{\pi}{2}(F_1 + F_2 - 2)}, \qquad K = \frac{\pi}{2}(F_1 + F_2 - 2),$$

where every operator is identified with its natural extension. Roughly speaking, when a quadruple of qubits 3 belonging to the continuous flow interacts with the couple of interest $\mathscr{H}$, the first two qubits of the quadruple $3_1 \otimes 3_2$ exchange their joint state with $\mathscr{H}$, while the other two qubits $3'_1 \otimes 3'_2$ are simple witnesses. Then

$$S_{(xx')(yy')} = \mathrm{Tr}_3\left[\left(|\beta_y \otimes \beta_{y'}\rangle\langle\beta_x \otimes \beta_{x'}| \otimes \mathbb{1}_{\mathscr{H}}\right)S\right] = |\beta_y\rangle\langle\beta_x|\,\delta_{x'y'}$$

and the Hudson-Parthasaraty equation is

$$dV(t) = \sum_{xyx'}\left(|\beta_y\rangle\langle\beta_x| - \delta_{xy}\right)V(t)\,d\Lambda_{(xx')(yx')}(t).$$

Therefore, there is no direct interaction between $\mathscr{H}_1$ and $\mathscr{H}_2$ as $K = \frac{\pi}{2}(F_1 + F_2 - 2)$ with F_1 involving only $\mathscr{H}_1$ and F_2 involving only $\mathscr{H}_2$. Let us remark that this is just one of those cases where the whole interaction is encoded in the domain of the global Hamiltonian H_T. Indeed,[25] H_T is just an extension of the free field Hamiltonian E_0, re-restricted to the domain of the "regular vectors" $\Phi \in \mathscr{K} \otimes \mathscr{H}$ such that $a_{xx'}(0^-)\Phi = \sum_{yy'} S_{(xx')(yy')}\,a_{yy'}(0^+)\Phi$ for all x, x'.

For the environment we choose the initial pure coherent state $|e(v)\rangle\langle e(v)|$ with argument

$$v(t) = \sqrt{\frac{\nu}{4}}\sum_{x=0}^3 |\beta_x\rangle \otimes |\beta_x\rangle \in 3 = \left(3_1 \otimes 3_2\right) \otimes \left(3'_1 \otimes 3'_2\right), \qquad \forall 0 \le t \le T,$$

where ν is a positive parameter and $T > 0$ is our arbitrary time horizon. Roughly speaking, even if the qubits $3'_1$ and $3'_2$ are not involved in the interaction with $\mathscr{H}_1$ and $\mathscr{H}_2$, they are initially entangled with the qubits 3_1 and 3_2 which exchange their state with $\mathscr{H}_1$ and $\mathscr{H}_2$.

Then, if ρ_0 is the system initial state, its reduced state at time t is

$$\eta(t) = \text{Tr}_{\mathscr{K}} \left[U_t \, |e(v)\rangle\langle e(v)| \otimes \rho_0 \, U_t^* \right] = e^{\mathscr{L}t} \rho_0,$$

where

$$\mathscr{L}\eta = \nu \, \frac{\text{Tr}\,\eta}{4} \, \mathbb{1} - \nu \, \eta,$$

so that

$$\eta(t) = \rho_0 e^{-\nu t} + \frac{\mathbb{1}}{4}(1 - e^{-\nu t}) \to \frac{\mathbb{1}}{4}, \qquad \text{for } t \to \infty,$$

and the state of $\mathscr{H}$ becomes maximally chaotic and any entanglement between $\mathscr{H}_1$ and $\mathscr{H}_2$ is destroyed by the interaction with the common bath.

The a priori concurrence goes to 0 at least exponentially,

$$C_{\eta(t)} \le C_{\rho_0} \, e^{-\nu t} \to 0, \qquad \text{for } t \to \infty,$$

and, depending on the system initial state ρ_0, we can even assist to entanglement sudden death.

This can be verified by considering an X state as initial state. Indeed, if η_{ij} are the matrix elements of ρ_0 with respect to the computational basis (1), we find

$$C_{\eta(t)} = 2 \max \left\{ 0, \, C_1(t), \, C_2(t) \right\},$$

$$C_1(t) = |\eta_{23}| \, e^{-\nu t} - \sqrt{\left(\eta_{11} e^{-\nu t} + \frac{1}{4}(1 - e^{-\nu t}) \right)\left(\eta_{44} e^{-\nu t} + \frac{1}{4}(1 - e^{-\nu t}) \right)},$$

$$C_2(t) = |\eta_{14}| \, e^{-\nu t} - \sqrt{\left(\eta_{22} e^{-\nu t} + \frac{1}{4}(1 - e^{-\nu t}) \right)\left(\eta_{33} e^{-\nu t} + \frac{1}{4}(1 - e^{-\nu t}) \right)}.$$

By the fact that

$$\lim_{t \to +\infty} C_1(t) = \lim_{t \to +\infty} C_2(t) = -\frac{1}{4},$$

if the initial X state ρ_0 has positive concurrence, it exists a finite time $t_D > 0$ for which $C_{\eta(t_D)} = 0$. The death time t_D can be explicitly computed. For example, if $\rho_0 = |\beta_1\rangle\langle\beta_1|$, then $t_D = \frac{\ln 3}{\nu}$.

Let us introduce now the continuous measurement. As a preliminary step, let us suppose we observe all the sixteen compatible processes of observables $\Lambda_{(xx')(xx')}(t)$. Roughly speaking, we count the quadruples of kinds (xx') that have been through an interaction with the couple $\mathscr{H}$ between time 0 and time t. Then the corresponding linear stochastic master equation for the non-normalized a posteriori state $\tilde{\sigma}(t)$ is

$$d\tilde{\sigma}(t) = \mathcal{L}[\tilde{\sigma}(t_-)]dt$$

$$+ \sum_{x,x'=0}^{3} \left(\frac{4\nu}{\lambda} |\beta_{x'}\rangle\langle\beta_x|\tilde{\sigma}(t_-)|\beta_x\rangle\langle\beta_{x'}| - \tilde{\sigma}(t_-) \right) \left(dN_{xx'}(t) - \frac{\lambda}{16}dt \right)$$

in a probability space $(\Omega, \mathscr{F}, \mathscr{F}_t, N_{xx'}(t), \mathbb{Q})$ where $N_{xx'}(t)$, $x, x' = 0, \ldots, 3$, are sixteen independent Poisson processes of rates $\lambda/16$ under $\mathbb{Q}$.

The definitive step is to consider the measurement of the (non-maximal) family of the four compatible processes of observables

$$\Lambda_{x'}(t) = \sum_{x=0}^{3} \Lambda_{(xx')(xx')}(t).$$

Roughly speaking, we count the quadruples of qubits, with the second couple of kind x', which have been through an interaction with the couple $\mathscr{H}$ between time 0 and time t. Then, by conditioning, we get the linear stochastic master equation for the non-normalized a posteriori state $\sigma(t)$,

$$d\sigma(t) = \mathcal{L}[\sigma(t_-)]dt$$

$$+ \sum_{x'=0}^{3} \left(\frac{\nu}{\lambda} \left(\text{Tr}\,\sigma(t_-) \right) |\beta_{x'}\rangle\langle\beta_{x'}| - \sigma(t_-) \right) \left(dN_{x'}(t) - \frac{\lambda}{4}dt \right),$$

in a probability space $(\Omega, \mathscr{F}, \mathscr{F}_t, N_{x'}(t), \mathbb{Q})$ where $N_{x'}(t)$, $x' = 0, \ldots, 3$, are four independent Poisson processes of rates $\lambda/4$ under $\mathbb{Q}$.

If $N(t) = \sum_{x'=0}^{3} N_{x'}(t)$ denotes the total counts up to time t, T_n denotes the arrival time of the count n and if X'_n denotes the mark of count n, the solution is

$$\sigma(t) = \begin{cases} \rho_0\, e^{-\nu t + \lambda t}, & \text{if } 0 \leq t < T_1, \\ |\beta_{X'_{N(t)}}\rangle\langle\beta_{X'_{N(t)}}|\, e^{-\nu t + \lambda t} \left(\frac{\nu}{\lambda} \right)^{N(t)}, & \text{if } t \geq T_1. \end{cases}$$

Then, under the physical probability $\mathbb{P}_T(d\omega) = \text{Tr}\,\{\sigma(T)\}\,\mathbb{Q}(d\omega)$, the four counting processes $N_{x'}(t)$ are four independent Poisson processes of rates $\nu/4$, which depend on the environment initial state, and the a posteriori state is

$$\rho(t) = \begin{cases} \rho_0, & \text{if } 0 \leq t < T_1, \\ |\beta_{X'_{N(t)}}\rangle\langle\beta_{X'_{N(t)}}|, & \text{if } t \geq T_1. \end{cases}$$

Roughly summarizing, a flow of quadruples of qubits $\mathfrak{Z}$ interacts with the two qubits $\mathscr{H}$. Actually only the couple $\mathfrak{Z}_1 \otimes \mathfrak{Z}_2$ interacts by exchanging its state with $\mathscr{H}$, while $3'_1 \otimes 3'_2$ is a simple witness which is, nevertheless, initially entangled with $3_1 \otimes 3_2$. As a result, the couple $\mathscr{H}$ becomes entangled with the couple $3'_1 \otimes 3'_2$ of the last interaction. By counting the quadruples gone through an interaction with $\mathscr{H}$ and measuring the projection valued measure $\{|\beta_{x'}\rangle\langle\beta_{x'}|\}_{x'=0}^3$ on $3'_1 \otimes 3'_2$, we get an output with the distribution of a marked Poisson process and, at every count, the a posteriori state of $\mathscr{H}$ jumps into the Bell state labelled by the corresponding mark X'.

We can also compute the random a posteriori concurrence

$$
C_{\rho(t)} = \begin{cases} C_{\rho_0}, & \text{if } 0 \le t < T_1, \\ 1, & \text{if } t \ge T_1, \end{cases}
$$

and we find that the a posteriori concurrence goes to 1, both almost surely and in the mean,

$$
C_{\rho(t)} \to 1, \qquad \text{for } t \to \infty, \qquad \mathbb{P}\text{-a.s.}, \qquad \forall \rho_0,
$$

$$
\mathbb{E}_{\mathbb{P}}[C_{\rho(t)}] = 1 - (1 - C_{\eta(0)})e^{-\nu t} \to 1, \qquad \text{for } t \to \infty, \qquad \forall \rho_0,
$$

while the a priori concurrence goes to 0,

$$
\mathbb{E}_{\mathbb{P}}[C_{\rho(t)}] \ge C_{\eta(t)} \to 0, \qquad \text{for } t \to \infty.
$$

Therefore, while any entanglement between the qubits is a priori destroyed by the interaction with the common bath, at the same time it is enough to monitor the bath in a proper way to get a maximal creation of the a posteriori entanglement, for any initial state of the qubits.

References

1. M.A. Nielsen, I.L. Chuang, *Quantum Computation and Quantum Information* (Cambridge University Press, Cambridge, 2000).
2. L. Accardi, K. Imafuku, *Control of quantum states by decoherence*, in L. Accardi, M. Ohya, N. Watanabe (eds.), *Quantum Information and Computing (Tokyo, 2003)*, QP-QP: Quantum Probability and White Noise Analysis Vol. 19, (World Scientific, Singapore, 2006) pp. 28–45.
3. Ting Yu, J.H. Eberly, *Sudden death of entanglement*, Science **323**(2009) 598–601.
4. M. Orszag, *Entanglement protection and generation in a two-atom system*, in R. Rebolledo, M. Orszag (eds.), *Quantum Probability and Related Topics*, QP-PQ: Quantum Probability and White Noise Analysis Vol. 27, (World Scientific, Singapore, 2011) pp. 227–260.
5. L. Diósi, *Progressive decoherence and total environmental disentanglement*, in F. Benatti, R. Floreanini *Irreversible Quantum Dynamics*, Lecture Notes in Physics, Vol. 622 (Springer, Berlin, 2003) pp. 157–163.

6. Ting Yu, J.H. Eberly, *Evolution from entanglement to decoherence of bipartite mixed "X" states*, QIC 7 (2007) 459–468.
7. F. Benatti, R. Floreanini, U. Marzolino, *Environment induced entanglement in a refined weak-coupling limit*, EPL 88 (2009) 20011.
8. F. Benatti, R. Floreanini, U. Marzolino, *Entangling two unequal atoms through a common bath*, Phys. Rev. A 81 (2010) 012105.
9. F. Benatti, A.M. Liguori, G. Paluzzano, *Entanglement and entropy rates in open quantum systems*, J. Phys. A: Math. Theor. 43 (2010) 045304.
10. F. Altintas, R. Eryigit, *Quantum correlations between identical and unidentical atoms in a dissipative environment*, J. Phys. B: At. Mol. Opt. Phys. 44 (2011) 125501.
11. A.R.R. Carvalho, J.J. Hope, *Stabilising entanglement by quantum jump-based feedback*, Phys. Rev. A 76 (2007) 010301(R).
12. A.R.R. Carvalho, A.J.S. Reid, J.J. Hope *Controlling entanglement by direct quantum feedback*, Phys. Rev. A 78 (2008) 012334.
13. E. Mascarenhas, B. Marques, D. Cavalcanti, M. Terra Cunha, M. França Santos, *Protection of quantum information and optimal singlet conversion through higher dimensional quantum systems and environment monitoring*, Phys. Rev. A 81 (2010) 032310.
14. S. Vogelsberger, D. Spehner, *Average entanglement for markovian quantum trajectories*, Phys. Rev. A 82 (2010) 052327.
15. C. Viviescas, I. Guevara, A.R.R. Carvalho, M. Busse, A. Buchleitner, *Entanglement dynamics in open two-qubit systems via diffusive quantum trajectories*, Phys. Rev. Lett. 105 (2010) 210502.
16. E. Mascarenhas, D. Cavalcanti, V. Vedral, M. França Santos, *Physically realizable entanglement by local continuous measurements*, Phys. Rev. A 83 (2011) 022311.
17. A. Barchielli, *Continual measurements in quantum mechanics and quantum stochastic calculus*. In S. Attal, A. Joye, C.-A. Pillet (eds.), *Open Quantum Systems III*, Lecture Notes in Mathematics **1882** (Springer, Berlin, 2006), pp. 207–291.
18. A. Barchielli, M. Gregoratti, *Quantum Trajectories and Measurements in Continuous Time — The diffusive case*, Lecture Notes in Physics **782** (Springer, Berlin, 2009).
19. W.K. Wootters, *Entanglement of formation of an arbitrary state of two qubits*, Phys. Rev. Lett. 80 (1998) 2245-2248.
20. D. Petz, *Quantum Information Theory and Quantum Statistics* (Springer, Berlin, 2008).
21. M. Horodecki, P. Horodecki, R. Horodecki, *Separability of mixed states: necessary and sufficient conditions*, Phys. Lett. A **223** (1996) 1–8.
22. K. R. Parthasarathy, *An Introduction to Quantum Stochastic Calculus* (Birkhäuser, Basel, 1992).
23. A. Barchielli, A. M. Paganoni, *Detection theory in quantum optics: stochastic representation*, Quantum Semiclass. Opt. **8** (1996) 133–156.
24. A. S. Holevo, *Exponential formulae in quantum stochastic calculus*, Proc. Roy. Soc. Edinburgh Sect. A **126** (1996) 375–389.

25. M. Gregoratti, *The Hamiltonian operator associated with some quantum stochastic evolutions*, Comm. Math. Phys. **222** (2001) 181–200.

COMPLETELY POSITIVE TRANSFORMATIONS OF QUANTUM OPERATIONS

GIULIO CHIRIBELLA

Perimeter Institute for Theoretical Physics,
31 Caroline St North, Waterloo, Ontario, N2L 2Y5, Canada
E-mail: gchiribella@perimeterinstitute.ca

ALESSANDRO TOIGO

Dipartimento di Matematica, Politecnico di Milano,
Piazza Leonardo da Vinci 32, I-20133 Milano,
and INFN, Sezione di Milano,
Via Celoria 16, I-20133 Milano, Italy
E-mail: alessandro.toigo@polimi.it

VERONICA UMANITÀ

Dipartimento di Matematica, Università di Genova,
Via Dodecaneso 35, I-16146 Genova, Italy
E-mail: umanita@dima.unige.it

Quantum supermaps are higher-order maps transforming quantum operations into quantum operations and satisfying suitable requirements of normality and complete positivity. Here we present the extension of the theory of quantum supermaps, originally formulated in the finite dimensional setting, to the case of higher-order maps transforming quantum operations on generic von Neumann algebras. In this setting, we provide two dilation theorems for quantum supermaps that are the analogues of the Stinespring and Radon-Nikodym theorems for quantum operations. A structure theorem for probability measures with values in the set of quantum supermaps is also illustrated. Finally, some applications are given, and in particular it is shown that all the supermaps defined in this paper can be implemented by connecting quantum devices in quantum circuits.

Keywords: Completely positive and completely bounded maps; dilation theorems; Stinespring representation; quantum information theory.

1. Introduction

Quantum supermaps[1,2] are the most general admissible transformations of quantum devices. Mathematically, the action of a quantum device is associated to a set of completely positive trace non-increasing maps, called *quantum operations*,[3,4] which transform the states of an input quantum system into states of an output quantum system. A quantum supermap is then a higher-order linear map that transforms quantum operations into quantum operations. The theory of quantum supermaps, developed in Refs. 1,2 for finite dimensional quantum systems, has proven to be a powerful tool for the treatment of many advanced topics in quantum information theory,[5-10] including in particular the optimal cloning and the optimal learning of unitary transformations[11,12] and quantum measurements.[13,14] Moreoever, quantum supermaps are interesting for the foundations of Quantum Mechanics as they are the possible dynamics in a toy model of non-causal theory,[15] which has a quartic relation between the number of distinguishable states and the number of parameters needed to specify a state.[16] Quantum supermaps also attracted interest in the mathematical physics literature, as they suggested the study of a general class of positive maps between convex subsets of the state space.[17]

Originally, the definition and the main theorems on quantum supermaps were presented in the context of full matrix algebras describing finite dimensional quantum systems.[1,2] In this paper we will present their extension to the case where the input of the quantum operations is allowed to be a generic von Neumann algebra and the output is the C^*-algebra of the bounded operators on an arbitrary Hilbert space. This generalization is useful for applications, because on the one hand it allows to treat quantum systems with infinite dimension, on the other hand it includes transformations of quantum measuring devices, i.e. maps from the commutative algebra of functions on the outcome space to the full operator algebra on the Hilbert space of the measured system (in the Heisenberg picture).

Quantum supermaps on finite dimensional quantum systems are defined axiomatically as completely positive linear maps transforming quantum operations into quantum operations (see Refs. 1,2 for the physical motivation of linearity and complete positivity). A quantum supermap is *deterministic* if it transforms quantum channels (i.e. unital completely positive maps[18]) into quantum channels. The following dilation theorem can be proved for deterministic supermaps in finite dimension:[1,2] denoting by $\mathcal{L}(\mathcal{H})$ and $\mathcal{L}(\mathcal{K})$ the C^*-algebras of the linear operators on the finite dimensional Hilbert spaces $\mathcal{H}$ and $\mathcal{K}$, respectively, and writing CP $(\mathcal{L}(\mathcal{H}), \mathcal{L}(\mathcal{K}))$ for the set of completely positive maps sending $\mathcal{L}(\mathcal{H})$ into $\mathcal{L}(\mathcal{K})$, we have that any deterministic supermap S transforming quantum operations in

CP $(\mathcal{L}(\mathcal{H}_1), \mathcal{L}(\mathcal{K}_1))$ to quantum operations in CP $(\mathcal{L}(\mathcal{H}_2), \mathcal{L}(\mathcal{K}_2))$ has the following form:

$$[\mathsf{S}(\mathcal{E})](A) = V_1^* \left[(\mathcal{E} \otimes \mathcal{I}_{\mathcal{V}_1})(V_2^*(A \otimes I_{\mathcal{V}_2})V_2) \right] V_1 \quad \forall A \in \mathcal{L}(\mathcal{H}_2) \tag{1}$$

for all $\mathcal{E} \in$ CP $(\mathcal{L}(\mathcal{H}_1), \mathcal{L}(\mathcal{K}_1))$, where $\mathcal{V}_1$ and $\mathcal{V}_2$ are two ancillary separable Hilbert spaces, $V_1 : \mathcal{K}_2 \to \mathcal{K}_1 \otimes \mathcal{V}_1$ and $V_2 : \mathcal{H}_1 \otimes \mathcal{V}_1 \to \mathcal{H}_2 \otimes \mathcal{V}_2$ are isometries, $\mathcal{I}_{\mathcal{V}_1}$ is the identity channel on $\mathcal{L}(\mathcal{V}_1)$ and $I_{\mathcal{V}_2}$ is the identity operator on $\mathcal{V}_2$. In the Schrödinger (or predual) picture, eq. (1) can be rewritten

$$[\mathsf{S}(\mathcal{E})]_*(\rho) = \mathrm{tr}_{\mathcal{V}_2} \left[V_2(\mathcal{E}_* \otimes \mathcal{I}_{\mathcal{V}_1})(V_1 \rho V_1^*) V_2^* \right] \quad \text{for all states } \rho \text{ on } \mathcal{K}_2 ,$$

and thus shows that the most general way to transform a quantum operation consists in

(1) applying an invertible transformation (corresponding to the isometry V_1), which transforms the system $\mathcal{K}_2$ into the composite system $\mathcal{K}_1 \otimes \mathcal{V}_1$;
(2) using the input device $\mathcal{E}_*$ on system $\mathcal{K}_1$, thus transforming it into system $\mathcal{H}_1$, while doing nothing on $\mathcal{V}_1$;
(3) applying an invertible transformation (corresponding to the isometry V_2), which transforms the composite system $\mathcal{H}_1 \otimes \mathcal{V}_1$ into the composite system $\mathcal{H}_2 \otimes \mathcal{V}_2$;
(4) discarding system $\mathcal{V}_2$ (mathematically, taking the partial trace over $\mathcal{V}_2$).

In this paper we will first present an extension of eq. (1) to the case of quantum supermaps acting on quantum operations with generic von Neumann algebras in the input, in particular removing all requirements of finite dimensionality. It will turn out that the definition of a suitable notion of *normality* of supermaps is the key point in order to extend the main theorems in the infinite dimensional case.[19,20] Then, as a second step we will state a Radon-Nikodym theorem for probabilistic supermaps, namely supermaps that are dominated by deterministic supermaps. The class of probabilistic supermaps is particularly interesting for physical applications, as such maps naturally appear in the description of quantum circuits that are designed to test properties of physical devices.[1,2,21] Higher-order quantum measurements are indeed described by *quantum superinstruments*, which are the generalization of the quantum instruments of Davies and Lewis.[22] The third main result exposed in the paper will then be a dilation theorem for quantum superinstruments, in analogy with Ozawa's dilation theorem for ordinary instruments.[23]

The present paper is intended as a review of Ref. 19 and our recent still unpublished work Ref. 20. As such, it contains a survey of the main results and applications, but defers the interested reader to Refs. 19 and 20 for a more detailed and technical exposition and for the complete proofs of the statements. The material of the paper is organized as follows. In Section 2 we fix the elementary definitions and notations, and state or recall some basic facts needed in the rest of the paper. In particular, in Section 2.1 we extend the notion of increasing sequences from positive operators to normal completely positive maps, while Section 2.2 contains some elementary results about the tensor product of weak*-continuous maps. In Section 3 we define normal completely positive supermaps and provide some examples. In Section 4 we state and comment the two main results of the paper, i.e our dilation Theorem 4.1 for deterministic supermaps and its extension to probabilistic supermaps contained in Theorem 4.2 (Radon-Nikodym theorem for probabilistic supermaps). As an application of Theorem 4.1, in Section 5 we show that every deterministic supermap transforming measurements into quantum operations can be realized by connecting devices in a quantum circuit. We then define quantum superinstruments in Section 6 and state a dilation theorem for them which is the generalization of classical Ozawa's result for ordinary instruments (see in particular Proposition 4.2 in Ref. 23). Finally, in Section 7 we apply the dilation theorem for quantum superinstruments in order to show how every abstract superinstrument describing a measurement on a quantum measuring device can be realized in a circuit.

2. Notations and preliminary results

In this paper, we will always mean by *Hilbert space* a complex and separable Hilbert space, with scalar product $\langle \cdot, \cdot \rangle$ linear in the second entry. If $\mathcal{H}$, $\mathcal{K}$ are Hilbert spaces, we denote by $\mathcal{L}(\mathcal{H}, \mathcal{K})$ the Banach space of bounded linear operators from $\mathcal{H}$ to $\mathcal{K}$ endowed with the uniform norm $\|\cdot\|_\infty$. If $\mathcal{H} = \mathcal{K}$, the shortened notation $\mathcal{L}(\mathcal{H}) := \mathcal{L}(\mathcal{H}, \mathcal{H})$ will be used, and $I_\mathcal{H}$ will be the identity operator in $\mathcal{L}(\mathcal{H})$. The linear space $\mathcal{L}(\mathcal{H})$ is ordered in the usual way. We denote by $\leq$ the order relation in $\mathcal{L}(\mathcal{H})$, and by $\mathcal{L}(\mathcal{H})_+$ the cone of positive operators.

Following Ref. 24 (see Definition 3.2 p. 72), by *von Neumann algebra* we mean a *-subalgebra $\mathcal{M} \subset \mathcal{L}(\mathcal{H})$ such that $\mathcal{M} = (\mathcal{M}')'$, where $\mathcal{M}'$ denotes the commutant of $\mathcal{M}$ in $\mathcal{L}(\mathcal{H})$. Note that, since all Hilbert spaces considered in the paper are separable, the von Neumann algebras considered here are those that are sometimes called *separable* in the literature. When $\mathcal{M}$ is regarded as an abstract von Neumann algebra (i.e. without reference to the representing Hilbert space $\mathcal{H}$), we will write its identity element $I_\mathcal{M}$ instead of $I_\mathcal{H}$. As usual, we define $\mathcal{M}_+ := \mathcal{M} \cap \mathcal{L}(\mathcal{H})_+$. The identity map on $\mathcal{M}$

will be denoted by $\mathcal{I}_{\mathcal{M}}$, and, when $\mathcal{M} \equiv \mathcal{L}(\mathcal{H})$, the abbreviated notation $\mathcal{I}_{\mathcal{H}} := \mathcal{I}_{\mathcal{L}(\mathcal{H})}$ will be used.

The algebraic tensor product of linear spaces U, V will be written $U \hat{\otimes} V$, while the notation $\mathcal{H} \otimes \mathcal{K}$ will be reserved to denote the Hilbert space tensor product of the Hilbert spaces $\mathcal{H}$ and $\mathcal{K}$. The inclusion $\mathcal{H} \hat{\otimes} \mathcal{K} \subset \mathcal{H} \otimes \mathcal{K}$ holds, and it is actually an equality iff $\mathcal{H}$ or $\mathcal{K}$ is finite dimensional.

If $A \in \mathcal{L}(\mathcal{H})$ and $B \in \mathcal{L}(\mathcal{K})$, their tensor product $A \otimes B$, which is well defined as a linear map on $\mathcal{H} \hat{\otimes} \mathcal{K}$, uniquely extends to a bounded operator $A \otimes B \in \mathcal{L}(\mathcal{H} \otimes \mathcal{K})$ in the usual way (see e.g. p. 183 in Ref. 24). Thus, the algebraic tensor product $\mathcal{L}(\mathcal{H}) \hat{\otimes} \mathcal{L}(\mathcal{K})$ can be regarded as a linear subspace of $\mathcal{L}(\mathcal{H} \otimes \mathcal{K})$. Even in this case, the equality $\mathcal{L}(\mathcal{H}) \hat{\otimes} \mathcal{L}(\mathcal{K}) = \mathcal{L}(\mathcal{H} \otimes \mathcal{K})$ holds iff $\mathcal{H}$ or $\mathcal{K}$ is finite dimensional. More generally, let $\mathcal{M} \subset \mathcal{L}(\mathcal{H})$ and $\mathcal{N} \subset \mathcal{L}(\mathcal{K})$ be two von Neumann algebras. Then, $\mathcal{M} \hat{\otimes} \mathcal{N}$ is a linear subspace of $\mathcal{L}(\mathcal{H} \otimes \mathcal{K})$. Its weak*-closure is the von Neumann algebra $\mathcal{M} \bar{\otimes} \mathcal{N} \subset \mathcal{L}(\mathcal{H} \otimes \mathcal{K})$ (see Definition 1.3 p. 183 in Ref. 24). Clearly, $\mathcal{M} \hat{\otimes} \mathcal{N} = \mathcal{M} \bar{\otimes} \mathcal{N}$ iff $\mathcal{M}$ or $\mathcal{N}$ is finite dimensional. It is a standard fact that $\mathcal{L}(\mathcal{H}) \bar{\otimes} \mathcal{L}(\mathcal{K}) = \mathcal{L}(\mathcal{H} \otimes \mathcal{K})$ (see eq. 10, p. 185 in Ref. 24).

We denote by $M_n(\mathbb{C})$ the linear space of square $n \times n$ complex matrices, which we identify as usual with the space $\mathcal{L}(\mathbb{C}^n)$. If $\mathcal{M}$ is a von Neumann algebra, we write $\mathcal{M}^{(n)} := M_n(\mathbb{C}) \bar{\otimes} \mathcal{M}$, which is a von Neumann algebra contained in $\mathcal{L}(\mathbb{C}^n \otimes \mathcal{H})$. As remarked above, $\mathcal{M}^{(n)}$ coincides with the algebraic tensor product $M_n(\mathbb{C}) \hat{\otimes} \mathcal{M}$. If $\mathcal{E} : M_m(\mathbb{C}) \to M_n(\mathbb{C})$ and $\mathcal{F} : \mathcal{M} \to \mathcal{N}$ are linear operators, we then see that their algebraic tensor product can be regarded as a linear map $\mathcal{E} \otimes \mathcal{F} : \mathcal{M}^{(m)} \to \mathcal{N}^{(n)}$. Since both $\mathcal{M}^{(m)}$ and $\mathcal{N}^{(n)}$ are von Neumann algebras, it makes sense to speak about positivity and boundedness of $\mathcal{E} \otimes \mathcal{F}$. This fact is at the heart of the following two very well known definitions. In them, we use $\mathcal{I}_n$ to denote the identity map on $M_n(\mathbb{C})$, i.e. $\mathcal{I}_n := \mathcal{I}_{M_n(\mathbb{C})}$.

Definition 2.1. Let $\mathcal{M}$, $\mathcal{N}$ be two von Neumann algebras. Then a linear map $\mathcal{E} : \mathcal{M} \to \mathcal{N}$ is

- *completely positive (CP)* if the linear map $\mathcal{I}_n \otimes \mathcal{E}$ is positive, i.e. maps $\mathcal{M}^{(n)}{}_+$ into $\mathcal{N}^{(n)}{}_+$, for all $n \in \mathbb{N}$;
- *completely bounded (CB)* if there exists $C > 0$ such that, for all $n \in \mathbb{N}$,

$$\|(\mathcal{I}_n \otimes \mathcal{E})(\tilde{A})\|_\infty \leq C \|\tilde{A}\|_\infty \quad \forall \tilde{A} \in \mathcal{M}^{(n)},$$

i.e. if the linear map $\mathcal{I}_n \otimes \mathcal{E}$ is bounded from the Banach space $\mathcal{M}^{(n)}$ into the Banach space $\mathcal{N}^{(n)}$ for all $n \in \mathbb{N}$, and the uniform norms of all the maps $\{\mathcal{I}_n \otimes \mathcal{E}\}_{n \in \mathbb{N}}$ are majorized by a constant independent of n.

We recall that a positive linear map $\mathcal{E} : \mathcal{M} \to \mathcal{N}$ is *normal* if it preserves the limits of increasing and bounded sequences, i.e. $\mathcal{E}(A_n) \uparrow \mathcal{E}(A)$ in $\mathcal{N}$ for all increasing sequences $\{A_n\}_{n\in\mathbb{N}}$ and A in $\mathcal{M}_+$ such that $A_n \uparrow A$ (as usual, the notation $A_n \uparrow A$ means that A is the *lower upper bound* of the sequence $\{A_n\}_{n\in\mathbb{N}}$ in $\mathcal{M}$, see e.g. Lemma 1.7.4 in Ref. 25). It is a standard fact that a positive linear map $\mathcal{E} : \mathcal{M} \to \mathcal{N}$ is normal if and only if it is weak*-continuous (Theorem 1.13.2 in Ref. 25).

We introduce the following notations

- CB $(\mathcal{M}, \mathcal{N})$ is the linear space of completely bounded *and weak*-continuous* maps from $\mathcal{M}$ to $\mathcal{N}$;
- CP $(\mathcal{M}, \mathcal{N})$ is the set of *normal* completely positive maps from $\mathcal{M}$ to $\mathcal{N}$;
- CP$_1$ $(\mathcal{M}, \mathcal{N})$ is the set of *quantum channels* from $\mathcal{M}$ to $\mathcal{N}$, i.e. the subset of elements $\mathcal{E} \in$ CP $(\mathcal{M}, \mathcal{N})$ such that $\mathcal{E}(I_\mathcal{M}) = I_\mathcal{N}$.

Remark 2.1. Suppose $\mathcal{M} \subset M_n(\mathbb{C})$ and $\mathcal{N} \subset M_n(\mathbb{C})$. Then the set CB $(\mathcal{M}, \mathcal{N})$ coincides with the space of all linear maps from $\mathcal{M}$ to $\mathcal{N}$ (see e.g. Exercise 3.11 in Ref. 26).

It is well known that each CP map is CB (Proposition 3.6 in Ref. 26) and that each CB map is in the the linear span of four CP maps (Theorem 8.5 in Ref. 26). This fact still holds true when one restricts to weak*-continuous maps, as the next theorem shows (see also Ref. 27 and Theorem 2 in Ref. 19 for the particular case $\mathcal{M} = \mathcal{L}(\mathcal{H})$ and $\mathcal{N} = \mathcal{L}(\mathcal{K})$).

Theorem 2.1. *The inclusion* CP $(\mathcal{M}, \mathcal{N}) \subset$ CB $(\mathcal{M}, \mathcal{N})$ *holds, and* CP $(\mathcal{M}, \mathcal{N})$ *is a cone in the linear space* CB $(\mathcal{M}, \mathcal{N})$. *For* $\mathcal{N} \equiv \mathcal{L}(\mathcal{K})$, *the linear space spanned by* CP $(\mathcal{M}, \mathcal{L}(\mathcal{K}))$ *coincides with* CB $(\mathcal{M}, \mathcal{L}(\mathcal{K}))$. *More precisely, if* $\mathcal{E} \in$ CB $(\mathcal{M}, \mathcal{L}(\mathcal{K}))$, *then there exists four maps* $\mathcal{E}_k \in$ CP $(\mathcal{M}, \mathcal{L}(\mathcal{K}))$ *$(k = 0, 1, 2, 3)$ such that* $\mathcal{E} = \sum_{k=0}^{3} i^k \mathcal{E}_k$.

The cone CP $(\mathcal{M}, \mathcal{N})$ induces a linear ordering in the space CB $(\mathcal{M}, \mathcal{N})$, that we will denote by $\preceq$. Namely, given two maps $\mathcal{E}, \mathcal{F} \in$ CB $(\mathcal{M}, \mathcal{N})$, we will write $\mathcal{E} \preceq \mathcal{F}$ whenever $\mathcal{F} - \mathcal{E} \in$ CP $(\mathcal{M}, \mathcal{N})$.

Two of the main features of CB weak*-continuous maps which we will need in the rest of the paper are the following:

- a notion of limit can be defined for a particular class of sequences in CB $(\mathcal{M}, \mathcal{N})$, which is the analogue of the lower upper bound for increasing bounded sequences of operators;

- if $\mathcal{M}_1$, $\mathcal{M}_2$, $\mathcal{N}_1$, $\mathcal{N}_2$ are von Neumann algebras, the maps in $\mathrm{CB}\,(\mathcal{M}_1,\mathcal{N}_1)$ and $\mathrm{CB}\,(\mathcal{M}_2,\mathcal{N}_2)$ can be tensored in order to obtain elements of $\mathrm{CB}\,(\mathcal{M}_1\bar{\otimes}\mathcal{M}_2,\mathcal{N}_1\bar{\otimes}\mathcal{N}_2)$.

As these concepts are the two main ingredients in our definition of supermaps and in the proof of a dilation theorem for them, we devote the next two sections to their explanation.

2.1. *Increasing sequences of normal CP maps*

We now introduce two definitions for sequences of maps in $\mathrm{CP}\,(\mathcal{M},\mathcal{N})$ that are analogous to the notion of increasing and bounded sequences of operators. We say that a sequence $\{\mathcal{E}_n\}_{n\in\mathbb{N}}$ of elements in $\mathrm{CP}\,(\mathcal{M},\mathcal{N})$ is

- *CP-increasing* if $\mathcal{E}_m \preceq \mathcal{E}_n$ whenever $m \leq n$,
- *CP-bounded* if there exists a map $\mathcal{E} \in \mathrm{CP}\,(\mathcal{M},\mathcal{N})$ such that $\mathcal{E}_n \preceq \mathcal{E}$ for all $n \in \mathbb{N}$.

The following result now shows that the notion of lower upper bound can be extended to CP-incresing and CP-bounded sequences in $\mathrm{CP}\,(\mathcal{M},\mathcal{N})$.

Proposition 2.1. *If* $\{\mathcal{E}_n\}_{n\in\mathbb{N}}$ *is a sequence in* $\mathrm{CP}\,(\mathcal{M},\mathcal{N})$ *which is CP-increasing and CP-bounded, then there exists a unique* $\mathcal{E} \in \mathrm{CP}\,(\mathcal{M},\mathcal{N})$ *such that*

$$\mathrm{wk}^*\text{-}\lim_n \mathcal{E}_n(A) = \mathcal{E}(A) \quad \forall A \in \mathcal{M}\,.$$

$\mathcal{E}$ *has the following property:* $\mathcal{E}_n \preceq \mathcal{E}$ *for all* $n \in \mathbb{N}$, *and, if* $\mathcal{E}' \in \mathrm{CP}\,(\mathcal{M},\mathcal{N})$ *is such that* $\mathcal{E}_n \preceq \mathcal{E}'$ *for all* $n \in \mathbb{N}$, *then* $\mathcal{E} \preceq \mathcal{E}'$.

If $\{\mathcal{E}_n\}_{n\in\mathbb{N}}$ and $\mathcal{E}$ are as in the statement of the above proposition, then we write $\mathcal{E}_n \Uparrow \mathcal{E}$.

We now give the most useful application of the previous result. First, note that the simplest example of maps in $\mathrm{CB}\,(\mathcal{M},\mathcal{L}(\mathcal{K}))$ is constructed in the following way. Suppose that $\mathcal{M}$ is a von Neumann algebra contained in $\mathcal{L}(\mathcal{H})$. For $E \in \mathcal{L}(\mathcal{H},\mathcal{K})$, $F \in \mathcal{L}(\mathcal{K},\mathcal{H})$, denote by $E \odot_\mathcal{M} F$ the linear operator

$$E \odot_\mathcal{M} F : \mathcal{M} \to \mathcal{L}(\mathcal{K})\,, \qquad (E \odot_\mathcal{M} F)(A) = EAF \quad \forall A \in \mathcal{M}\,.$$

Then it is easy to show that $E \odot_\mathcal{M} F \in \mathrm{CB}\,(\mathcal{M},\mathcal{L}(\mathcal{K}))$, and, if $E = F^*$, actually $F^* \odot_\mathcal{M} F \in \mathrm{CP}\,(\mathcal{M},\mathcal{L}(\mathcal{K}))$. The importance of the elementary maps $E \odot_\mathcal{M} F$'s is made clear by the next theorem.

Theorem 2.2 (Kraus theorem). *Suppose* $\mathcal{M} \subset \mathcal{L}(\mathcal{H})$ *is a von Neumann algebra. Then,* $\mathcal{E} \in \mathrm{CP}\,(\mathcal{M}, \mathcal{L}(\mathcal{K}))$ *if and only if there exists a finite or countable set* $I \subset \mathbb{N}$ *and a sequence* $\{E_i\}_{i \in I}$ *of elements in* $\mathcal{L}(\mathcal{K}, \mathcal{H})$ *such that the sequence of partial sums* $\{\sum_{i \leq n} E_i^* \odot_\mathcal{M} E_i\}_{n \in \mathbb{N}}$ *converges to* $\mathcal{E}$ *in* $\mathrm{CP}\,(\mathcal{M}, \mathcal{L}(\mathcal{K}))$ *in the sense of Proposition 2.1.*

If $\mathcal{E}$ and $\{E_i\}_{i \in I}$ are as in item (2) of the above theorem, the expression $\sum_{i \in I} E_i^* \odot_\mathcal{M} E_i$ is the *Kraus form* of $\mathcal{E}$.

Kraus theorem is very important, as it shows that every map in $\mathrm{CB}\,(\mathcal{M}, \mathcal{L}(\mathcal{K}))$ can be decomposed into a (possibly infinite) sum of elementary maps $E_i \odot_\mathcal{M} F_i$. Indeed, by Theorem 2.1 we can choose four elements $\mathcal{E}_k \in \mathrm{CP}\,(\mathcal{M}, \mathcal{L}(\mathcal{K}))$ ($k = 0, 1, 2, 3$) such that $\mathcal{E} = \sum_{k=0}^3 i^k \mathcal{E}_k$, and by Theorem 2.2 each $\mathcal{E}_k$ can be written in the Kraus form $\mathcal{E}_k = \sum_{i \in I_k} E_i^{(k)*} \odot_\mathcal{M} E_i^{(k)}$. It is clear, however, that such decomposition is not unique even if $\mathcal{E} \in \mathrm{CP}\,(\mathcal{M}, \mathcal{L}(\mathcal{K}))$ itself.

2.2. Tensor product of weak*-continuous CB maps

If $\mathcal{E} : \mathcal{L}(\mathcal{H}_1) \to \mathcal{L}(\mathcal{K}_1)$ and $\mathcal{F} : \mathcal{L}(\mathcal{H}_2) \to \mathcal{L}(\mathcal{K}_2)$ are linear *bounded* maps, their tensor product $\mathcal{E} \otimes \mathcal{F}$ is well defined as a linear map $\mathcal{L}(\mathcal{H}_1) \hat{\otimes} \mathcal{L}(\mathcal{H}_2) \to \mathcal{L}(\mathcal{K}_1) \hat{\otimes} \mathcal{L}(\mathcal{K}_2)$. However, unless $\mathcal{H}_1$ and $\mathcal{K}_1$, or alternatively $\mathcal{H}_2$ and $\mathcal{K}_2$, are finite dimensional, in general one can not extend $\mathcal{E} \otimes \mathcal{F}$ to an operator $\mathcal{L}(\mathcal{H}_1 \otimes \mathcal{H}_2) \to \mathcal{L}(\mathcal{K}_1 \otimes \mathcal{K}_2)$. Weak*-continuous CB maps constitute an important exception to this obstruction, as it is made clear by the following proposition.

Proposition 2.2. *Let* $\mathcal{M}_1, \mathcal{M}_2, \mathcal{N}_1, \mathcal{N}_2$ *be von Neumann algebras. Given two maps* $\mathcal{E} \in \mathrm{CB}\,(\mathcal{M}_1, \mathcal{N}_1)$ *and* $\mathcal{F} \in \mathrm{CB}\,(\mathcal{M}_2, \mathcal{N}_2)$, *there exists a unique map* $\mathcal{E} \otimes \mathcal{F} \in \mathrm{CB}\,(\mathcal{M}_1 \bar{\otimes} \mathcal{M}_2, \mathcal{N}_1 \bar{\otimes} \mathcal{N}_2)$ *such that*

$$(\mathcal{E} \otimes \mathcal{F})(A \otimes B) = \mathcal{E}(A) \otimes \mathcal{F}(B) \quad \forall A \in \mathcal{M}_1, B \in \mathcal{M}_2.$$

If $\mathcal{E}$ *and* $\mathcal{F}$ *are CP, then* $\mathcal{E} \otimes \mathcal{F} \in \mathrm{CP}\,(\mathcal{M}_1 \bar{\otimes} \mathcal{M}_2, \mathcal{N}_1 \bar{\otimes} \mathcal{N}_2)$.

Note that, if $\mathcal{M}_1$ and $\mathcal{N}_1$ (or, equivalently, $\mathcal{M}_2$ and $\mathcal{N}_2$) are finite dimensional, we had already a notion of tensor product at our disposal, i.e. the algebraic tensor product. Indeed, when $\mathcal{M}_1 = M_m(\mathbb{C})$ and $\mathcal{N}_1 = M_n(\mathbb{C})$, the product $\mathcal{E} \otimes \mathcal{F}$ defined in Proposition 2.2 clearly coincides with the algebraic product that we already encountered in the definition of CB and CP maps (see Definition 2.1, where $\mathcal{E} = \mathcal{I}_n$). Moreover, in this case we actually have the equality

$$\mathrm{CB}\,(M_m(\mathbb{C}), M_n(\mathbb{C})) \,\hat{\otimes}\, \mathrm{CB}\,(\mathcal{M}, \mathcal{N}) = \mathrm{CB}\left(\mathcal{M}^{(m)}, \mathcal{N}^{(n)}\right). \tag{2}$$

It is easy to check that the tensor product defined above preserves ordering (i.e. $\mathcal{E}_1 \preceq \mathcal{E}_2$ and $\mathcal{F}_1 \preceq \mathcal{F}_2$ imply $\mathcal{E}_1 \otimes \mathcal{F}_1 \preceq \mathcal{E}_2 \otimes \mathcal{F}_2$) and lower upper bounds (i.e., if $\mathcal{E}_\lambda$, $\mathcal{E}$ and $\mathcal{F}$ are normal CP maps such that $\mathcal{E}_\lambda \Uparrow \mathcal{E}$, then $\mathcal{E}_\lambda \otimes \mathcal{F} \Uparrow \mathcal{E} \otimes \mathcal{F}$).

3. Quantum supermaps

In this section, we introduce the central objects in our study, i.e. the set of linear maps $\mathsf{S} : \mathrm{CB}\,(\mathcal{M}_1, \mathcal{N}_1) \to \mathrm{CB}\,(\mathcal{M}_2, \mathcal{N}_2)$ which mathematically describe the physically admissible transformations of quantum channels. Before giving the precise definition, we need to fix the following terminology.

Definition 3.1. Suppose $\mathcal{M}_1$, $\mathcal{M}_2$, $\mathcal{N}_1$, $\mathcal{N}_2$ are von Neumann algebras. A linear map $\mathsf{S} : \mathrm{CB}\,(\mathcal{M}_1, \mathcal{N}_1) \to \mathrm{CB}\,(\mathcal{M}_2, \mathcal{N}_2)$ is

- *positive* if $\mathsf{S}(\mathcal{E}) \succeq 0$ for all $\mathcal{E} \succeq 0$;
- *completely positive (CP)* if the map

$$\mathsf{I}_n \otimes \mathsf{S} : \mathrm{CB}\left(\mathcal{M}_1^{(n)}, \mathcal{N}_1^{(n)}\right) \to \mathrm{CB}\left(\mathcal{M}_2^{(n)}, \mathcal{N}_2^{(n)}\right)$$

is positive for every $n \in \mathbb{N}$, where I_n is the identity map on the space $\mathrm{CB}\,(M_n(\mathbb{C}), M_n(\mathbb{C}))$;
- *normal* if $\mathsf{S}(\mathcal{E}_n) \Uparrow \mathsf{S}(\mathcal{E})$ for all sequences $\{\mathcal{E}_n\}_{n \in \mathbb{N}}$ in $\mathrm{CP}\,(\mathcal{M}_1, \mathcal{N}_1)$ such that $\mathcal{E}_n \Uparrow \mathcal{E}$.

Note that in the above definition of complete positivity we used the identification $\mathrm{CB}\left(\mathcal{M}^{(n)}, \mathcal{N}^{(n)}\right) = \mathrm{CB}\,(M_n(\mathbb{C}), M_n(\mathbb{C})) \,\hat{\otimes}\, \mathrm{CB}\,(\mathcal{M}, \mathcal{N})$ of eq. (2).

We are now in position to define quantum supermaps.

Definition 3.2. A *quantum supermap* (or simply, *supermap*) is a normal completely positive linear map $\mathsf{S} : \mathrm{CB}\,(\mathcal{M}_1, \mathcal{N}_1) \to \mathrm{CB}\,(\mathcal{M}_2, \mathcal{N}_2)$.

The convex set of quantum supermaps from $\mathrm{CB}\,(\mathcal{M}_1, \mathcal{N}_1)$ to $\mathrm{CB}\,(\mathcal{M}_2, \mathcal{N}_2)$ will be denoted by $\mathrm{SCP}\,(\mathcal{M}_1, \mathcal{N}_1; \mathcal{M}_2, \mathcal{N}_2)$. A partial order $\ll$ can be introduced in it as follows: given two maps $\mathsf{S}_1, \mathsf{S}_2 \in \mathrm{SCP}\,(\mathcal{M}_1, \mathcal{N}_1; \mathcal{M}_2, \mathcal{N}_2)$, we write $\mathsf{S}_1 \ll \mathsf{S}_2$ if $\mathsf{S}_2 - \mathsf{S}_1 \in \mathrm{SCP}\,(\mathcal{M}_1, \mathcal{N}_1; \mathcal{M}_2, \mathcal{N}_2)$.

We now specialize the definition of quantum supermaps to the following two main cases of interest.

Definition 3.3. A quantum supermap $\mathsf{S} \in \mathrm{SCP}\,(\mathcal{M}_1, \mathcal{N}_1; \mathcal{M}_2, \mathcal{N}_2)$ is

- *deterministic* if it preserves the set of quantum channels, that is, if $\mathsf{S}(\mathcal{E}) \in \mathrm{CP}_1\,(\mathcal{M}_2, \mathcal{N}_2)$ for all $\mathcal{E} \in \mathrm{CP}_1\,(\mathcal{M}_1, \mathcal{N}_1)$;

- *probabilistic* if a deterministic supermap $\mathsf{T} \in \mathrm{SCP}\,(\mathcal{M}_1, \mathcal{N}_1; \mathcal{M}_2, \mathcal{N}_2)$ exists, such that $\mathsf{S} \ll \mathsf{T}$.

Deterministic supermaps are clearly probabilistic. The subset of deterministic supermaps in $\mathrm{SCP}\,(\mathcal{M}_1, \mathcal{N}_1; \mathcal{M}_2, \mathcal{N}_2)$ will be labeled by $\mathrm{SCP}_1\,(\mathcal{M}_1, \mathcal{N}_1; \mathcal{M}_2, \mathcal{N}_2)$.

Obviously, the composition of two quantum supermaps is a supermap: for every $\mathsf{S}_1 \in \mathrm{SCP}\,(\mathcal{M}_1, \mathcal{N}_1; \mathcal{M}_2, \mathcal{N}_2)$ and $\mathsf{S}_2 \in \mathrm{SCP}\,(\mathcal{M}_2, \mathcal{N}_2; \mathcal{M}_3, \mathcal{N}_3)$, we have $\mathsf{S}_2 \mathsf{S}_1 \in \mathrm{SCP}\,(\mathcal{M}_1, \mathcal{N}_1; \mathcal{M}_3, \mathcal{N}_3)$. Similarly, the composition of two probabilistic [resp. deterministic] supermaps is a probabilistic [resp. deterministic] supermap.

We now introduce two examples of supermaps which will play a very important role in the next section.

Example 3.1 (Concatenation). *Given two maps* $\mathcal{A} \in \mathrm{CP}\,(\mathcal{N}_1, \mathcal{N}_2)$ *and* $\mathcal{B} \in \mathrm{CP}\,(\mathcal{M}_2, \mathcal{M}_1)$, *define the linear map*

$$\mathsf{C}_{\mathcal{A},\mathcal{B}} : \mathrm{CB}\,(\mathcal{M}_1, \mathcal{N}_1) \to \mathrm{CB}\,(\mathcal{M}_2, \mathcal{N}_2) \,,$$
$$\mathsf{C}_{\mathcal{A},\mathcal{B}}(\mathcal{E}) = \mathcal{A}\mathcal{E}\mathcal{B} \quad \forall \mathcal{E} \in \mathrm{CB}\,(\mathcal{M}_1, \mathcal{N}_1) \,.$$

Then $\mathsf{C}_{\mathcal{A},\mathcal{B}} \in \mathrm{SCP}\,(\mathcal{M}_1, \mathcal{N}_1; \mathcal{M}_2, \mathcal{N}_2)$. *Moreover, if* $\mathcal{A}$ *and* $\mathcal{B}$ *are quantum channels, then* $\mathsf{C}_{\mathcal{A},\mathcal{B}}$ *is deterministic.*

Example 3.2 (Amplification). *Suppose* $\mathcal{V}$ *is a Hilbert space, and define the linear map*

$$\Pi_{\mathcal{V}} : \mathrm{CB}\,(\mathcal{M}, \mathcal{N}) \to \mathrm{CB}\,(\mathcal{M}\bar{\otimes}\mathcal{L}(\mathcal{V}), \mathcal{N}\bar{\otimes}\mathcal{L}(\mathcal{V})) \,,$$
$$\Pi_{\mathcal{V}}(\mathcal{E}) = \mathcal{E} \otimes \mathcal{I}_{\mathcal{V}} \quad \forall \mathcal{E} \in \mathrm{CB}\,(\mathcal{M}, \mathcal{N}) \,,$$

where we recall that $\mathcal{I}_{\mathcal{V}} := \mathcal{I}_{\mathcal{L}(\mathcal{V})}$ *(cf. Proposition 2.2 for the definition of the tensor product). Then the map* $\Pi_{\mathcal{V}}$ *is a deterministic supermap, that is,* $\Pi_{\mathcal{V}} \in \mathrm{SCP}_1\,(\mathcal{M}, \mathcal{N}; \mathcal{M}\bar{\otimes}\mathcal{L}(\mathcal{V}), \mathcal{N}\bar{\otimes}\mathcal{L}(\mathcal{V}))$.

The main result in the next two sections is that every deterministic or probabilistic supermap in $\mathrm{SCP}\,(\mathcal{M}_1, \mathcal{L}(\mathcal{K}_1); \mathcal{M}_2, \mathcal{L}(\mathcal{K}_2))$ is the composition of an amplification followed by a concatenation (Theorems 4.1 and 4.2 below).

4. Dilation of deterministic and probabilistic supermaps

Our central result is the following dilation theorem for deterministic supermaps.

Theorem 4.1 (Dilation of deterministic supermaps). *Suppose* $\mathcal{M}_1$, $\mathcal{M}_2$ *are von Neumann algebras. A linear map* $S : \mathrm{CB}\,(\mathcal{M}_1, \mathcal{L}(\mathcal{K}_1)) \to \mathrm{CB}\,(\mathcal{M}_2, \mathcal{L}(\mathcal{K}_2))$ *is a deterministic supermap if and only if there exists a triple* $(\mathcal{V}, V, \mathcal{F})$, *where*

- $\mathcal{V}$ *is a Hilbert space*
- $V : \mathcal{K}_2 \to \mathcal{K}_1 \otimes \mathcal{V}$ *is an isometry*
- $\mathcal{F}$ *is a quantum channel in* $\mathrm{CP}_1\,(\mathcal{M}_2, \mathcal{M}_1 \bar{\otimes} \mathcal{L}(\mathcal{V}))$

such that

$$[S(\mathcal{E})](A) = V^* \left[(\mathcal{E} \otimes \mathcal{I}_{\mathcal{V}}) \mathcal{F}(A) \right] V \quad \forall \mathcal{E} \in \mathrm{CB}\,(\mathcal{M}_1, \mathcal{L}(\mathcal{K}_1))\,,\, A \in \mathcal{M}_2\,. \tag{3}$$

The triple $(\mathcal{V}, V, \mathcal{F})$ *can always be chosen in a way that*

$$\mathcal{V} = \overline{\mathrm{span}}\, \{ (u^* \otimes I_{\mathcal{V}}) V v \mid u \in \mathcal{K}_1\,,\, v \in \mathcal{K}_2 \}\,. \tag{4}$$

We remark that in eq. (4) the adjoint u^* of $u \in \mathcal{K}_1$ is the linear functional $u^* : w \mapsto \langle u, w \rangle$ on $\mathcal{K}_1$.

Note that, if we define the quantum channel $\mathcal{A} := V^* \odot_{\mathcal{L}(\mathcal{K}_1 \otimes \mathcal{V})} V$, then eq. (3) is equivalent to

$$S = C_{\mathcal{A}, \mathcal{F}} \Pi_{\mathcal{V}}\,,$$

where $C_{\mathcal{A}, \mathcal{F}}$ and $\Pi_{\mathcal{V}}$ are the concatenation and amplification supermaps defined in Examples 3.1 and 3.2. In particular, we see that, if a linear map $S : \mathrm{CB}\,(\mathcal{M}_1, \mathcal{L}(\mathcal{K}_1)) \to \mathrm{CB}\,(\mathcal{M}_2, \mathcal{L}(\mathcal{K}_2))$ is defined as in eq. (3), then $S \in \mathrm{SCP}_1\,(\mathcal{M}_1, \mathcal{L}(\mathcal{K}_1); \mathcal{M}_2, \mathcal{L}(\mathcal{K}_2))$ by the composition property of deterministic supermaps. The converse statement is more difficult to be shown, and a sketch of its proof will be provided in the next subsection.

Definition 4.1. If a Hilbert space $\mathcal{V}$, an isometry $V : \mathcal{K}_2 \to \mathcal{K}_1 \otimes \mathcal{V}$, and a quantum channel $\mathcal{F} \in \mathrm{CP}_1\,(\mathcal{M}_2, \mathcal{M}_1 \bar{\otimes} \mathcal{L}(\mathcal{V}))$ are such that eq. (3) holds, then we say that the triple $(\mathcal{V}, V, \mathcal{F})$ is a *dilation* of the supermap S. If also eq. (4) holds, then we say that the dilation $(\mathcal{V}, V, \mathcal{F})$ is *minimal*.

The importance of the minimality property is highlighted by the following fact.

Proposition 4.1. *Let* $(\mathcal{V}, V, \mathcal{F})$ *and* $(\mathcal{V}', V', \mathcal{F}')$ *be two dilations of the deterministic supermap* $S \in \mathrm{SCP}_1\,(\mathcal{M}_1, \mathcal{L}(\mathcal{K}_1); \mathcal{M}_2, \mathcal{L}(\mathcal{K}_2))$. *If* $(\mathcal{V}, V, \mathcal{F})$ *is minimal, then there exists a unique isometry* $W : \mathcal{V} \to \mathcal{V}'$ *such that* $V' = (I_{\mathcal{K}_1} \otimes W) V$ *and* $\mathcal{F}(A) = (I_{\mathcal{M}_1} \otimes W^*) \mathcal{F}'(A)(I_{\mathcal{M}_1} \otimes W)$ *for all* $A \in \mathcal{M}_2$. *Moreover, if also the dilation* $(\mathcal{V}', V', \mathcal{F}')$ *is minimal, then the isometry* W *is actually unitary.*

Remark 4.1. Suppose $\mathcal{M}_1 = \mathcal{L}(\mathcal{H}_1)$ and $\mathcal{M}_2 \subset \mathcal{L}(\mathcal{H}_2)$. In this case, a linear map $\mathsf{S} : \mathrm{CB}\,(\mathcal{L}(\mathcal{H}_1), \mathcal{L}(\mathcal{K}_1)) \to \mathrm{CB}\,(\mathcal{M}_2, \mathcal{L}(\mathcal{K}_2))$ is a deterministic supermap if and only if there exist two separable Hilbert spaces $\mathcal{V}, \mathcal{U}$ and two isometries $V : \mathcal{K}_2 \to \mathcal{K}_1 \otimes \mathcal{V}$, $U : \mathcal{H}_1 \otimes \mathcal{V} \to \mathcal{H}_2 \otimes \mathcal{U}$ such that

$$[\mathsf{S}(\mathcal{E})](A) = V^* \left[(\mathcal{E} \otimes \mathcal{I}_\mathcal{V})(U^*(A \otimes I_\mathcal{U})U) \right] V \tag{5}$$

for all $\mathcal{E} \in \mathrm{CB}\,(\mathcal{L}(\mathcal{H}_1), \mathcal{L}(\mathcal{K}_1))$ and $A \in \mathcal{M}_2$. Indeed, by Stinespring theorem (Theorem 4.3 p. 165 in Ref. 28 and the discussion following it) every quantum channel $\mathcal{F} \in \mathrm{CP}_1\,(\mathcal{M}_2, \mathcal{L}(\mathcal{H}_1) \bar{\otimes} \mathcal{L}(\mathcal{V})) = \mathrm{CP}_1\,(\mathcal{M}_2, \mathcal{L}(\mathcal{H}_1 \otimes \mathcal{V}))$ can be written as

$$\mathcal{F}(A) = U^*(A \otimes I_\mathcal{U})U \quad \forall A \in \mathcal{M}_2$$

for some separable Hilbert space $\mathcal{U}$ and some isometry $U : \mathcal{H}_1 \otimes \mathcal{V} \to \mathcal{H}_2 \otimes \mathcal{U}$. Eq. (5) then follows by eq. (3). Note that in this way we recover Theorem 5 of Ref. 19 as a particular case of Theorem 4.1 above.

Remark 4.2. As anticipated in the Introduction, eq. (3) is the desired generalization of the analogous finite dimensional result in Refs. 1,2. The physical interpretation of the dilation of deterministic supermaps is clear in the Schrödinger picture: indeed, turning eq. (3) into its predual, we obtain

$$[\mathsf{S}(\mathcal{E})]_*(\rho) = \mathcal{F}_* \left[(\mathcal{E} \otimes \mathcal{I}_\mathcal{V})_*(V \rho V^*) \right]$$

for all ρ in the space $\mathcal{T}(\mathcal{K}_2)$ of trace class operators on $\mathcal{K}_2$ and $\mathcal{E} \in \mathrm{CB}\,(\mathcal{M}_1, \mathcal{L}(\mathcal{K}_1))$. If $\mathcal{M}_i = \mathcal{L}(\mathcal{H}_i)$, take the Stinespring dilation $\mathcal{F}(A) = U^*(A \otimes I_\mathcal{U})U$ of $\mathcal{F}$. The last equation then rewrites

$$[\mathsf{S}(\mathcal{E})]_*(\rho) = \mathrm{tr}_\mathcal{U} \left\{ U \left[(\mathcal{E} \otimes \mathcal{I}_\mathcal{V})_*(V \rho V^*) \right] U^* \right\}$$

where $\mathrm{tr}_\mathcal{U}$ denotes the partial trace over $\mathcal{U}$. If ρ is a quantum state (i.e. $\rho \geq 0$ and $\mathrm{tr}\,(\rho) = 1$), this means that the quantum system with Hilbert space $\mathcal{K}_2$ first undergoes the invertible evolution V, then the dilated quantum channel $(\mathcal{E} \otimes \mathcal{I}_\mathcal{V})_*$, and finally the invertible evolution U, after which the ancillary system with Hilbert space $\mathcal{U}$ is discarded. It is interesting to note that the same kind of sequential composition of invertible evolutions also appears in a very different context: the reconstruction of quantum stochastic processes from correlation kernels.[29–31] That context is very different from the present framework of higher-order maps, and it is a remarkable feature of Theorem 4.1 that any deterministic supermap on the space of quantum operations can be achieved through a two-step sequence of invertible evolutions.

Theorem 4.1 contains as a special case the Stinespring dilation of quantum channels. This fact is illustrated in the following example.

Example 4.1 (Stinespring theorem). *Suppose that $\mathcal{M}_1 = \mathcal{M}_2 = \mathbb{C}$, the trivial von Neumann algebra. In this case we have the identification* $\mathrm{CB}\,(\mathbb{C}, \mathcal{L}(\mathcal{K}_i)) = \mathcal{L}(\mathcal{K}_i)$. *Precisely, the element* $\mathcal{E} \in \mathrm{CB}\,(\mathbb{C}, \mathcal{K}_i)$ *is identified with the operator* $A_{\mathcal{E}} = \mathcal{E}(1) \in \mathcal{L}(\mathcal{K}_i)$. *Moreover, we clearly have* $\mathrm{CP}_1\,(\mathcal{M}_2, \mathcal{M}_1 \bar{\otimes} \mathcal{L}(\mathcal{V})) = \{I_{\mathcal{V}}\}$, *hence eq. (3) reads*

$$[\mathsf{S}(\mathcal{E})](1) = V^*(A_{\mathcal{E}} \otimes I_{\mathcal{V}})V\,,$$

which is just Stinespring dilation for normal CP maps. A linear map $\mathsf{S} : \mathcal{L}(\mathcal{K}_1) \to \mathcal{L}(\mathcal{K}_2)$ *is thus in* $\mathrm{SCP}_1\,(\mathbb{C}, \mathcal{L}(\mathcal{K}_1); \mathbb{C}, \mathcal{L}(\mathcal{K}_2))$ *if and only if it is a unital normal CP map, i.e. a quantum channel.*

The dilation theorem for deterministic supermaps can be generalized to probabilistic supermaps. In this case, the following theorem provides an analog of the Radon-Nikodym theorem for CP maps (compare with Refs. 32, 33, and see also Ref. 34 for the particular case of quantum operations).

Theorem 4.2 (Radon-Nikodym theorem for supermaps). *Suppose* $\mathsf{S} \in \mathrm{SCP}_1\,(\mathcal{M}_1, \mathcal{L}(\mathcal{K}_1); \mathcal{M}_2, \mathcal{L}(\mathcal{K}_2))$ *and let* $(\mathcal{V}, V, \mathcal{F})$ *be its minimal dilation. If* $\mathsf{T} \in \mathrm{SCP}\,(\mathcal{M}_1, \mathcal{L}(\mathcal{K}_1); \mathcal{M}_2, \mathcal{L}(\mathcal{K}_2))$ *is such that* $\mathsf{T} \ll \mathsf{S}$, *then there exists a unique element* $\mathcal{G} \in \mathrm{CP}\,(\mathcal{M}_2, \mathcal{M}_1 \bar{\otimes} \mathcal{L}(\mathcal{V}))$ *with* $\mathcal{G} \preceq \mathcal{F}$ *and such that*

$$[\mathsf{T}(\mathcal{E})]\,(A) = V^*[(\mathcal{E} \otimes \mathcal{I}_{\mathcal{V}})\mathcal{G}(A)]V \quad \forall \mathcal{E} \in \mathrm{CB}\,(\mathcal{M}_1, \mathcal{L}(\mathcal{K}_1))\,, A \in \mathcal{M}_2\,. \quad (6)$$

Definition 4.2. With the notations of Theorem 4.2, the map $\mathcal{G} \in \mathrm{CP}\,(\mathcal{M}_2, \mathcal{M}_1 \bar{\otimes} \mathcal{L}(\mathcal{V}))$ defined by eq. (6) is the *Radon-Nikodym derivative* of the supermap T with respect to S.

4.1. *Sketch of the proof of Theorem 4.1*

Here we provide a sketch of the proof of our central dilation Theorem 4.1. The interested reader is referred to Refs. 19,20 for the details.

In the following, we will restrict ourselves to the simplified case in which the deterministic supermap S belongs to the set $\mathrm{SCP}\,(\mathcal{L}(\mathcal{H}), \mathcal{L}(\mathcal{H}); \mathcal{N}, \mathcal{L}(\mathcal{K}))$, i.e. assume $\mathcal{M}_1 = \mathcal{L}(\mathcal{K}_1)$ in the notations of Theorem 4.1. The proof can be divided into several steps.

(1) Each supermap $\mathsf{S} \in \mathrm{SCP}\,(\mathcal{L}(\mathcal{H}), \mathcal{L}(\mathcal{H}); \mathcal{N}, \mathcal{L}(\mathcal{K}))$ defines a sesquilinear form $\langle \cdot, \cdot \rangle_1$ on the algebraic tensor product $\mathcal{L}(\mathcal{H}) \hat{\otimes} \mathcal{N} \hat{\otimes} \mathcal{K}$ as follows

$$\langle E_1 \otimes A_1 \otimes v_1, E_2 \otimes A_2 \otimes v_2 \rangle_1 := \langle v_1, [\mathsf{S}\,(E_1^* \odot_{\mathcal{M}_1} E_2)]\,(A_1^* A_2)\,v_2 \rangle\,.$$

It is not difficult to show that complete positivity of S implies that the form $\langle \cdot, \cdot \rangle_1$ is positive semidefinite. If $\mathcal{R} \subset \mathcal{L}(\mathcal{H}) \hat{\otimes} \mathcal{N} \hat{\otimes} \mathcal{K}$ is the radical of the form $\langle \cdot, \cdot \rangle_1$, the quotient space $\mathcal{L}(\mathcal{H}) \hat{\otimes} \mathcal{N} \hat{\otimes} \mathcal{K} / \mathcal{R}$ can then be completed to a

Hilbert space, say $\hat{\mathcal{U}}_1$. We denote by $\langle \cdot, \cdot \rangle_1$ the resulting scalar product in $\hat{\mathcal{U}}_1$.

(2) We can also use $\langle \cdot, \cdot \rangle_1$ to introduce a second positive semidefinite sesquilinear form $\langle \cdot, \cdot \rangle_2$, this time defined on the algebraic tensor product $\mathcal{L}(\mathcal{H}) \hat{\otimes} \mathcal{K}$ and given by

$$\langle E_1 \otimes v_1, E_2 \otimes v_2 \rangle_2 := \langle E_1 \otimes I_{\mathcal{N}} \otimes v_1, E_2 \otimes I_{\mathcal{N}} \otimes v_2 \rangle_1 .$$

As before, if $\mathcal{R}' \subset \mathcal{L}(\mathcal{H}) \hat{\otimes} \mathcal{K}$ is the radical of $\langle \cdot, \cdot \rangle_2$, we denote by $\hat{\mathcal{U}}_2$ the Hilbert space completion of the quotient $\mathcal{L}(\mathcal{H}) \hat{\otimes} \mathcal{K} / \mathcal{R}'$, and let $\langle \cdot, \cdot \rangle_2$ be the scalar product extended to $\hat{\mathcal{U}}_2$. It can be proven that the Hilbert space $\hat{\mathcal{U}}_2$ is separable.

(3) Now, we define two linear maps

$$U_1 : \mathcal{L}(\mathcal{H}) \hat{\otimes} \mathcal{K} \to \mathcal{L}(\mathcal{H}) \hat{\otimes} \mathcal{N} \hat{\otimes} \mathcal{K} \qquad U_1(E \otimes v) = E \otimes I_{\mathcal{N}} \otimes v$$
$$U_2 : \mathcal{K} \to \mathcal{L}(\mathcal{H}) \hat{\otimes} \mathcal{K} \qquad U_2 v = I_{\mathcal{H}} \otimes v .$$

It is easy to verify by definitions that U_1 and U_2 extend to isometries $U_1 : \hat{\mathcal{U}}_2 \to \hat{\mathcal{U}}_1$ and $U_2 : \mathcal{K} \to \hat{\mathcal{U}}_2$, respectively.

(4) For all $B \in \mathcal{N}$, we introduce the linear operator $\pi_1(B)$ on $\mathcal{L}(\mathcal{H}) \hat{\otimes} \mathcal{N} \hat{\otimes} \mathcal{K}$, defined by

$$[\pi_1(B)](E \otimes A \otimes v) = E \otimes BA \otimes v$$

for all $E \in \mathcal{L}(\mathcal{H})$, $A \in \mathcal{N}$ and $v \in \mathcal{K}$. Using again the definitions, it is easy to show that π_1 extends to a normal unital $*$-homomorphism of $\mathcal{N}$ into $\mathcal{L}(\hat{\mathcal{U}}_1)$.

(5) For all $F \in \mathcal{L}(\mathcal{H})$, we introduce the linear operator $\pi_2(F)$ on $\mathcal{L}(\mathcal{H}) \hat{\otimes} \mathcal{K}$, defined by

$$[\pi_2(F)](E \otimes v) = FE \otimes v$$

for all $E \in \mathcal{L}(\mathcal{H})$ and $v \in \mathcal{K}$. It can be shown that also π_2 extends to a normal unital $*$-homomorphism of $\mathcal{L}(\mathcal{H})$ into $\mathcal{L}(\hat{\mathcal{U}}_2)$. However, we remark that in this case the proof is more involved than in step (4), and makes essential use of the fact that the supermap S is deterministic (see Lemma 4.3 in Ref. 20) and normal.

(6) By Lemma 2.2 p. 139 in Ref. 3, separability of $\hat{\mathcal{U}}_2$ and normality of π_2 imply that there exists a (separable) Hilbert space $\mathcal{V}$ such that $\hat{\mathcal{U}}_2 = \mathcal{H} \otimes \mathcal{V}$ and $\pi_2(F) = F \otimes I_{\mathcal{V}}$ for all $F \in \mathcal{L}(\mathcal{H})$. Note that, if $E \in \mathcal{L}(\mathcal{H})$, $A \in \mathcal{N}$ and $v \in \mathcal{K}$, then by an immediate application of definitions we have

$$\pi_1(A) U_1 \pi_2(E) U_2 v = E \otimes A \otimes v \quad \text{as an element of } \mathcal{L}(\mathcal{H}) \hat{\otimes} \mathcal{N} \hat{\otimes} \mathcal{K}.$$

(7) We define a linear map $\mathcal{F} : \mathcal{N} \to \mathcal{L}(\hat{\mathcal{U}}_2) = \mathcal{L}(\mathcal{H} \otimes \mathcal{V})$, given by

$$\mathcal{F}(A) := U_1^* \pi_1(A) U_1 \quad \forall A \in \mathcal{N}.$$

Clearly, $\mathcal{F}$ is a unital CP map. By normality of the representation π_1, it follows that actually $\mathcal{F} \in \mathrm{CP}_1 (\mathcal{N}, \mathcal{L}(\mathcal{H} \otimes \mathcal{V}))$.

(8) At this point, we are in position to prove eq. (3) for elementary CP maps. Indeed, if $E \in \mathcal{L}(\mathcal{H})$, $A \in \mathcal{N}$ and $v, w \in \mathcal{K}$, then we have, for $\mathcal{E} = E^* \odot_{\mathcal{L}(\mathcal{H})} E$,

$$\begin{aligned}
\langle v, [\mathsf{S}(\mathcal{E})] (A) w \rangle &= \langle E \otimes I_{\mathcal{N}} \otimes v, E \otimes A \otimes w \rangle_1 \\
&= \langle U_1 \pi_2(E) U_2 v, \pi_1(A) U_1 \pi_2(E) U_2 w \rangle_1 \\
&= \langle \pi_2(E) U_2 v, \mathcal{F}(A) \pi_2(E) U_2 w \rangle_2 \\
&= \langle v, U_2^*(E^* \otimes I_{\mathcal{V}}) \mathcal{F}(A)(E \otimes I_{\mathcal{V}}) U_2 w \rangle \\
&= \langle v, U_2^* [(\mathcal{E} \otimes \mathcal{I}_{\mathcal{V}}) \mathcal{F}(A)] U_2 w \rangle .
\end{aligned}$$

Setting $V := U_2$, we then obtain

$$[\mathsf{S}(\mathcal{E})] (A) = V^* [(\mathcal{E} \otimes \mathcal{I}_{\mathcal{V}}) \mathcal{F}(A)] V \quad \forall A \in \mathcal{N},$$

i.e. eq. (3) in the special case $\mathcal{E} = E^* \odot_{\mathcal{L}(\mathcal{H})} E$.

(9) By Kraus Theorem 2.2, eq. (3) for generic $\mathcal{E} \in \mathrm{CP}(\mathcal{L}(\mathcal{H}), \mathcal{L}(\mathcal{H}))$ then follows from step (8) using normality of S and of the amplification supermap $\Pi_{\mathcal{U}} : \mathcal{E} \mapsto \mathcal{E} \otimes \mathcal{I}_{\mathcal{U}}$. Finally, linearity and Theorem 2.1 extend the equality to all $\mathcal{E} \in \mathrm{CB}(\mathcal{L}(\mathcal{H}), \mathcal{L}(\mathcal{H}))$. This concludes the proof of Theorem 4.1.

5. An application of Theorem 4.1: Transforming a quantum measurement into a quantum channel

For simplicity we consider here quantum measurements with a countable set of outcomes, denoted by X. In the algebraic language, a measurement on the quantum system with Hilbert space $\mathcal{K}_1$ and with outcomes in X is described by a quantum channel $\mathcal{E} \in \mathrm{CP}_1 (\mathcal{M}_1, \mathcal{L}(\mathcal{K}_1))$, where $\mathcal{M}_1 \equiv \ell^\infty(X)$ is the von Neumann algebra of the bounded complex functions (i.e. sequences) on X with uniform norm $\|f\|_\infty := \sup_{i \in X} |f_i|$. The channel $\mathcal{E}$ maps the function $f \in \ell^\infty(X)$ into the operator

$$\mathcal{E}(f) = \sum_{i \in X} f_i P_i \in \mathcal{L}(\mathcal{K}_1), \tag{7}$$

where each P_i is a non-negative operator in $\mathcal{L}(\mathcal{K}_1)$ and $\sum_{i \in X} P_i = I_{\mathcal{K}_1}$. Note that the map $i \mapsto P_i$ is a normalized *positive operator valued measure (POVM)* based on the discrete space X and with values in $\mathcal{L}(\mathcal{K}_1)$.

Actually, eq. (7) allows us to identify the convex set of measurements $\mathrm{CP}_1\left(\ell^\infty(X), \mathcal{L}(\mathcal{K}_1)\right)$ with the set of *all* normalized $\mathcal{L}(\mathcal{K}_1)$-valued POVMs on X.[a]

The probability of obtaining the outcome $i \in X$ when the measurement is performed on a system prepared in the quantum state $\rho \in \mathcal{T}(\mathcal{K}_1) = \mathcal{L}(\mathcal{K}_1)_*$ is given by the Born rule

$$p_i = \mathrm{tr}\,(\rho P_i)\,,$$

and the expectation value of the function $f \in \ell^\infty(X)$ with respect to the probability distribution p is given by

$$\mathbb{E}_p(f) := \sum_{i \in X} p_i f_i = \mathrm{tr}\,[\rho \mathcal{E}(f)]\,.$$

The above equation allows us to interpret the channel $\mathcal{E}$ as an *operator valued expectation* (see e.g. Ref. 35).

Now, consider the deterministic supermaps sending quantum measurements in $\mathrm{CP}\left(\ell^\infty(X), \mathcal{L}(\mathcal{K}_1)\right)$ to quantum operations in $\mathrm{CP}\left(\mathcal{M}_2, \mathcal{L}(\mathcal{K}_2)\right)$, where $\mathcal{M}_2 \equiv \mathcal{L}(\mathcal{H}_2)$. In this case, our dilation Theorem 4.1 (in the predual form of Remark 4.2) states that every deterministic supermap $\mathsf{S} : \mathrm{CB}\left(\ell^\infty(X), \mathcal{L}(\mathcal{K}_1)\right) \to \mathrm{CB}\left(\mathcal{L}(\mathcal{H}_2), \mathcal{L}(\mathcal{K}_2)\right)$ is of the form

$$[\mathsf{S}(\mathcal{E})]_*(\rho) = \mathcal{F}_*[(\mathcal{E} \otimes \mathcal{I}_\mathcal{V})_*(V \rho V^*)] \quad \forall \mathcal{E} \in \mathrm{CB}\left(\ell^\infty(X), \mathcal{L}(\mathcal{K}_1)\right),\, \rho \in \mathcal{T}(\mathcal{K}_2) \tag{8}$$

where $\mathcal{V}$ is a Hilbert space, $V : \mathcal{K}_2 \to \mathcal{K}_1 \otimes \mathcal{V}$ is an isometry, and $\mathcal{F} \in \mathrm{CP}_1\left(\mathcal{L}(\mathcal{H}_2), \ell^\infty(X)\bar{\otimes}\mathcal{L}(\mathcal{V})\right)$ is a quantum channel. In our case, we have the identification

$$\ell^\infty(X)\bar{\otimes}\mathcal{L}(\mathcal{V}) \simeq \ell^\infty(X; \mathcal{L}(\mathcal{V}))\,,$$

where $\ell^\infty(X; \mathcal{L}(\mathcal{V}))$ is the von Neumann algebra of the bounded $\mathcal{L}(\mathcal{V})$-valued functions on X. Its predual space is

$$(\ell^\infty(X)\bar{\otimes}\mathcal{L}(\mathcal{V}))_* \simeq \ell^1(X; \mathcal{T}(\mathcal{V}))\,,$$

i.e. the space of norm-summable sequences with index in X and values in the Banach space of the trace class operators on $\mathcal{V}$ (see Theorem 1.22.13 in Ref. 25). In the Schrödinger picture, the channel $\mathcal{F}_*$ can be realized by first reading the classical information carried by the system with algebra

---

[a]Indeed, by commutativity of $\ell^\infty(X)$ the set $\mathrm{CP}_1\left(\ell^\infty(X), \mathcal{L}(\mathcal{K}_1)\right)$ coincides with the set of all normalized weak*-continuous *positive* maps from $\ell^\infty(X)$ into $\mathcal{L}(\mathcal{K}_1)$ (Theorem 3.11 in Ref. 26). The latter set is just the set of all normalized $\mathcal{L}(\mathcal{K}_1)$-valued POVMs on X, the identification being the one given in eq. (7).

$\ell^\infty(X)$ and, conditionally to the value $i \in X$, by performing the quantum channel $\mathcal{F}_{i*} : \mathcal{T}(\mathcal{V}) \to \mathcal{T}(\mathcal{H}_2)$ given by

$$\mathcal{F}_{i*}(\sigma) = \mathcal{F}_*(\delta_i \sigma) \quad \forall \sigma \in \mathcal{T}(\mathcal{V}) \, ,$$

where $\delta_i \sigma \in \ell^1(X; \mathcal{T}(\mathcal{V}))$ is the sequence $(\delta_i \sigma)_k = \delta_{ik} \sigma \; \forall k \in X$ (δ_{ik} is just Kronecker delta). Indeed, in this way eq. (8) can be rewritten

$$[\mathsf{S}(\mathcal{E})]_*(\rho) = \sum_{i \in X} \mathcal{F}_{i*}[(\mathcal{E} \otimes \mathcal{I}_\mathcal{V})_*(V \rho V^*)_i] \, .$$

In other words, Theorem 4.1 states that the most general transformation of a quantum measurement on $\mathcal{K}_1$ into a quantum channel from states on $\mathcal{K}_2$ to states on $\mathcal{H}_2$ can be realized by

(1) applying an invertible dynamics (the isometry V) that transforms the input system $\mathcal{K}_2$ into the composite system $\mathcal{K}_1 \otimes \mathcal{V}$, where $\mathcal{V}$ is an ancillary system;
(2) performing the given measurement $\mathcal{E}$ on $\mathcal{K}_1$, thus obtaining the outcome $i \in X$;
(3) conditionally to the outcome $i \in X$, applying a physical transformation (the channel $\mathcal{F}_{i*}$) on the ancillary system $\mathcal{V}$, thus converting it into the output system $\mathcal{H}_2$.

6. Superinstruments

Quantum superinstruments describe measurement processes where the measured object is not a quantum system, as in ordinary instruments, but rather a quantum device. While ordinary quantum instruments are defined as measures with values in the set of quantum operations (see Ref. 22, and also Ref. 3 for a more complete exposition), quantum superinstruments are defined as probability measures with values in the set of quantum supermaps.

Definition 6.1. Let Ω be a measurable space with σ-algebra $\sigma(\Omega)$ and let S be a map from $\sigma(\Omega)$ to $\mathrm{SCP}(\mathcal{M}_1, \mathcal{L}(\mathcal{K}_1); \mathcal{M}_2, \mathcal{L}(\mathcal{K}_2))$, sending the measurable subset $B \in \sigma(\Omega)$ to the quantum supermap $\mathsf{S}_B \in \mathrm{SCP}(\mathcal{M}_1, \mathcal{L}(\mathcal{K}_1); \mathcal{M}_2, \mathcal{L}(\mathcal{K}_2))$. We say that S is a *quantum superinstrument* if it satisfies the following properties:

(i) S_Ω is deterministic;
(ii) if $n \in \mathbb{N} \cup \{\infty\}$ and $B = \bigcup_{i=1}^n B_i$ with $B_i \cap B_j = \emptyset$ for $i \neq j$, then $\mathsf{S}_B = \sum_{i=1}^n \mathsf{S}_{B_i}$, where if $n = \infty$ convergence of the series is understood

in the following sense:

$$[\mathsf{S}_B(\mathcal{E})](A) = \text{wk}^*\text{-}\lim_k \sum_{i=1}^{k} [\mathsf{S}_{B_i}(\mathcal{E})](A)$$

for all $\mathcal{E} \in \text{CB}\,(\mathcal{M}_1, \mathcal{L}(\mathcal{K}_1))$ and $A \in \mathcal{M}_2$.

We will briefly see that every quantum superinstrument is associated to an ordinary quantum instrument in an unique way. Before giving the precise statement, we recall the notion of quantum instrument, which is central in the statistical description of quantum measurements:

Definition 6.2. A map $\mathcal{J} : \sigma(\Omega) \to \text{CP}\,(\mathcal{M}, \mathcal{N})$ is a *quantum instrument* if it satisfies the following properties:

(i) $\mathcal{J}_\Omega$ is a quantum channel;
(ii) if $n \in \mathbb{N} \cup \{\infty\}$ and $B = \bigcup_{i=1}^{n} B_i$ with $B_i \cap B_j = \emptyset$ for $i \neq j$, then $\mathcal{J}_B = \sum_{i=1}^{n} \mathcal{J}_{B_i}$, where if $n = \infty$ convergence of the series is understood in the following sense:

$$\mathcal{J}_B(A) = \text{wk}^*\text{-}\lim_k \sum_{i=1}^{k} \mathcal{J}_{B_i}(A) \quad \forall A \in \mathcal{M}\,.$$

With an easy application of Radon-Nikodym Theorem 4.2, one can then prove the following dilation theorem for quantum superinstruments.

Theorem 6.1 (Dilation of quantum superinstruments).
Suppose that $\mathsf{S} : \sigma(\Omega) \to \text{SCP}\,(\mathcal{M}_1, \mathcal{L}(\mathcal{K}_1); \mathcal{M}_2, \mathcal{L}(\mathcal{K}_2))$ *is a quantum superinstrument and let* $(\mathcal{V}, V, \mathcal{F})$ *be the minimal dilation of the deterministic supermap* S_Ω. *Then there exists a unique quantum instrument* $\mathcal{J} : \sigma(\Omega) \to \text{CP}\,(\mathcal{M}_2, \mathcal{M}_1 \bar{\otimes} \mathcal{L}(\mathcal{V}))$ *such that*

$$[\mathsf{S}_B(\mathcal{E})](A) = V^*[(\mathcal{E} \otimes \mathcal{I}_\mathcal{V})\mathcal{J}_B(A)]V \quad \forall \mathcal{E} \in \text{CB}\,(\mathcal{M}_1, \mathcal{L}(\mathcal{K}_1))\,,\ A \in \mathcal{M}_2 \quad (9)$$

for all $B \in \sigma(\Omega)$.

The physical interpretation of the dilation of quantum superinstruments is clear in the Schrödinger picture. Indeed, taking the predual of eq. (9), we have for all $\rho \in \mathcal{T}\,(\mathcal{K}_2)$ and $\mathcal{E} \in \text{CB}\,(\mathcal{M}_1, \mathcal{L}(\mathcal{K}_1))$

$$[\mathsf{S}_B(\mathcal{E})]_*(\rho) = \mathcal{J}_{B\,*}\,[(\mathcal{E} \otimes \mathcal{I}_\mathcal{V})_*(V \rho V^*)]\,.$$

This means that a quantum state ρ first is coupled with an ancillary system with Hilbert space $\mathcal{V}$ and the overall system undergoes the invertible evolution V; then the system is transformed by means of the quantum channel $\mathcal{E}$, while nothing is done on the ancilla; finally, the quantum measurement $\mathcal{J}$ is performed on the system + ancilla, and after that the ancilla is discarded.

7. Application of Theorem 6.1: Measuring a measurement

Suppose that we want to characterize some property of a quantum measuring device on a system with Hilbert space $\mathcal{K}_1$: for example, we may have a device performing a projective measurement on an unknown orthonormal basis, and want to find out the basis. In this case the set of possible answers to our question is thus the set of all orthonormal bases. In a more abstract setting, the possible outcomes will constitute a measure space Ω with σ-algebra $\sigma(\Omega)$. This includes also the case of full tomography of the measurement device,[36–39] in which the outcomes in Ω label all possible measurements.

The mathematical object describing our task will be a superinstrument taking the given measurement as input and yielding an outcome in the set $B \in \sigma(\Omega)$ with some probability. In the algebraic framework, we will describe the input measurement as a quantum channel $\mathcal{E} \in \mathrm{CP}\,(\mathcal{M}_1, \mathcal{L}(\mathcal{K}_1))$, where $\mathcal{M}_1 \equiv \ell^\infty(X)$ is the algebra of the complex bounded functions on X (see the discussion in Section 5).

7.1. *Outcome statistics for a measurement on a measuring device*

If we only care about the outcomes in Ω and their statistical distribution, then the output of the superinstrument will be trivial, that is $\mathcal{M}_2 \equiv \mathcal{L}(\mathcal{K}_2) \equiv \mathbb{C}$. In this case, Theorem 6.1 states that every superinstrument $\mathsf{S} : \sigma(\Omega) \to \mathrm{SCP}\,(\ell^\infty(X), \mathcal{L}(\mathcal{K}_1); \mathbb{C}, \mathbb{C})$ will be of the form

$$\mathsf{S}_B(\mathcal{E}) = \langle v, (\mathcal{E} \otimes \mathcal{I}_\mathcal{V})(\mathcal{J}_B)v \rangle \quad \forall \mathcal{E} \in \mathrm{CB}\,(\ell^\infty(X), \mathcal{L}(\mathcal{K}_1))\,, \, B \in \sigma(\Omega)\,,$$

where $\mathcal{V}$ is an ancillary Hilbert space, $v \in \mathcal{K}_1 \otimes \mathcal{V}$ is a unit vector, and $\mathcal{J} : \sigma(\Omega) \to \mathrm{CP}\,(\mathbb{C}, \mathcal{M}_1 \bar{\otimes} \mathcal{L}(\mathcal{V})) \simeq \ell^\infty(X; \mathcal{L}(\mathcal{V}))$ is just a weak*-countably additive positive measure on Ω with values in $\ell^\infty(X; \mathcal{L}(\mathcal{V}))$, satisfying $(\mathcal{J}_\Omega)_i = I_\mathcal{V} \, \forall i \in X$. Note that in this case each supermap S_B is actually a linear map $\mathrm{CB}\,(\ell^\infty(X), \mathcal{L}(\mathcal{K}_1)) \to \mathbb{C}$, and, if $\mathcal{E}$ is a quantum channel, the map $B \mapsto \mathsf{S}_B(\mathcal{E})$ is a probability measure on Ω. In the Schrödinger picture

$$\mathsf{S}_B(\mathcal{E}) = [\mathcal{J}_{B*}(\mathcal{E} \otimes \mathcal{I}_\mathcal{V})_*](\omega_v)\,, \tag{10}$$

where ω_v is the state in $\mathcal{T}\,(\mathcal{K}_1 \otimes \mathcal{V})$ given by $\omega_v(A) := \langle v, Av \rangle \, \forall A \in \mathcal{L}(\mathcal{K}_1 \otimes \mathcal{V})$. Note that $\mathcal{J}_{B*} : \ell^1(X; \mathcal{T}\,(\mathcal{V})) \to \mathbb{C}$. Thus, if for all $i \in X$ we define the following normalized POVM on Ω

$$Q_i : \sigma(\Omega) \to \mathcal{L}(\mathcal{V})\,, \qquad Q_{i,B} := (\mathcal{J}_B)_i\,,$$

then we have

$$\mathcal{J}_{B*}(\delta_i\,\sigma) = \operatorname{tr}(\sigma Q_{i,B}) \quad \forall \sigma \in \mathcal{T}(\mathcal{V})$$

and eq. (10) becomes

$$\mathsf{S}_B(\mathcal{E}) = \sum_{i \in X} \operatorname{tr}\left[Q_{i,B}(\mathcal{E} \otimes \mathcal{I}_\mathcal{V})_*(\omega_v)_i\right],$$

which shows that, conditionally on the classical information $i \in X$, we just perform a measurement of the normalized POVM Q_i on the states in $\mathcal{T}(\mathcal{V})$. In other words, Theorem 6.1 claims that the most general way to extract information about a measuring device on system $\mathcal{K}_1$ consists in

(1) preparing a pure bipartite state ω_v in $\mathcal{K}_1 \otimes \mathcal{V}$;
(2) performing the given measurement $\mathcal{E}$ on $\mathcal{K}_1$, thus obtaining the outcome $i \in X$;
(3) conditionally on the outcome $i \in X$, performing a measurement (the POVM Q_i) on the ancillary system $\mathcal{V}$, thus obtaining an outcome in Ω.

Note that the choice of the POVM Q_i depends in general on the outcome of the first measurement $\mathcal{E}$.

7.2. *Tranformations of measuring devices induced by a higher-order measurement*

In a quantum measurement it is often interesting to consider not only the statistics of the outcomes, but also how the measured object changes due to the measurement process. For example, in the case of ordinary quantum measurements, one is interested in studying the state reduction due to the occurrence of particular measurement outcomes We can ask the same question in the case of higher-order measurements on quantum devices: for example, we can imagine a measurement process where a measuring device is tested, producing outcomes in Ω, and transformed into a new measuring device. This situation is described mathematically by a quantum superinstrument with outcomes in Ω, sending measurements in CP $(\mathcal{M}_1, \mathcal{L}(\mathcal{K}_1))$ to measurements in CP $(\mathcal{M}_2, \mathcal{L}(\mathcal{K}_2))$, where $\mathcal{M}_1 \equiv \ell^\infty(X)$ and $\mathcal{M}_2 \equiv \ell^\infty(Y)$ for some countable sets X and Y.

In this case, Theorem 6.1 states that every superinstrument $\mathsf{S} : \sigma(\Omega) \to$ SCP $(\ell^\infty(X), \mathcal{L}(\mathcal{K}_1); \ell^\infty(Y), \mathcal{L}(\mathcal{K}_2))$ is of the form

$$[\mathsf{S}_B(\mathcal{E})](f) = V^*[(\mathcal{E} \otimes \mathcal{I}_\mathcal{V})\mathcal{J}_B(f)]V \quad \forall \mathcal{E} \in \mathrm{CB}\left(\ell^\infty(X), \mathcal{L}(\mathcal{K}_1)\right), f \in \ell^\infty(Y)$$

for all $B \in \sigma(\Omega)$, where $\mathcal{V}$ is an ancillary Hilbert space, $V \in \mathcal{L}(\mathcal{K}_2, \mathcal{K}_1 \otimes \mathcal{V})$ is an isometry, and $\mathcal{J} : \sigma(\Omega) \to \mathrm{CP}\left(\ell^\infty(Y), \ell^\infty(X; \mathcal{L}(\mathcal{V}))\right)$ is an instrument.

Note that, by commutativity of $\ell^\infty(Y)$, the set $\mathrm{CP}\left(\ell^\infty(Y), \ell^\infty(X; \mathcal{L}(\mathcal{V}))\right)$ is actually the set of weak*-continuous *positive* maps from $\ell^\infty(Y)$ into $\ell^\infty(X; \mathcal{L}(\mathcal{V}))$. If for all $i \in X$ we define the positive map

$$\mathcal{J}_{i,B} : \ell^\infty(Y) \to \mathcal{L}(\mathcal{V}), \qquad \mathcal{J}_{i,B}(f) := \mathcal{J}_B(f)_i,$$

then the mapping $\mathcal{J}_i : \sigma(\Omega) \to \mathrm{CP}\left(\ell^\infty(Y), \mathcal{L}(\mathcal{V})\right)$ is an instrument, with preduals

$$\mathcal{J}_{i,B*} : \mathcal{T}(\mathcal{V}) \to \ell^1(Y), \qquad \mathcal{J}_{i,B*}(\sigma) = \mathcal{J}_{B*}(\delta_i\,\sigma)$$

for all $B \in \sigma(\Omega)$. From the relation

$$[\mathsf{S}_B(\mathcal{E})]_*(\rho) = [\mathcal{J}_{B*}(\mathcal{E} \otimes \mathcal{I}_\mathcal{V})_*](V\rho V^*) = \sum_{i \in X} \mathcal{J}_{i,B*}[(\mathcal{E} \otimes \mathcal{I}_\mathcal{V})_*(V\rho V^*)_i],$$

holding for all states $\rho \in \mathcal{T}(\mathcal{K}_2)$, we then see that the most general measurement on a quantum measuring device can be implemented by

(1) applying an invertible dynamics (the isometry V) that transforms the input system $\mathcal{K}_2$ into the composite system $\mathcal{K}_1 \otimes \mathcal{V}$, where $\mathcal{V}$ is an ancillary system;

(2) performing the given measurement $\mathcal{E}$ on $\mathcal{K}_1$, thus obtaining the outcome $i \in X$;

(3) conditionally to the outcome $i \in X$, performing a quantum measurement (the predual instrument $\mathcal{J}_{i*}$), thus obtaining an outcome in Ω and transforming the ancillary system $\mathcal{V}$ into the classical system described by the commutative algebra $\ell^\infty(Y)$.

When Ω is a countable set, we have that the instrument $\mathcal{J} : \sigma(\Omega) \to \mathrm{CP}\left(\ell^\infty(Y), \ell^\infty(X, \mathcal{L}(\mathcal{V}))\right)$ is completely specified by its action on singleton sets, that is, by the quantum operations $\{\mathcal{J}_\omega \in \mathrm{CP}\left(\ell^\infty(Y), \ell^\infty(X, \mathcal{L}(\mathcal{V}))\right) \mid \omega \in \Omega\}$. In this case, if for all $i \in X$ we set

$$Q^{(i)}_{\omega,j} := \mathcal{J}_\omega(\delta_j)_i = \mathcal{J}_{i,\omega}(\delta_j) \quad \forall(\omega, j) \in \Omega \times Y,$$

then the map $(\omega, j) \mapsto Q^{(i)}_{\omega,j}$ is a normalized POVM on the product set $\Omega \times Y$ and with values in $\mathcal{L}(\mathcal{V})$. Note that, in terms of the POVM $Q^{(i)}$, we can express each $\mathcal{J}_{i,\omega}$ as

$$\mathcal{J}_{i,\omega}(f) = \sum_{j \in Y} f_j\, Q^{(i)}_{\omega,j} \quad \forall f \in \ell^\infty(Y)$$

or, equivalently,

$$(\mathcal{J}_{i,\omega*}(\sigma))_j = \mathrm{tr}\left(\sigma Q^{(i)}_{\omega,j}\right) \quad \forall \sigma \in \mathcal{T}(\mathcal{V}).$$

In other words, the step (3) in the measurement process can be interpreted as a quantum measurement with outcome $(\omega, j) \in \Omega \times Y$, where only the classical information concerning the index $j \in Y$ is encoded in a physical system available for future experiments, whereas the information concerning index $\omega \in \Omega$ becomes unavailable after being red out by the experimenter.

Acknowledgements

Research at Perimeter Institute for Theoretical Physics is supported in part by Canada through NSERC and by Ontario through MRI. This work was partly supported by the National Basic Research Program of China (973) 2011CBA00300 (2011CBA00302).

References

1. Chiribella, G., D'Ariano, G. M., and Perinotti, P., Transforming quantum operations: quantum supermaps, Europhys. Lett. **83**, 30004 (2008)
2. Chiribella, G., D'Ariano, G. M., and Perinotti, P., A theoretical framework for quantum networks, Phys. Rev. A **80**, 022339 (2009)
3. Davies, E. B., *Quantum Theory of Open Systems*, Academic Press, London (1976)
4. Kraus, K., General state changes in quantum theory, Ann. Phys. **64**, 311-335 (1971)
5. Bisio, A., Chiribella, G., D'Ariano, G. M., Facchini, S., and Perinotti, P., Optimal quantum tomography for states, measurements, and transformations, Phys. Rev. Lett. **102**, 010404 (2009)
6. Bisio, A., Chiribella, G., D'Ariano, G. M., and Perinotti, P., Information-disturbance tradeoff in estimating a unitary transformation, Phys. Rev. A **82**, 062305 (2010)
7. Chiribella, G., D'Ariano, G. M., and Perinotti, P., Memory effects in quantum channel discrimination, Phys. Rev. Lett. **101**, 180501 (2008)
8. Chiribella, G., D'Ariano, G. M., and Perinotti, P., Optimal covariant quantum networks, AIP Conf. Proc. **1110**, 47-56 (2009)
9. Sedlak, M., and Ziman, M., Unambiguous comparison of unitary channels, Phys. Rev. A **79**, 012303 (2009)
10. Ziman, M., Process POVM: A mathematical framework for the description of process tomography experiments, Phys. Rev. A **77**, 062112 (2008)
11. Bisio, A., Chiribella, G., D'Ariano, G. M., Facchini, S., and Perinotti, P., Optimal quantum learning of a unitary transformation, Phys. Rev. A **81**, 032324 (2010)
12. Chiribella, G., D'Ariano, G. M., and Perinotti, P., Optimal cloning of a unitary transformation, Phys. Rev. Lett. **101**, 180504 (2008)
13. Bisio, A., D'Ariano, G. .M., Perinotti, P., and Sedlak, M., Cloning of a quantum measurement, Phys. Rev. A **84**, 042330 (2011)

14. Bisio, A., D'Ariano, G. .M., Perinotti, P., and Sedlak, M., Quantum learning algorithms for quantum measurements, Phys. Lett. A **375**, 3425-3434 (2011)
15. Chiribella, G., D'Ariano, G. M., and Perinotti, P., Probabilistic theories with purification, Phys. Rev. A **81**, 062348 (2010)
16. Zyczkowski, K., Quartic quantum theory: an extension of the standard quantum mechanics, J. Phys. A **41**, 355302 (2008)
17. Jencova, A., Generalized channels: channels for convex subsets of the state space, arXiv:1105.1899 (2011)
18. Holevo, A. S., *Statistical Structure of Quantum Theory*, Springer, Berlin (2001)
19. Chiribella, G., Toigo, A., and Umanità, V., Normal completely positive maps on the space of quantum operations, arXiv:1012.3197 (2010)
20. Chiribella, G., Toigo, A., and Umanità, V., Quantum supermaps transforming quantum operations on von Neumann algebras, preprint (2011)
21. Chiribella, G., D'Ariano, G. M., and Perinotti, P., Quantum circuits architecture, Phys. Rev. Lett. **101**, 060401 (2008)
22. Davies, E. B., and Lewis, J. T., An operational approach to quantum probability, Comm. Math. Phys. **17**, 239-260 (1970)
23. Ozawa, M., Quantum measuring processes of continuous observables, J. Math. Phys. **25**, 79-87 (1984)
24. Takesaki, M., *Theory of Operator Algebras, Volume I*, Springer, Berlin (1979)
25. Sakai, S., *C*-algebras and W*-algebras*, Springer, Berlin (1971)
26. Paulsen, V., *Completely bounded maps and operator algebras*, Cambridge University Press, Cambridge, UK (2002)
27. Haagerup, U., Decomposition of completely bounded maps on operator algebras, unpublished preprint
28. Attal, S., Joye, A., and Pillet, C.-A., Open Quantum Systems II - The Markovian Approach, Lecture Notes in Math., 1881, Springer, Berlin (2006)
29. Belavkin, V. P., Reconstruction theorem for a quantum stochastic process, Theor. Math. Phys. **62**, 275-289 (1985)
30. Lindblad, G., Non-Markovian quantum stochastic processes and their entropy, Comm. Math. Phys. **65**, 281-294 (1979)
31. Parthasarathy, K. R., A continuous time version of Stinespring's theorem on completely positive maps, *Quantum probability and applications, V* (Heidelberg, 1988), 296-300, Lecture Notes in Math., 1442, Springer, Berlin (1990)
32. Arveson, W. B., Subalgebras of C*-algebras, Acta Math. **123**, 141-224 (1969)
33. Belavkin, V. P., and Staszewski, P., A Radon-Nikodym theorem for completely positive maps, Rep. Math. Phys. **24**, 49-55 (1986)
34. Raginsky, M., Radon-Nikodym derivatives of quantum operations, J. Math. Phys. **44**, 5003-5019 (2003)
35. Chiribella, G., D'Ariano, G. M., and Schlingemann, D. M., Barycentric decomposition of quantum measurements in finite dimensions, J. Math. Phys. **51**, 022111 (2010)
36. D'Ariano, G. M., Maccone, L., and Presti, P. L., Quantum calibration of measurement instrumentation, Phys. Rev. Lett. **93**, 250407 (2004)
37. Fiurasek, J. Maximum-likelihood estimation of quantum measurement,

Phys. Rev. A **64**, 024102 (2001)

38. Luis, A., and Sanchez-Soto, L. L., Complete characterization of arbitrary quantum measurement processes, Phys. Rev. Lett. **83**, 3573-3576 (1999)

39. Lundeen, J. S., Feito, A., Coldenstrodt-Ronge, H., Pregnell, K .L., Silberhorn, C., Ralph, T. C., Eisert, J., Plenio, M. B., and Walmsley, I. A., Tomography of quantum detectors, Nature Physics **5**, 27-30 (2009)

INVARIANT OPERATORS IN SCHRÖDINGER SETTING

VLADIMIR K. DOBREV

Institute for Nuclear Research and Nuclear Energy,
Bulgarian Academy of Sciences,
72 Tsarigradsko Chaussee, 1784 Sofia, Bulgaria
* E-mail: dobrev@inrne.bas.bg

We give a brief overview of some group-theoretic results related to the Schrödinger equation and the Schrödinger algebra. We first recall the interpretation of non-relativistic holography as equivalence between representations of the Schrödinger algebra describing bulk fields and boundary fields. One important result is the explicit construction of the boundary-to-bulk operators in the framework of representation theory, and that these operators and the bulk-to-boundary operators are intertwining operators. In analogy to the relativistic case each bulk field has two boundary fields with conjugated conformal weights. These fields are related by another intertwining operator given by a two-point function on the boundary. Further, we recall the fact that there is a hierarchy of equations on the boundary, invariant w.r.t. Schrödinger algebra. The new results are the explicit construction of an analogous hierarchy of invariant equations in the bulk, and that the two hierarchies are equivalent via the bulk-to-boundary intertwining operators.

Keywords: Schrödinger equation, Schrödinger algebra, invariant operators

1. Introduction

The role of nonrelativistic symmetries in theoretical physics was always important. One of the popular fields in theoretical physics - string theory, pretending to be a universal theory - encompasses together relativistic quantum field theory, classical gravity, and certainly, nonrelativistic quantum mechanics, in such a way that it is not even necessary to separate these components, cf., e.g.,[1–4]. The cornerstone of quantum mechanics - the Schrödinger equation - appears naturally here at the conference of Quantum Probability in almost every talk.

Altogether, it is not a surprise that the Schrödinger group - the group that is the maximal group of symmetry of the Schrödinger equation - is playing more and more a prominent role in theoretical physics in general, and quantum probability, in particular, cf., e.g.,[5–17]. Indeed, this symmetry appears in this conference series not for the first time - earlier it was brought by contributions of, e.g., P. Feinsilver, M. Henkel, R. Schott,[18–20].

Originally, the Schrödinger group, actually the Schrödinger algebra, was introduced by Niederer[21] and Hagen[22] as nonrelativistic limit of the vector-field realization of the conformal algebra.

Recently, Son[6] proposed another method of identifying the Schrödinger algebra in d+1 space-time. Namely, Son started from anti de Sitter (AdS) space in d+3 dimensional space-time with metric that is invariant under the corresponding conformal algebra so(d+1,2) and then deformed the AdS metric to reduce the symmetry to the Schrödinger algebra.

In view of the relation of the conformal and Schrödinger algebra there arises the natural question. Is there a nonrelativistic analogue of the AdS/CFT correspondence (CFT stands for Conformal Field Theory), in which the conformal symmetry is replaced by Schrödinger symmetry. Indeed, this is to be expected since the Schrödinger equation should play a role both in the bulk and on the boundary.

Thus, we study the nonrelativistic analogue of the AdS/CFT correspondence. First let us remind the two ingredients of the AdS/CFT correspondence:[23–25]

1. the holography principle, which is very old, and means the reconstruction of some objects in the bulk (that may be classical or quantum) from some objects on the boundary;

2. the reconstruction of quantum objects, like 2-point functions on the boundary, from appropriate actions on the bulk.

Here the main focus is on the first ingredient and we consider the simplest case of the (3+1)-dimensional bulk. It is shown that the holography principle is established using representation theory only, that is, we do not specify any action. (Such representation-theoretic intertwining operator realization of the AdS/CFT correspondence in the conformal case was given in[26].)

For the implementation of the first ingredient in the Schrödinger algebra context we used in[27] a method that is standard in the mathematical literature for the construction of discrete series representations of real semisimple Lie groups,[28,29] and which method was applied in the physics literature first in[30] in exactly an AdS/CFT setting, though that term was not used then.

The method utilizes the fact that in the bulk the Casimir operators are not fixed numerically. Thus, when a vector-field realization of the algebra in consideration is substituted in the Casimir it turns into a differential operator. In contrast, the boundary Casimir operators are fixed by the quantum numbers of the fields under consideration. Then the bulk/boundary correspondence forces an eigenvalue equation involving the Casimir differential operator. That eigenvalue equation is used to find the two-point Green function in the bulk which is then used to construct the boundary-to-bulk integral operator. This operator maps a boundary field to a bulk field similarly to what was done in the conformal context by Witten[25] (see also[26]). This is the first main result of[27].

The second main result of[27] is that this operator is an intertwining operator, namely, it intertwines the two representations of the Schrödinger algebra acting in the bulk and on the boundary. This also helps us to establish that each bulk field has actually two bulk-to-boundary limits. The two boundary fields have conjugated conformal weights Δ, $3 - \Delta$, and they are related by a boundary two-point function.

In the present paper we discuss also the Schrödinger equation as an invariant differential equation. On the boundary this was done long time ago in[31] (extending the approach in the semi-simple group setting[32]), constructing actually an infinite hierarchy of invariant differential equations (see also[33]). Here we extend this construction to the bulk combining techniques from[27] and[31].

2. Preliminaries

The Schrödinger algebra $\mathcal{S}(d)$ in $(d+1)$-dimensional space-time is generated by:
- time translation P_t
- space translations P_k (which generate a subalgebra $t(d)$), $k = 1, \cdots, d$
- Galilei boosts G_k (which generate a subalgebra $g(d)$), $k = 1, \cdots, d$
- rotations $J_{k\ell} = -J_{\ell k}$ (which generate the subalgebra $so(d)$), $k, \ell = 1, \cdots, d$
- dilatation D
- conformal transformation K

The non-trivial commutation relations are:[34]

$$[P_t, D] = 2P_t, \quad [D, K] = 2K, \quad [P_t, K] = D, \tag{1}$$
$$[P_t, G_k] = P_k, \quad [P_k, D] = P_k,$$
$$[P_k, K] = G_k, \quad [D, G_k] = G_k,$$
$$[P_i, J_{k\ell}] = \delta_{i\ell} P_k - \delta_{ik} P_\ell,$$

$$[G_i, J_{k\ell}] = \delta_{i\ell}G_k - \delta_{ik}G_\ell, \tag{2}$$

$$[J_{ij}, J_{k\ell}] = \delta_{ik}J_{j\ell} + \delta_{j\ell}J_{ik} - \delta_{i\ell}J_{jk} - \delta_{jk}J_{i\ell},$$

Actually, we shall work with the central extension of the Schrödinger algebra: $\hat{\mathcal{S}}(d)$, obtained by adding the central element M to $\mathcal{S}(d)$ which enters the additional commutation relations:

$$[P_k, G_\ell] = \delta_{k\ell}M . \tag{3}$$

The generators J_{ij}, P_i form the $((d+1)d/2)$–dimensional Euclidean subalgebra $\mathcal{E}(d)$.

The generators J_{ij}, P_i, D form the $((d+1)d/2 + 1)$–dimensional Euclidean Weyl subalgebra $\mathcal{W}(d)$.

The subalgebras $\tilde{\mathcal{E}}(d)$ and $\tilde{\mathcal{W}}(d)$ generated by J_{ij}, G_i and by J_{ij}, G_i, D, resp., are isomorphic to $\mathcal{E}(d)$, $\mathcal{W}(d)$, resp.

The generators J_{ij}, P_i, G_i, P_t form the $((d+1)(d+2)/2)$–dimensional Galilei subalgebra $\mathcal{G}(d)$. The generators J_{ij}, P_i, G_i, K form another $((d+1)(d+2)/2)$–dimensional subalgebra $\tilde{\mathcal{G}}(d)$ which is isomorphic to the Galilei subalgebra.

The isomorphic pairs mentioned above are conjugated to each other.

For the structure of $\hat{\mathcal{S}}(d)$ it is also important to note that the generators D, K, P_t form an $sl(2, \mathbb{R})$ subalgebra - cf. the first line of (1).

Obviously $\mathcal{S}(d)$ is not semisimple and has the following Levi–Malcev decomposition (for $d \neq 2$):

$$\begin{aligned}
\mathcal{S}(d) &= \mathcal{N}(d) \rtimes \mathcal{M}(d) \tag{4} \\
\mathcal{N}(d) &= t(d) \oplus g(d) \\
\mathcal{M}(d) &= sl(2, \mathbb{R}) \oplus so(d)
\end{aligned}$$

with $\mathcal{M}(d)$ acting on $\mathcal{N}(d)$, where the maximal solvable ideal $\mathcal{N}(d)$ is abelian, while the semisimple subalgebra (the Levi factor) is $\mathcal{M}(d)$.

[For $d = 2$ the maximal solvable ideal $t(d) \oplus g(d) \oplus so(2)$ is not abelian, while the Levi factor $sl(2, \mathbb{R})$ is simple.]

We recall the triangular decomposition of the algebra $sl(2, \mathbb{R}) = sl(2, \mathbb{R})^+ \oplus sl(2, \mathbb{R})^0 \oplus sl(2, \mathbb{R})^-$, where $sl(2, \mathbb{R})^+$ is spanned by K, the Cartan subalgebra $sl(2, \mathbb{R})^0$ is spanned by D, and $sl(2, \mathbb{R})^-$ is spanned by P_t. Taking into account also the triangular decomposition: $so(d) = so(d)^+ \oplus so(d)^0 \oplus so(d)^-$, (more precisely of its complexification

$so(d, \mathbb{C})$), we can introduce the following triangular decomposition:

$$\hat{\mathcal{S}}(d) = \hat{\mathcal{S}}(d)^+ \oplus \hat{\mathcal{S}}(d)^0 \oplus \hat{\mathcal{S}}(d)^- \tag{5}$$
$$\hat{\mathcal{S}}(d)^+ = g(d) \oplus sl(2, \mathbb{R})^+ \oplus so(d)^+$$
$$\hat{\mathcal{S}}(d)^0 = sl(2, \mathbb{R})^0 \oplus so(d)^0 \oplus \text{lin.span } M ,$$
$$\hat{\mathcal{S}}(d)^- = t(d) \oplus sl(2, \mathbb{R})^- \oplus so(d)^-$$

(Clearly, for $d = 1$ the $so(d)$ factors are missing, while for $d = 2$ only the Cartan subalgebra $so(d)^0$ survives.)

Now we restrict to the 1+1 dimensional case, $d = 1$. In this case the centrally extended Schrödinger algebra has six generators:

- time translation: H
- space translation: P
- Galilei boost: G
- dilatation: D
- conformal transformation: K
- mass: M

with the following non-vanishing commutation relations:

$$[H, D] = 2H, \quad [D, K] = 2K, \quad [H, K] = D,$$
$$[H, G] = P, \quad [P, D] = P,$$
$$[P, K] = G, \quad [D, G] = G, \tag{6}$$
$$[P, G] = M,$$

Further we need also the Casimir operator. It turns out that the lowest order nontrivial Casimir operator is the 4-th order one:[35]

$$\tilde{C}_4 = (2MD - \{P, G\})^2 - 2\{2MK - G^2, 2MH - P^2\} \tag{7}$$

In fact, there are many cancellations, and the central generator M is a common linear multiple.

3. Choice of bulk and boundary

We would like to select as bulk space the four-dimensional space (t, x, x_-, z) introduced in[6]:

$$ds^2 = -\frac{2(dt)^2}{z^4} + \frac{-2dtdx_- + (dx)^2 + dz^2}{z^2} \tag{8}$$

We require that the Schrödinger algebra is an isometry of the above metric. We also need to replace the central element M by the derivative of the variable x_- which is chosen so that $\frac{\partial}{\partial x_-}$ continues to be central. Thus, a vector-field realization of the Schrödinger algebra is given by:

$$H = \frac{\partial}{\partial t}, \qquad P = \frac{\partial}{\partial x}, \qquad M = \frac{\partial}{\partial x_-},$$

$$G = t\frac{\partial}{\partial x} + xM, \tag{9}$$

$$D = x\frac{\partial}{\partial x} + z\frac{\partial}{\partial z} + 2t\frac{\partial}{\partial t},$$

$$K = t\left(x\frac{\partial}{\partial x} + z\frac{\partial}{\partial z} + t\frac{\partial}{\partial t}\right)$$
$$+ \frac{1}{2}(x^2 + z^2)M$$

and it generates an isometry of (8). This vector-field realization of the Schrödinger algebra acts on the bulk fields $\phi(t, x, x_-, z)$.

In this realization the Casimir becomes:

$$\tilde{C}_4 = M^2 C_4,$$

$$C_4 = \hat{Z}^2 - 4\hat{Z} - 4z^2\hat{S} \;=\; 4z^2\partial_z^2 - 8z\partial_z + 5 - 4z^2\hat{S}\ , \tag{10}$$

$$\hat{S} \equiv 2\partial_t\partial_- - \partial_x^2\ , \tag{11}$$

$$\hat{Z} \equiv 2z\partial_z - 1$$

Note that (11) is the pro-Schrödinger operator.

Next we consider a well known vector-field realization of the Schrödinger algebra:[34]

$$H = \frac{\partial}{\partial t}, \qquad P = \frac{\partial}{\partial x},$$

$$G = t\frac{\partial}{\partial x} + xM, \tag{12}$$

$$D = x\frac{\partial}{\partial x} + \Delta + 2t\frac{\partial}{\partial t},$$

$$K = t\left(x\frac{\partial}{\partial x} + \Delta + t\frac{\partial}{\partial t}\right) + \frac{1}{2}x^2 M$$

where Δ is the conformal weight.

We would like to treat the realization (12) as vector-field realization on the boundary of the chosen bulk. Clearly, this is natural if we also write the generator M as $M = \frac{\partial}{\partial x_-}$.

Obviously, the variable z is the variable distinguishing the bulk, namely, the boundary is obtained when $z = 0$. (The exact map will be displayed below. Heuristically, passing from (9) to (12) one first replaces $z\frac{\partial}{\partial z}$ with Δ and then sets $z = 0$.)

Thus, the vector-field realization of the Schrödinger algebra (12) acts on the boundary field $\phi(t, x, x_-)$ with fixed conformal weight Δ. In this realization the Casimir becomes:

$$\tilde{C}_4^0 = M^2 C_4^0, \qquad C_4^0 = (2\Delta - 1)(2\Delta - 5) \tag{13}$$

As expected C_4^0 is a constant which has the same value if we replace Δ by $3 - \Delta$:

$$C_4^0(\Delta) = C_4^0(3 - \Delta) \tag{14}$$

This already means that the two boundary fields with conformal weights Δ and $3 - \Delta$ are related, or in mathematical language, that the corresponding representations are (partially) equivalent.

4. Boundary-to-bulk correspondence

In this Section we review the paper.[27] As we explained in the Introduction we concentrate on one aspect of AdS/CFT,[24,25] namely, the holography principle, or boundary-to-bulk correspondence, which means to have an operator which maps a boundary field φ to a bulk field ϕ, cf.,[25] also[26]. Mathematically, this means the following. We treat both the boundary fields and the bulk fields as representation spaces of the Schrödinger algebra. The action of the Schrödinger algebra in the boundary, resp. bulk, representation spaces is given by formulae (12), resp. by formulae (9). The boundary-to-bulk operator maps the boundary representation space to the bulk representation space.

The fields on the boundary are fixed by the value of the conformal weight Δ, correspondingly, as we saw, the Casimir has the eigenvalue determined by Δ:

$$C_4^0 \varphi(t, x, x_-) = \lambda \varphi(t, x, x_-) , \tag{15}$$
$$\lambda = (2\Delta - 1)(2\Delta - 5)$$

Thus, the first requirement for the corresponding field on the bulk $\phi(t, x, x_-, z)$ is to satisfy the same eigenvalue equation, namely, we require:

$$C_4 \phi(t, x, x_-, z) = \lambda \phi(t, x, x_-, z) , \tag{16}$$
$$\lambda = (2\Delta - 1)(2\Delta - 5)$$

where C_4 is the differential operator given in (10). Thus, in the bulk the eigenvalue condition is a differential equation.

The other condition is the behaviour of the bulk field when we approach the boundary:

$$\phi(t, x, x_-, z) \;\rightarrow\; z^\alpha \varphi(t, x, x_-)\,, \tag{17}$$

$$\alpha = \Delta, 3 - \Delta$$

Let us denote by $\hat{C}^\alpha$ the space of bulk functions $\phi(t, x, x_-, z)$ satisfying (16) and (17).

To find the boundary-to-bulk operator we first find the two-point Green function in the bulk solving the differential equation:

$$(C_4 - \lambda)\, G(\chi, z\,;\, \chi', z') = z'^4\, \delta^3(\chi - \chi')\, \delta(z - z') \tag{18}$$

where $\chi = (t, x_-, x)$.

It is important to use an invariant variable which here is:

$$u = \frac{4zz'}{(x - x')^2 - 2(t - t')(x_- - x'_-) + (z + z')^2} \tag{19}$$

The normalization is chosen so that for coinciding points we have $u = 1$.

In terms of u the Casimir becomes:

$$C_4 = 4u^2(1 - u)\frac{d^2}{du^2} - 8u\frac{d}{du} + 5 \tag{20}$$

The eigenvalue equation can be reduced to the hypergeometric equation by the substitution:

$$G(\chi, z; \chi', z') \;=\; G(u) \;=\; u^\alpha \hat{G}(u) \tag{21}$$

and the two solutions are:

$$\hat{G}(u) = F(\alpha, \alpha - 1; 2(\alpha - 1); u)\,, \quad \alpha = \Delta, 3 - \Delta \tag{22}$$

where $F = {}_2F_1$ is the standard hypergeometric function.

As expected at $u = 1$ both solutions are singular: by,[36] they can be recast into:

$$G(u) = \frac{u^\Delta}{1 - u}F(\Delta - 2, \Delta - 1; 2(\Delta - 1); u), \quad \alpha = \Delta,$$

$$G(u) = \frac{u^{3-\Delta}}{1 - u}F(1 - \Delta, 2 - \Delta; 2(2 - \Delta); u), \quad \alpha = 3 - \Delta\,.$$

Now the boundary-to-bulk operator is obtained from the two-point bulk Green function by bringing one of the points to the boundary, however, one has to take into account all info from the field on the boundary.

More precisely, we express the function in the bulk with boundary behaviour (17) through the function on the boundary by the formula:

$$\phi(\chi, z) = \int d^3\chi' \, S_\alpha(\chi - \chi', z) \, \varphi(\chi'), \tag{23}$$

where $d^3\chi' = dx'_+ dx'_- dx'$ and $S_\alpha(\chi - \chi', z)$ is defined by

$$S_\alpha(\chi - \chi', z) = \lim_{z' \to 0} z'^{-\alpha} G(u) = \left[\frac{4z}{(x - x')^2 - 2(t - t')(x_- - x'_-) + z^2} \right]^\alpha \tag{24}$$

An important ingredient of this approach is that the bulk-to-boundary and boundary-to-bulk operators are actually intertwining operators. To see this we need some more notation.

Let us denote by L_α the bulk-to-boundary operator :

$$(L_\alpha \, \phi)(\chi) \doteq \lim_{z \to 0} z^{-\alpha} \phi(\chi, z), \tag{25}$$

where $\alpha = \Delta, 3 - \Delta$ consistently with (17). The intertwining property is:

$$L_\alpha \circ \hat{X} = \tilde{X}_\alpha \circ L_\alpha, \qquad X \in \hat{\mathcal{S}}(1), \tag{26}$$

where $\tilde{X}_\alpha$ denotes the action of the generator X on the boundary (12) (with Δ replaced by α from (17)), $\hat{X}$ denotes the action of the generator X in the bulk (9).

Let us denote by $\tilde{L}_\alpha$ the boundary-to-bulk operator in (23):

$$\phi(\chi, z) = (\tilde{L}_\alpha \varphi)(\chi, z) \doteq \int d^3\chi' \, S_\alpha(\chi - \chi', z) \, \varphi(\chi') \tag{27}$$

The intertwining property now is:

$$\tilde{L}_\alpha \circ \tilde{X}_{3-\alpha} = \hat{X} \circ \tilde{L}_\alpha, \qquad X \in \hat{\mathcal{S}}(1). \tag{28}$$

Next we check consistency of the bulk-to-boundary and boundary-to-bulk operators, namely, their consecutive application in both orders should be the identity map:

$$L_{3-\alpha} \circ \tilde{L}_\alpha = 1_{\text{boundary}}, \tag{29}$$

$$\tilde{L}_\alpha \circ L_{3-\alpha} = 1_{\text{bulk}}. \tag{30}$$

Checking (29) in[27] was obtained:

$$(L_{3-\alpha} \circ \tilde{L}_\alpha \, \varphi)(\chi) = 2^{2\alpha} \pi^{3/2} \frac{\Gamma(\alpha - \frac{3}{2})}{\Gamma(\alpha)} \, \varphi(\chi) \tag{31}$$

Thus, in order to obtain (29) exactly, we have to normalize, e.g., $\tilde{L}_\alpha$.

We note the excluded values $\alpha - 3/2 \notin \mathbb{Z}_-$ for which the two intertwining operators are not inverse to each other. This means that at least one of the representations is reducible. This reducibility was established[31] for the associated Verma modules with lowest weight determined by the conformal weight Δ. (For more information on the representation theory and related hierarchies of invariant differential operators and equations based on the approach of,[32] cf.[33].)

Checking (30) is now straightforward, but also fails for the excluded values.

Note that checking (29) we used (25) for $\alpha \to 3 - \alpha$, i.e., we used one possible limit of the bulk field (23). But it is important to note that this bulk field has also the boundary as given in (25). Namely, we can consider the field:

$$\varphi_0(\chi) \doteq (L_\alpha \ \phi)(\chi) = \lim_{z \to 0} z^{-\alpha} \phi(\chi, z), \tag{32}$$

where $\phi(\chi, z)$ is given by (23). We obtain immediately:

$$\varphi_0(\chi) = \int d^3\chi' \, G_\alpha(\chi - \chi') \, \varphi(\chi'), \tag{33}$$

where

$$G_\alpha(\chi) = \left[\frac{4}{x^2 - 2tx_-} \right]^\alpha. \tag{34}$$

If we denote by G_α the operator in (33) then we have the intertwining property:

$$\tilde{X}_\alpha \circ G_\alpha = G_\alpha \circ \tilde{X}_{3-\alpha} \ . \tag{35}$$

Thus, the two boundary fields corresponding to the two limits of the bulk field are equivalent (partially equivalent for $\alpha \in \mathbb{Z} + 3/2$). The intertwining kernel has the properties of the conformal two-point function.

Thus, for generic Δ the bulk fields obtained for the two values of α are not only equivalent - they coincide, since both have the two fields φ_0 and φ as boundaries.

Remark: For the relativistic AdS/CFT correspondence the above analysis relating the two fields in (33) was given in[26]. An alternative treatment relating these two fields via the Legendre transform was given later in[37].

As in the relativistic case there is a range of dimensions when both fields $\Delta, 3 - \Delta$ are physical:

$$\Delta_-^0 \equiv 1/2 < \Delta < 5/2 \equiv \Delta_+^0 \ . \tag{36}$$

At these bounds the Casimir eigenvalue $\lambda = (2\Delta - 1)(2\Delta - 5)$ becomes zero.

5. Singular vectors and invariant differential equations

5.1. *Singular vectors*

In this subsection we follow[31]. We consider lowest weight modules (LWM) of $\hat{\mathcal{S}}(d)$, in particular, Verma modules, which are standard for semisimple Lie algebras (SSLA) and their q–deformations. A lowest weight module is characterized by its lowest weight vector v_0 and its lowest weight. The lowest weight vector is characterized by the property of being annihilated by $\hat{\mathcal{S}}^+$ and by being an eigenvector of the Cartan generators. The lowest weight is given by the eigenvalues of the Cartan generators on v_0.

Here the Cartan generators are D, M and we can write all those properties as:

$$D\, v_0 \;=\; \Delta\, v_0\,, \quad M\, v_0 \;=\; M\, v_0\,, \tag{37}$$
$$P_x\, v_0 \;=\; 0, \quad P_t\, v_0 \;=\; 0$$

where $\Delta \in \mathbb{R}$ will be called the (conformal) weight.

We denote by $\mathcal{B}$ the nonpositively graded subalgebra generated by D, M, P_x, P_t. (This is an analogue of a Borel subalgebra.) A Verma module V^Δ is defined as the LWM with lowest weight Δ, induced from a one–dimensional representation of $\mathcal{B}$ spanned by v_0, on which the generators of $\mathcal{B}$ act as in (37). The Verma module is given explicitly by $V^\Delta = U(\hat{\mathcal{S}}^+) \otimes v_0$, where $U(\hat{\mathcal{S}}^+)$ is the universal enveloping algebra of $\hat{\mathcal{S}}^+$. Clearly, $U(\hat{\mathcal{S}}^+)$ is abelian and has basis elements $p_{k,\ell} = G^k K^\ell$. The basis vectors of the Verma module are $v_{k,\ell} = p_{k,\ell} \otimes v_0$, (with $v_{0,0} = v_0$). The action of $\hat{\mathcal{S}}$ on this basis is derived easily from (6):

$$
\begin{aligned}
D\, v_{k,\ell} &= (k + 2\ell + \Delta)\, v_{k,\ell} \\
G\, v_{k,\ell} &= v_{k+1,\ell} \\
K\, v_{k,\ell} &= v_{k,\ell+1} \\
P_x\, v_{k,\ell} &= \ell\, v_{k+1,\ell-1} + mk\, v_{k-1,\ell} \\
P_t\, v_{k,\ell} &= \ell(k + \ell - 1 + \Delta)\, v_{k,\ell-1} + m\frac{k(k-1)}{2}\, v_{k-2,\ell}
\end{aligned}
\tag{38}
$$

Because of (38) we notice that the Verma module V^Δ can be decomposed in homogeneous (w.r.t. D) subspaces as follows:

$$
\begin{aligned}
V^\Delta &= \oplus_{n=0}^\infty\, V_n^\Delta \\
V_n^\Delta &= \text{lin.span.}\,\{v_{k,\ell} \mid k + 2\ell = n\} \\
\dim V_n^\Delta &= 1 + \left[\frac{n}{2}\right]
\end{aligned}
\tag{39}
$$

Next we analyze the reducibility of V^Δ through the so-called singular vectors. In analogy to the SSLA situation (cf., e.g.,[32]) a singular vector

v_s here is a homogeneous element of V^Δ , such that $v_s \notin \mathfrak{C}v_0$, and

$$P_x\, v_s \;\; = \;\; 0\; , \quad P_t\, v_s \;\; = \;\; 0 \tag{40}$$

All possible singular vectors were given explicitly in,[31] where was proved:

Proposition 1.   The singular vectors of the Verma module V^Δ over $\hat{\mathcal{S}}$ are given as follows:

$$v_s^p \;\; = \;\; a_0 \sum_{\ell=0}^{p/2} (-2m)^\ell \binom{p/2}{\ell}\, v_{p-2\ell,\ell} \tag{41}$$

$$= \;\; a_0 \Big(G^2 \,-\, 2MK \Big)^{p/2} \otimes v_0\; , \quad \Delta \;=\; \frac{3-p}{2}\; , \quad p \in 2I\!N,\; Ma_0 \neq 0$$

$$v_s^p \;\; = \;\; a_0 v_{p0} \;\; = \;\; a_0 G^p \otimes v_0\; , \quad \Delta \text{ arbitrary},\; p \in I\!N,\; M = 0,\; a_0 \neq 0\; . \quad \Diamond$$

Further in[31] was proved:

Theorem 1.   The list of the irreducible lowest weight modules over the (centrally extended) Schrödinger algebra is given by:

- V^Δ , when $\Delta \neq (3-p)/2$, $p \in 2I\!N$ and $M \neq 0$;
- $\mathcal{L}^{(p-3)/2}$, when $\Delta = (3-p)/2$, $p \in 2I\!N$ and $M \neq 0$;
- $\tilde{\mathcal{L}}_0^\Delta$, when $\Delta \notin \mathbb{Z}_+$ and $M = 0$;
- $\mathcal{L}_0^\Delta$, when $\Delta \in \mathbb{Z}_+$ and $M = 0$,

where $\mathcal{L}^{(p-3)/2}$, $\tilde{\mathcal{L}}_0^\Delta$, $\mathcal{L}_0^\Delta$, are the irreducible factor-modules of the Verma modules $V^{(p-3)/2}$ (with $M \neq 0$), V^Δ (with $\Delta \notin \mathbb{Z}_+$, $M = 0$), V^Δ (with $\Delta \in \mathbb{Z}_+$, $M = 0$), respectively, over the corresponding maximal submodules. $\Diamond$

5.2.   *Generalized Schrödinger equations from a vector–field realization of the Schrödinger algebra*

Now we shall employ vector–field representation (12) as in.[31] This realization was used to construct a polynomial realization of the irreducible lowest weight modules considered in the previous Subsection. For this realization we represent the lowest weight vector by the function 1. Indeed, the constants in (12) are chosen so that (37) is satisfied:

$$D\, 1 \;\; = \;\; \Delta\; , \quad M\, 1 \;\; = \;\; M\; , \quad P_x\, 1 \;\; = \;\; 0, \quad P_t\, 1 \;\; = \;\; 0 \tag{42}$$

Applying the basis elements $p_{k,\ell} = G^k K^\ell$ of the universal enveloping algebra $U(\hat{\mathcal{S}}^+)$ to 1 we get polynomials in x, t. Let us introduce notation for these polynomials by $f_{k,\ell} \equiv p_{k,\ell}\, 1$. (In partial cases we have explicit expressions for $f_{k,\ell}$ from[31] but we shall not need them here.)

Let us denote by C^Δ the spaces spanned by the elements $f_{k,\ell}$, and by L^Δ the irreducible subspace of C^Δ. Now in[31] was shown:

Theorem 2. The irreducible spaces L^Δ give a realization of the irreducible lowest weight representations of $\hat{\mathcal{S}}(1)$ given in Theorem 1. $\Diamond$

We consider now in more detail the most interesting cases of the representations $L^{(3-p)/2}$ with $M \neq 0$ and $p \in 2N$.

We first introduce an operator by the polynomial $G^2 - 2MK \in U(\hat{\mathcal{S}}^+)$ expressed this polynomial in the vector–field realization:

$$S \doteq G^2 - 2MK = t^2\left(\partial_x^2 - 2M\partial_t\right) + 2Mt\left(\frac{1}{2} - \Delta\right) \tag{43}$$

In these case we have:[31]

Proposition 2. Each basis polynomial $f_{k,\ell}$ of $L^{(p-3)/2}$ satisfies:

$$S^{p/2} f_{k,\ell} = \left(t^2(\partial_x^2 - 2M\partial_t) + (p-2)Mt\right)^{p/2} f_{k,\ell} = \tag{44}$$

$$= t^p \left(\partial_x^2 - 2M\partial_t\right)^{p/2} f_{k,\ell} = 0, \quad \Delta = \frac{3-p}{2}. \qquad \Diamond$$

Thus, there is a hierarchy of equations:

$$t^p \left(\partial_x^2 - 2M\partial_t\right)^{p/2} f = 0 \tag{45}$$

In the case of function spaces with elements which are polynomials in t (as our representation spaces) or singular at most as $t^{-p/2}$ for $t \to 0$, the hierarchy is:

$$\left(\partial_x^2 - 2M\partial_t\right)^{p/2} f = 0 \tag{46}$$

The above Proposition also shows that our representation spaces are comprised from solutions of the corresponding equations (46). The case $p = 2$ and M real is the ordinary heat or diffusion equation and for $p = 2$ and m purely imaginary we get the free Schrödinger equation. So the members of our hierarchy of equations which are invariant under the Schrödinger group have generically higher orders of derivatives in t. This shows that the Schrödinger symmetry is not necessarily connected with first order (in t) differential operators.

We can further extend[32] to our non-semisimple situation by considering equations with non-zero RHS. However, invariance w.r.t. the Schrödinger algebra requires that the RHS is an element of the irreducible representation space $C^{(p+3)/2}$, while the functions in the LHS are not restricted to the solution subspace of (46). Thus, using the operator in (45) we obtained the following hierarchy of generalized heat/Schrödinger equations :

$$t^p \left(\partial_x^2 - 2M\partial_t\right)^{p/2} f = j, \quad f \in C^{(3-p)/2}, \quad j \in C^{(3+p)/2} \tag{47}$$

5.3. Generalized Schrödinger equations in the bulk

Now we shall employ the bulk vector–field representation (9) trying similarly to the previous subsection to construct generalized Schrödinger equations in the bulk. We start with the operator (distinguishing bulk operators by hats):

$$\hat{S} \;\doteq\; \hat{G}^2 - 2\partial_-\hat{K} \;=\; t^2\left(\partial_x^2 - 2\partial_-\partial_t\right) + 2t(\frac{1}{2} - z\partial_z)\partial_- \;-\; z^2\partial_-^2 \quad (48)$$

We could use the one-point invariant variable obtained from u by setting in (19) $t' = 0$, $x' = 0$, $x'_- = 0$, $z' = 1$, i.e., we use

$$\tilde{u} = \frac{4z}{x^2 - 2tx_- + (z+1)^2} \quad (49)$$

Substituting this change in (48) we obtain:

$$\hat{S} \;=\; \frac{t^2}{z}\left(\tilde{u}^3\partial_{\tilde{u}} + \frac{\tilde{u}^4}{2}\partial_{\tilde{u}}^2\right) . \quad (50)$$

We shall elaborate on the use of (50) elsewhere.

Now we set an Ansatz for the fields in the bulk: $\phi(t, x, x_-, z) = e^{Mx_-}\phi(t, x, z)$ which leads to the identification $\partial_- = M$ both in the bulk and on the boundary. Thus, we shall use:

$$\hat{S}_0 \;\doteq\; \hat{G}_0^2 - 2M\hat{K}_0 \;=\; t^2\left(\partial_x^2 - 2M\partial_t\right) + 2tM(\frac{1}{2} - z\partial_z) - z^2M^2 \quad (51)$$

Thus, we obtain the following *Schrödinger-like equation* in the bulk:

$$\hat{S}_0\,\phi \;=\; \phi' , \qquad \phi \in \hat{C}^{1/2} , \quad \phi' \in \hat{C}^{5/2} . \quad (52)$$

The relation to the Schrödinger equation on the boundary is seen by the following commutative diagram:

$$\hat{C}^{1/2} \;\underset{\hat{S}_0}{\longrightarrow}\; \hat{C}^{5/2}$$

$$\downarrow L_{1/2} \qquad \downarrow L_{1/2} \qquad\qquad (53)$$

$$C^{1/2} \;\underset{S}{\longrightarrow}\; C^{5/2}$$

where $L_{1/2}$ is the bulk-to-boundary operator defined in (25), and (53) may be re-written as the intertwining relation:

$$S \circ L_{1/2} = L_{1/2} \circ \hat{S}_0 , \quad \text{acting as operator} \quad \hat{C}^{1/2} \longrightarrow C^{5/2} \quad (54)$$

The relation (54) (and so (53)) follows by substitution of the definitions.

As expected, we have a *Schrödinger-like hierarchy of equations* in the bulk:

$$(\hat{S}_0)^{p/2} \phi = \phi' , \quad \phi \in \hat{C}^{(3-p)/2} , \quad \phi' \in \hat{C}^{(3+p)/2} , \quad p \in 2I\!N \quad (55)$$

They are equivalent to the Schrödinger hierarchy of equations on the boundary (47) which is proved by showing the analogues of (53) and (54):

$$
\begin{array}{ccc}
\hat{C}^\Delta & \xrightarrow{\ (\hat{S}_0)^{p/2}\ } & \hat{C}^{3-\Delta} \\[2mm]
\downarrow L_\Delta & \downarrow L_\Delta & \\[2mm]
C^\Delta & \xrightarrow{\ S^{p/2}\ } & C^{3-\Delta}
\end{array}
\quad (56)
$$

$$S^{p/2} \circ L_\Delta = L_\Delta \circ (\hat{S}_0)^{p/2} , \quad \text{acting as operator} \quad \hat{C}^\Delta \longrightarrow C^{3-\Delta}, (57)$$
$$\Delta = (3 - p)/2, \quad p \in 2I\!N$$

Acknowledgments

The author would like to thank Professor Luigi Accardi for the kind invitation to speak at the 32nd International Conference on Quantum Probability and Related Topics (Trento, 2011).

The author was supported in part by Bulgarian NSF grant *DO 02-257*.

References

1. J. Maldacena, D. Martelli and Y. Tachikawa, JHEP **0810** (2008) 072; arXiv:0807.1100 [hep-th].
2. H. Aoki, J. Nishimura and Y. Susaki, JHEP **0904** (2009) 055; arXiv:0810.5234 [hep-th].
3. M. Schvellinger, JHEP **0812** (2008) 004; arXiv:0810.3011 [hep-th].
4. A. Akhavan, M. Alishahiha, A. Davody and A. Vahedi, JHEP **0903** (2009) 053; arXiv:0811.3067 [hep-th]; Fermions in non-relativistic AdS/CFT correspondence, arXiv:0902.0276 [hep-th].
5. Y. Nishida and D.T. Son, Phys. Rev. D **76** (2007) 086004; arXiv:0706.3746 [hep-th].
6. D.T. Son, Phys. Rev. **D78** (2008) 046003; arXiv:0804.3972 [hep-th].
7. K. Balasubramanian and J. McGreevy, Phys. Rev. Lett. **101** (2008) 061601; arXiv:0804.4053 [hep-th].
8. W.D. Goldberger, JHEP **0903** (2009) 069; arXiv:0806.2867 [hep-th].

9. D. Minic and M. Pleimling, Phys. Rev. **E78** (2008) 061108.
10. C. Duval, M. Hassaine and P.A. Horvathy, Annals Phys. **324** (2009) 1158-1167.
11. M. Taylor, Non-relativistic holography, `arXiv:0812.0530 [hep-th]`.
12. A. Adams, A. Maloney, A. Sinha and S.E. Vazquez, JHEP **0903** (2009) 097; `arXiv:0812.0166 [hep-th]`.
13. A. Donos and J.P. Gauntlett, JHEP **0903** (2009) 138; `arXiv:0901.0818 [hep-th]`; Solutions of type IIB and D=11 supergravity with Schrodinger(z) symmetry, `arXiv:0905.1098 [hep-th]`.
14. A. Bagchi and R. Gopakumar, Galilean Conformal Algebras and AdS/CFT, `arXiv:0902.1385 [hep-th]`.
15. E.O. Colgain and H. Yavartanoo, NR CFT_3 duals in M-theory, `arXiv:0904.0588 [hep-th]`.
16. N. Bobev, A. Kundu and K. Pilch, Supersymmetric IIB Solutions with Schrödinger Symmetry, `arXiv:0905.0673 [hep-th]`.
17. H. Ooguri and C.-S. Park, Supersymmetric non-relativistic geometries in M-theory, `arXiv:0905.1954 [hep-th]`.
18. M. Henkel, J. Stat. Phys. **75** (1994) 1023-1061.
19. Ph. Feinsilver, J. Kocik and R. Schott, Representations of the Schrodinger algebra and Appell systems, Fortschr. d. Physik, **52** (2004) 343-359.
20. S. Stoimenov and M. Henkel, J. Phys. Conf. Ser. **40** (2006) 144.
21. U. Niederer, Helv. Phys. Acta **45** (1972) 802-810.
22. C.R. Hagen, Phys. Rev. **D5** (1972) 377-388.
23. J. Maldacena, Adv. Theor. Math. Phys. **2** (1998) 231-252; hep-th/9711200.
24. S.S. Gubser, I.R. Klebanov and A.M. Polyakov, Phys. Lett. **428B** (1998) 105-114; hep-th/9802109.
25. E. Witten, Adv. Theor. Math. Phys. **2** (1998) 253-291; hep-th/9802150.
26. V.K. Dobrev, Nucl. Phys. **B553** [PM] (1999) 559-582; hep-th/9812194.
27. N. Aizawa and V.K. Dobrev, Nucl. Phys. **B828** [PM] (2010) 581593.
28. R. Hotta, J. Math. Soc. Japan, **23** (1971) 384-407.
29. W. Schmid, Rive Univ. Studies, **56** (1970) 99-108.
30. V.K. Dobrev, G. Mack, V.B. Petkova, S.G. Petrova and I.T. Todorov, *Harmonic Analysis on the n-Dimensional Lorentz Group and Its Applications to Conformal Quantum Field Theory*, Lecture Notes in Physics, Vol. 63 (Springer, 1977).
31. V.K. Dobrev, H.-D. Doebner and C. Mrugalla, Rep. Math. Phys. **39** (1997) 201-218.
32. V.K. Dobrev, Reports Math. Phys. **25** (1988) 159-181.
33. V.K. Dobrev, H.-D. Doebner and C. Mrugalla, J. Phys. **A29** (1996) 5909-5918; Mod. Phys. Lett. **A14** (1999) 1113-1122;   N. Aizawa, V.K. Dobrev and H.-D. Doebner, in: 'Quantum Theory and Symmetries II', Proceedings of the 2nd QTS Symposium, (Cracow, 2001), (World Sci, Singapore, 2002) pp. 222-227;   N. Aizawa, V.K. Dobrev, H.-D. Doebner and S. Stoimenov, Proceedings of the VII International Workshop "Lie Theory and Its Applications in Physics", (Varna, 2007), eds. H.-D. Doebner et al, (Heron Press, Sofia, 2008) pp. 372-399;   V.K. Dobrev and S. Stoimenov, Physics of Atomic

Nuclei, **73**, No. 11, 19161924 (2010).
34. A.O. Barut and R. Rączka, *Theory of Group Representations and Applications*, (PWN, Warszawa, 1980).
35. M. Perroud, Helv. Phys. Acta, **50** (1977) 233-252.
36. H. Bateman and A. Erdelyi, *Higher Transcendental Functions*, Vol. 1 (New-York, McGraw-Hill, 1953).
37. I.R. Klebanov and E. Witten, Nucl.Phys. B556 (1999) 89-114; hep-th/9905104.

GENERATION OF SEMIGROUPS BY DEGENERATE ELLIPTIC OPERATORS ARISING IN OPEN QUANTUM SYSTEMS

FRANCO FAGNOLA

Dipartimento di Matematica "F. Brioschi", Politecnico di Milano,
I-20133 Milano, Italy
E-mail: franco.fagnola@polimi.it
www.mate.polimi.it/qp

LEOPOLDO PANTALEÓN MARTÍNEZ

ENP UNAM,
Mexico City, Mexico
E-mail: polo_paml@yahoo.com.mx

We show that certain second order, degenerate elliptic, differential operators on $L^2(\mathbb{R}^d; \mathbb{C})$ with complex and possibly singular coefficients, arising in models of open quantum systems, generate strongly continuous contraction semigroups.

Keywords: Open quantum systems, quantum Markov semigroups, minimal semigroup, degenerate elliptic operators.

1. Introduction

The evolution of states of an open quantum system, under the Markov approximation, is described by a master equation in the Gorini-Kossakowski-Sudarshan-Lindblad (GKSL) form

$$\frac{d\rho_t}{dt} = G\rho_t + \sum_{\ell \geq 1} L_\ell\, \rho_t L_\ell^* + \rho_t\, G^* \tag{1}$$

where ρ_t, the state of the system at time t, is a positive operator on h with unit trace on a complex separable Hilbert space h and G, L_ℓ are possibly unbounded operators on h satisfying the trace-preservation condition $G^* + \sum_{\ell \geq 1} L_\ell^* L_\ell + G = 0$.

In order to construct the quantum Markov semigroup solving the equation (1) by the minimal semigroup method the operator G must be the generator of a strongly continuous contraction semigroup on h. By the trace-preservation condition, formally, we have

$$G = -\frac{1}{2} \sum_{\ell \geq 1} L_\ell^* L_\ell - \mathrm{i}H$$

where H is a symmetric operator on h; therefore both G and the adjoint G^* are formally dissipative.

The operator G is unbounded in many interesting situations (see e.g. Refs. 1–4,6,7,9,10,12,14,15,18). When dealing with systems of bosons in particular (see e.g. Refs. 2–4,9,10,12,18 and the references therein) the Hilbert space h is typically $L^2(\mathbb{R}^d; \mathbb{C})$, the operators L_ℓ are first order differential operators with complex valued coefficients and H is a modified system Hamiltonian of the form $-\Delta + V$ with a potential V that can be singular (e.g. Coulomb potential) or unbounded. As a consequence G is a second order differential operator on $L^2(\mathbb{R}^d; \mathbb{C})$, which is elliptic, possibly degenerate when there are only few non-zero L_ℓ, with complex-valued coefficients that can be unbounded or singular.

The strongly elliptic case has been extensively studied in the literature (see Ref. 5 and the references therein). There exists also several results in the degenerate elliptic case however: (1) the Hilbert space is often $L^2(\Omega; \mathbb{C})$ with Ω a bounded subset of $\mathbb{R}^d$, (2) coefficients are often real-valued and regular, (3) several authors are interested in generation of analytic semigroups and this is not our case, for instance, when many L_ℓ are zero.

This paper aims at filling this gap verifying that an important class of operators G of the above form generates a strongly continuous contraction semigroup on $L^2(\mathbb{R}^d; \mathbb{C})$ and establishing some regularity property needed later in the study of the quantum Markov semigroup associated with G and L_ℓ.

We start describing our class of models. Then we prove in Sect.3 a generation result for degenerate elliptic operators with complex-valued, regular and bounded coefficients. This, in the case of real-valued coefficients, was proved by B. Wong-Dzung[19] in $L^p(\mathbb{R}^d; \mathbb{C})$ spaces. Our method is based on approximations with strongly elliptic operators and a result by Cannarsa and Vespri[5] for second order, strongly elliptic operators with complex-valued coefficients. Finally we extend our results to include Coulomb-type potentials by perturbation methods in Sect. 4.

We would like to stress here that our result is the first step towards proving conservativity, namely well-definiteness, of a class of quantum Markov semigroups including those in Alicki.[3] Approximating the Coulomb potential by a regular one, this was done in Chebotarev and Fagnola.[8]

2. Open quantum system models

In this section we describe the class of operators G arising from models of open quantum systems, as, for instance, the heavy-ion collision model introduced by R. Alicki,[4] that we shall consider in this paper.

The Hilbert space is $h = L^2(\mathbb{R}^d; \mathbb{C})$ and the Hamiltonian H is

$$H_0 := -\frac{1}{2m}\Delta, \qquad H = H_0 + V$$

where the potential Δ is the Laplacian, V is a function on $\mathbb{R}^d$ that can be written in the form $V_1 + V_2$ with $V_1 \in L^2(\mathbb{R}^d; \mathbb{R})$ and $V_2 \in L^\infty(\mathbb{R}^d; \mathbb{R})$.

Kraus' operators L_ℓ in the dissipative part of (1) ($\ell = 1, \ldots, n$, $n \in \mathbb{N}$) are defined on the domain $\mathrm{Dom}(\Delta)$ by

$$L_\ell = \sum_{k=1}^{d} \sigma_{\ell k}(x)\partial_k + \eta_\ell(x), \tag{2}$$

where $\sigma_{\ell k}$ and η_ℓ are complex-valued bounded regular functions on $\mathbb{R}^d$.

Define the operators on $\mathrm{Dom}(\Delta)$

$$G_0 := -\frac{1}{2}\sum_{\ell=1}^{n} L_\ell^* L_\ell, \qquad G_1 := G_0 - iH_0, \qquad G := G_0 - iH \tag{3}$$

namely, G_1 is the differential operator,

$$\frac{1}{2}\sum_{h,k}\partial_h\left((\sigma^*\sigma)_{hk} + \frac{i\delta_{hk}}{m}\right)\partial_k + \frac{1}{2}\sum_{h,\ell}\partial_h\overline{\sigma}_{\ell h}\eta_\ell - \frac{1}{2}\sum_{k,\ell}\overline{\eta}_\ell\sigma_{\ell k}\partial_k - \frac{1}{2}\sum_\ell|\eta_\ell|^2$$

Note that, putting $A = \sigma^*\sigma$, we have $A_{kh} = \overline{A}_{hk}$, then

$$\sum_{h,k}\partial_h A_{hk}\partial_k = \frac{1}{2}\sum_{h,k}\partial_h\left(A_{hk} + A_{kh}\right)\partial_k + \frac{1}{2}\sum_{h,k}\partial_h\left(A_{hk} - A_{kh}\right)\partial_k.$$

Thus, by a change of indices and commutation of ∂_k and ∂_h with A_{kh}, we can write $\sum_{h,k} \partial_h (A_{hk} - A_{kh}) \partial_k$ as

$$\sum_{h,k} \partial_k A_{kh} \partial_h - \sum_{h,k} \partial_h A_{kh} \partial_k$$

$$= \sum_{h,k} (A_{kh} \partial_h \partial_k + M(\partial_k A_{kh}) \partial_h) - \sum_{h,k} (A_{kh} \partial_h \partial_k + M(\partial_h A_{kh}) \partial_k)$$

$$= \sum_{h,k} \left(M(\partial_k A_{kh}) \partial_h - M(\partial_h A_{kh}) \partial_k \right)$$

$$= \sum_{h,k} M(\partial_h (A_{hk} - A_{kh})) \partial_k$$

which is a first order differential operator. Summing up we can write G_1 as

$$G_1 = G_2 - iH_0 + G_3 \tag{4}$$

where, denoting by $a = (\sigma^* \sigma + (\sigma^* \sigma)^T)/2$ the *real symmetric* part of the hermitian matrix $\sigma^* \sigma$, G_2 and G_3 are given by

$$G_2 = \frac{1}{2} \sum_{h,k} \partial_h \, a_{hk} \, \partial_k \tag{5}$$

$$G_3 = \frac{1}{2} \sum_{h,k} M(\partial_h ((\sigma^* \sigma)_{hk} - (\sigma^* \sigma)_{kh})) \partial_k \tag{6}$$

$$+ \frac{1}{2} \sum_{h,\ell} \partial_h \overline{\sigma}_{\ell h} \eta_\ell - \frac{1}{2} \sum_{k,\ell} \overline{\eta}_\ell \sigma_{\ell k} \partial_k - \frac{1}{2} \sum_\ell |\eta_\ell|^2 . \tag{7}$$

We denote by $C_b^k(\mathbb{R}^d, \mathbb{C})$ the vector space of bounded continuous functions with bounded continuous derivatives up to the k-th order. Throughout the paper we assume the following regularity condition

R *For all $1 \leq \ell \leq n$, $1 \leq k \leq d$, functions $\sigma_{\ell k}$ and η_ℓ belong to $C_b^3(\mathbb{R}^d, \mathbb{C})$.*

Note that, under the above condition, G_3 is a first order differential operator with coefficients in $C_b^2(\mathbb{R}^d, \mathbb{C})$.

We prove first that the operator G_1 defined on the dense domain $\mathrm{Dom}(\Delta)$ generates a strongly continuous contraction semigroup on h applying the Lumer-Phillips theorem and then apply a perturbation theorem to include the potential V. To this end we start proving two inequalities allowing us to control the graph norm of G_1 and establish closedness.

Lemma 2.1. *There exists a constant $c(\sigma, \eta)$, depending on σ and η such that, for all $u \in \mathrm{Dom}(\Delta)$, we have*

$$|\Re \langle i\Delta u, G_0 u \rangle| \leq c(\sigma, \eta) (\|\Delta u\| + \|u\|) \cdot \|\Delta u\|^{1/2} \cdot \|u\|^{1/2} .$$

Proof. A straightforward computation yields

$$2\Re\langle i\Delta u, G_0 u\rangle = \langle i\partial_j u, \partial_j L_\ell^* L_\ell u\rangle + \langle \partial_j L_\ell^* L_\ell u, i\partial_j u\rangle$$
$$= -i\langle \partial_j u, L_\ell^* L_\ell \partial_j u\rangle + i\langle L_\ell^* L_\ell \partial_j u, \partial_j u\rangle$$
$$- i\langle \partial_j u, [\partial_j, L_\ell^* L_\ell] u\rangle + i\langle [\partial_j, L_\ell^* L_\ell] u, \partial_j u\rangle.$$

The first and second term cancel; thus we find the inequality

$$2\left|\Re\langle i\Delta u, G_0 u\rangle\right| \le 2\sum_{j=1}^{d} \left\|[\partial_j, L_\ell^* L_\ell] u\right\| \cdot \left\|\partial_j u\right\|$$

$$\le 2\left(\sum_{j=1}^{d} \left\|\partial_j u\right\|^2\right)^{1/2} \left(\sum_{j=1}^{d} \left\|[\partial_j, L_\ell^* L_\ell] u\right\|^2\right)^{1/2}.$$

The commutator $[\partial_j, L_\ell^* L_\ell]$ is given by

$$L_\ell^* [\partial_j, L_\ell] + [\partial_j, L_\ell^*] L_\ell$$
$$= L_\ell^* (M(\partial_j\sigma_{\ell k})\partial_k + M(\partial_j\eta_\ell)) + (-\partial_h M(\partial_j\overline{\sigma}_{\ell h}) + M(\partial_j\overline{\eta}_\ell)) L_\ell$$
$$= -\partial_h\overline{\sigma}_{\ell h}M(\partial_j\sigma_{\ell k})\partial_k - \partial_h M(\partial_j\overline{\sigma}_{\ell h})\sigma_{\ell k}\partial_k + \overline{\eta}_\ell M(\partial_j\sigma_{\ell k})\partial_k$$
$$- \partial_h M(\partial_j\overline{\sigma}_{\ell h})\eta_\ell - \partial_h\overline{\sigma}_{\ell h}M(\partial_j\eta_\ell) + M(\partial_j\overline{\eta}_\ell)\sigma_{\ell k}\partial_k + M(\partial_j(\overline{\eta}_\ell\eta_\ell))$$
$$= -\partial_h M(\partial_j(\overline{\sigma}_{\ell h}\sigma_{\ell k}))\partial_k - \partial_h M(\partial_j(\overline{\sigma}_{\ell h}\eta_\ell))$$
$$+ M(\partial_j(\overline{\eta}_\ell\sigma_{\ell k}))\partial_k + M(\partial_j(\overline{\eta}_\ell\eta_\ell)).$$

This is a second order differential operator $\sum_{|\alpha|\le 2}\varphi_\alpha\partial_\alpha$ where φ_α are polynomial functions of $\sigma_{\ell k}$, η_ℓ and their partial derivatives up to the second order. Therefore there exists a constant $c(\sigma,\eta)$, depending on $\sigma_{\ell k}$, η_ℓ and d such that

$$\sum_{j=1}^{d} \left\|[\partial_j, L_\ell^* L_\ell] u\right\|^2 \le c(\sigma,\eta)^2 \left(\|\Delta u\| + \|u\|\right)^2.$$

Moreover,

$$\sum_{j=1}^{d} \left\|\partial_j u\right\|^2 = \langle u, -\Delta u\rangle \le \|\Delta u\| \cdot \|u\|$$

and the proof is complete. □

Proposition 2.1. *For all $\varepsilon \in\,]0,1[$ there exists a constant $c(\sigma,\eta,m,\varepsilon)$, depending on σ, η, m and ε such that, for all $u \in \mathrm{Dom}(\Delta)$, we have*

$$\|G_1 u\|^2 \le \|G_0 u\|^2 + (1+\varepsilon)\|H_0 u\|^2 + c(\sigma,\eta,m,\varepsilon)\|u\|^2, \tag{8}$$
$$\|G_1 u\|^2 \ge \|G_0 u\|^2 + (1-\varepsilon)\|H_0 u\|^2 - c(\sigma,\eta,m,\varepsilon)\|u\|^2. \tag{9}$$

Proof. We prove only (9); the proof of (8) is similar. For all $u \in \text{Dom}(\Delta)$, by Lemma 2.1, we have

$$\|Gu\|^2 = \|G_0 u\|^2 + \|H_0 u\|^2 + 2\Re \langle -iH_0 u, G_0 u \rangle$$
$$\geq \|G_0 u\|^2 + \|H_0 u\|^2 - m^{-1} c(\sigma, \eta) \left(\|\Delta u\| + \|u\| \right) \|\Delta u\|^{1/2} \|u\|^{1/2}.$$

Note that, by the convexity inequality

$$x^\lambda y^{1-\lambda} \leq \lambda x + (1 - \lambda) y$$

for all $x, y > 0$ and $\lambda \in [0, 1]$ we have

$$x^{3/4} y^{1/4} + x^{1/4} y^{3/4} = (\varepsilon x)^{3/4} (\varepsilon^{-3} y)^{1/4} + (\varepsilon x)^{1/4} (\varepsilon^{-1/3} y)^{3/4}$$
$$\leq \frac{3\varepsilon x}{4} + \frac{y}{4\varepsilon^3} + \frac{\varepsilon x}{4} + \frac{3y}{4\varepsilon^{1/3}}$$
$$= \varepsilon x + \frac{\left(\varepsilon^{-3} + 3\varepsilon^{-1/3} \right) y}{4}.$$

An application of this inequality with $x = \|\Delta u\|^2$, $y = \|u\|^2$, replacing ε by $\varepsilon / 4mc(\sigma, \eta)$ yields

$$m^{-1} c(\sigma, \eta) \left(\|\Delta u\| + \|u\| \right) \|\Delta u\|^{1/2} \|u\|^{1/2}$$
$$\leq \frac{\varepsilon}{(2m)^2} \|\Delta u\|^2 + \frac{c(\sigma, \eta)}{4m} \left(\frac{(4mc(\sigma, \eta))^3}{\varepsilon^3} + \frac{3(4mc(\sigma, \eta))^{1/3}}{\varepsilon^{1/3}} \right) \|u\|^2.$$

This proves (9) with the explicit form of $c(\sigma, \eta, m, \varepsilon)$. □

Proposition 2.2. *The operator G_1 defined on the dense domain* $\text{Dom}(\Delta)$ *is closed and dissipative. Moreover the graph norms of G_1 and Δ are equivalent.*

Proof. Let $(u_k)_{k \geq 1}$ be a sequence in $\text{Dom}(\Delta)$ converging to $u \in \mathsf{h}$ such that $(G_1 u_k)_{k \geq 1}$ converges to some $v \in \mathsf{h}$. By the inequality (9), for any $\varepsilon \in]0, 1[$, we have

$$\frac{1 - \varepsilon}{4m^2} \|\Delta(u_k - u_j)\|^2 \leq \|G_1(u_k - u_j)\|^2 + c(\sigma, \eta, m, \varepsilon) \|u_k - u_j\|^2.$$

Therefore the sequence $(\Delta u_k)_{k \geq 1}$ is also convergent and, since the operator Δ is closed, u belongs to $\text{Dom}(\Delta) = \text{Dom}(G_1)$ and, by **R**, $(G_1 u_k)_{k \geq 1}$ converges to $G_1 u$.

Dissipativity is clear because, for all $u \in \text{Dom}(\Delta) = \text{Dom}(G_1)$, we have $2\Re \langle u, G_1 u \rangle = - \sum_\ell \|L_\ell u\|^2 \leq 0$.

The equivalence of graph norms follows directly from (9) and (8) together with an inequality of the type $\|G_0 u\|^2 \leq c_1(\sigma, \eta) \|\Delta u\|^2$ because G_0 is a second order differential operator with bounded coefficients. □

3. G_1 generates a semigroup

We shall prove that G_1 generates a strongly continuous contraction semi-group on h applying the following version of Lumer-Phillips theorem (see Ref. 11 Th. 3.15 p. 76).

Theorem 3.1. *For a densely defined, dissipative operator A on a Banach space $\mathcal{X}$ the following statements are equivalent:*

(a) *the closure $\overline{A}$ of A generates a strongly continuous contraction semi-group on $\mathcal{X}$,*
(b) *the range of $\lambda - A$ is dense in $\mathcal{X}$ for some (hence all) $\lambda > 0$.*

In order to check the range condition we take a $\nu > 0$ and approximate G_1 by the operator

$$E := G_1 + \frac{\nu}{2}\Delta = G_2 + \frac{\nu}{2}\Delta + G_3$$

with domain $\text{Dom}(\Delta)$. This is an elliptic operator since it is a partial differential operator of second order with highest order coefficients

$$\frac{1}{2}\left(a_{hk} + \nu\delta_{hk}\right),$$

where δ_{hk} is the Dirac delta, for all $1 \leq h, k \leq d$, $a = (\sigma^*\sigma + (\sigma^*\sigma)^T)/2$ is the real symmetric part of the hermitian matrix $\sigma^*\sigma$ and, for all $z \in \mathbb{C}^d$,

$$2\Re \langle z, (a + \nu 1)z \rangle = \|\sigma z\|^2 + \|\sigma^T z\|^2 + 2\nu\|z\|^2 \geq 2\nu\|z\|^2.$$

We refer to[5] Theorem 1.2 p.859 for the proof of the following result

Theorem 3.2. *There exists $\lambda_0 > 0$ such that, if λ is a complex number with $\Re\lambda > \lambda_0$, then the equation $\lambda u - Eu = f$ has a unique solution $u \in H^2(\mathbb{R}^d; \mathbb{C})$ for all $f \in L^2(\mathbb{R}^d; \mathbb{C})$.*

In our proof, however, we need a regularity property of solutions: if $f \in H^2(\mathbb{R}^d; \mathbb{C})$ then $u \in H^4(\mathbb{R}^d; \mathbb{C})$. This follows essentially by an elliptic regularity theorem. However, since we could not find a full proof under our hypotheses, we also check this fact here.

Definition 3.1. Let $f \in L^2(\mathbb{R}^d; \mathbb{C})$. A weak solution of the equation $\lambda u - Eu = f$ is a $u \in H^1(\mathbb{R}^d; \mathbb{C})$ such that, for all $\varphi \in C_c^\infty(\mathbb{R}^d; \mathbb{C})$

$$\lambda \langle \varphi, u \rangle + \left(\frac{\nu}{2} + \frac{i}{2m}\right)\sum_{j=1}^d \langle \partial_j\varphi, \partial_j u \rangle + \frac{1}{2}\sum_{\ell=1}^n \langle L_\ell\varphi, L_\ell u \rangle = \langle \varphi, f \rangle \quad (10)$$

Proposition 3.1. *For all λ with $\Re(\lambda) > 0$ weak solutions of $\lambda u - Eu = f$, $f \in L^2(\mathbb{R}^d; \mathbb{C})$ are unique.*

Proof. It suffices to prove that any weak solution $u \in H^1(\mathbb{R}^d; \mathbb{C})$ of $\lambda u - Eu = 0$ is zero. To this end let $(\varphi_k)_{k \geq 1}$ be a sequence in $C_c^\infty(\mathbb{R}^d; \mathbb{C})$ converging to $u \in H^1(\mathbb{R}^d; \mathbb{C})$, namely

$$\lim_{k \to \infty} \|\varphi_k - u\| = 0, \qquad \lim_{k \to \infty} \|\partial_j \varphi_k - \partial_j u\| = 0,$$

for all $1 \leq j \leq d$. Writing (10) for $f = 0$ and $\varphi = \varphi_k$ and letting k tend to infinity we find

$$\lambda \|u\|^2 + \left(\frac{\nu}{2} + \frac{i}{2m}\right) \sum_{j=1}^d \|\partial_j u\|^2 + \frac{1}{2} \sum_{\ell=1}^n \|L_\ell u\|^2 = 0.$$

Taking the real part we find $\Re(\lambda) \|u\|^2 = 0$, i.e. $u = 0$. $\qquad\square$

Theorem 3.2 and uniqueness of weak solutions yield the following regularity result.

Proposition 3.2. *There exists $\lambda_1 > \lambda_0$ such that, if λ is a complex number with $\Re(\lambda) > \lambda_1$, then the equation $\lambda u - Eu = f$ has a unique solution $u \in H^4(\mathbb{R}^d; \mathbb{C})$ for all $f \in H^2(\mathbb{R}^d; \mathbb{C})$.*

Proof. A solution $u \in H^2(\mathbb{R}^d; \mathbb{C})$ of $\lambda u - Eu = f$ is also a weak solution, therefore writing (10) for $\partial_k \varphi$ and integrating by parts we find

$$\lambda \langle \varphi, \partial_k u \rangle + \left(\frac{\nu}{2} + \frac{i}{2m}\right) \sum_{j=1}^d \langle \partial_j \varphi, \partial_k \partial_j u \rangle + \frac{1}{2} \sum_{\ell=1}^n \langle L_\ell \varphi, \partial_k L_\ell u \rangle$$

$$= \frac{1}{2} \sum_{\ell=1}^n \langle [L_\ell, \partial_k] \varphi, L_\ell u \rangle + \langle \varphi, \partial_k f \rangle .$$

Writing $\partial_k L_\ell u = L_\ell \partial_k u + [\partial_k, L_\ell] u$ we find that $\partial_k u$ is a weak solution to (10) with f replaced by $\tilde{f} := \partial_k f + D_{2,k} u$ where $D_{2,k}$ is the second order differential operator with coefficients in $C_b^1(\mathbb{R}^d; \mathbb{C})$ given by

$$2D_{2,k} = [L_\ell^*, \partial_k] L_\ell + L_\ell^* [L_\ell, \partial_k]$$
$$= \partial_j \left(\lceil \partial_k \overline{\sigma}_{\ell j} \rceil \sigma_{\ell h}\right) \partial_h + \partial_j \left(\overline{\sigma}_{\ell j} \lceil \partial_k \sigma_{\ell h} \rceil\right) \partial_h - \lceil \partial_k \overline{\eta}_\ell \rceil \eta_\ell - \overline{\eta}_\ell \lceil \partial_k \eta_\ell \rceil$$
$$+ \partial_j \left(\lceil \partial_k \overline{\sigma}_{\ell j} \rceil \eta_\ell + \overline{\sigma}_{\ell j} \lceil \partial_k \eta_\ell \rceil\right) - \left(\lceil \partial_k \overline{\eta}_\ell \rceil \sigma_{\ell h} + \overline{\eta}_\ell \lceil \partial_k \sigma_{\ell h} \rceil\right) \partial_h$$
$$= \partial_j (\sigma^* \sigma)_{jk} \partial_k + \partial_j \lceil \partial_k (\overline{\sigma}_{\ell j} \eta_\ell) \rceil - \lceil \partial_k (\overline{\eta}_\ell \sigma_{\ell h}) \rceil \partial_h - \lceil \partial_k (\overline{\eta}_\ell \eta_\ell) \rceil .$$

By Theorem 3.2, there exists a unique $v \in H^2(\mathbb{R}^d; \mathbb{C})$ solving $\lambda v - Ev = \tilde{f}$. Uniqueness of the weak solution implies that $v = \partial_k u$; thus $\partial_k u \in H^2(\mathbb{R}^d; \mathbb{C})$, namely $u \in H^3(\mathbb{R}^d; \mathbb{C})$.

Writing (10) for $\partial_k \partial_h \varphi$, in a similar way we can show that $\partial_k \partial_h u$ is a weak solution of

$$\lambda v - Ev = \partial_h \partial_k f + (\partial_h D_{2,k} + D_{2,h} \partial_k) u.$$

Since f belongs to $H^2(\mathbb{R}^d; \mathbb{C})$ and $\partial_h D_{2,k} + D_{2,h} \partial_k$ is a third order differential operator with bounded continuous coefficients, it follows from Theorem 3.2 and uniqueness of weak solutions that $\partial_k \partial_h u$ belongs to $H^2(\mathbb{R}^d; \mathbb{C})$ and the proof is complete.   □

We shall now prove an inequality allowing us to find an a priori estimate.

Proposition 3.3. *For all $u \in \mathrm{Dom}(\Delta^2)$ we have*

$$2\Re \langle \Delta u, \Delta G_1 u \rangle \leq b \, \|\Delta u\| \, (\|\Delta u\| + \|u\|) \tag{11}$$

where b is a positive constant depending only on functions $\sigma_{\ell k}$ and η_ℓ.

Proof. Note that

$$2\Re \langle \Delta u, \Delta G_1 u \rangle = 2\Re \langle \Delta u, G_1 \Delta u \rangle + 2\Re \langle \Delta u, [\Delta, G_1] u \rangle \leq 2\Re \langle \Delta u, [\Delta, G_1] u \rangle$$

because G_1 is dissipative. Writing G_1 as in (4) and noting that $[\Delta, G_1] = [\Delta, G_2] + [\Delta, G_3]$, we have

$$2\Re \langle \Delta u, \Delta G_1 u \rangle \leq 2\Re \langle \Delta u, [\Delta, G_2] u \rangle + 2\Re \langle \Delta u, [\Delta, G_3] u \rangle$$
$$\leq 2\Re \langle \Delta u, [\Delta, G_2] u \rangle + 2 \|\Delta u\| \cdot \|[\Delta, G_3] u\|. \tag{12}$$

We shall now prove

$$2[\Delta, G_2] = \partial_j [\partial_j, \partial_h a_{hk} \partial_k] + [\partial_j, \partial_h a_{hk} \partial_k] \partial_j$$
$$= \partial_j \partial_h M (\partial_j a_{hk}) \partial_k + \partial_h M (\partial_j a_{hk}) \partial_k \partial_j$$

therefore, putting $M_{hk}^j := \partial_j a_{hk}$ and noting that $M_{hk}^j = M_{kh}^j$ because a is real symmetric,

$$2\Re \langle \Delta u, [\Delta, G_2] u \rangle = \Re \left\langle \partial_s \partial_s u, \left(\partial_j \partial_h M_{hk}^j \partial_k + \partial_h M_{hk}^j \partial_k \partial_j \right) u \right\rangle$$
$$= \Re \left\langle \partial_h \partial_s u, \partial_s \left(\partial_j M_{hk}^j + M_{hk}^j \partial_j \right) \partial_k u \right\rangle.$$

Commuting partial derivative ∂_s with $\partial_j M_{hk}^j + M_{hk}^j \partial_j$ we find then

$$2\Re \langle \Delta u, [\Delta, G_2] u \rangle = \Re \left\langle \partial_h \partial_s u, \left(\partial_j M_{hk}^j + M_{hk}^j \partial_j \right) \partial_s \partial_k u \right\rangle$$
$$+ \Re \langle \partial_h \partial_s u, (\partial_j M (\partial_j \partial_s a_{hk}) + M (\partial_j \partial_s a_{hk}) \partial_j) \partial_k u \rangle$$

The first term in the right-hand side vanishes because, exchanging h with k, it turns out to be equal to to one-half of

$$\left\langle \partial_h \partial_s u, \left(\partial_j M_{hk}^j + M_{hk}^j \partial_j\right) \partial_s \partial_k u\right\rangle + \left\langle \left(\partial_j M_{hk}^j + M_{hk}^j \partial_j\right) \partial_s \partial_k u, \partial_h \partial_s u\right\rangle$$

$$= \left\langle \partial_h \partial_s u, \left(\partial_j M_{hk}^j + M_{hk}^j \partial_j\right) \partial_s \partial_k u\right\rangle - \left\langle \partial_s \partial_k u, \left(\partial_j M_{hk}^j + M_{hk}^j \partial_j\right) \partial_h \partial_s u\right\rangle$$

$$= \left\langle \partial_h \partial_s u, \left(\partial_j M_{hk}^j + M_{hk}^j \partial_j\right) \partial_s \partial_k u\right\rangle - \left\langle \partial_s \partial_h u, \left(\partial_j M_{kh}^j + M_{kh}^j \partial_j\right) \partial_k \partial_s u\right\rangle$$

and $M_{kh}^j = M_{hk}^j$ since a is real symmetric.

It follows that $2\Re \left\langle \Delta u, [\Delta, G_2]u\right\rangle$ can be written as

$$\Re \left\langle \partial_h \partial_s u, (\partial_j M(\partial_j \partial_s a_{hk}) + M(\partial_j \partial_s a_{hk})\partial_j) \partial_k u\right\rangle$$

$$= 2\Re \left\langle \partial_h \partial_s u, M(\partial_j \partial_s a_{hk})\partial_j \partial_k u\right\rangle + \Re \left\langle \partial_h \partial_s u, M(\partial_s \Delta a_{hk})\partial_k u\right\rangle$$

Letting

$$b_2 = \max_{j,s,k,h} \|\partial_j \partial_s a_{hk}\|_\infty , \qquad b_3 = \max_{s,h,k} \|\partial_s \Delta a_{hk}\|_\infty$$

we have

$$2\Re \left\langle \partial_h \partial_s u, M(\partial_j \partial_s a_{hk})\partial_j \partial_k u\right\rangle \le 2b_2 \sum_{j,s,k,h} \|\partial_h \partial_s u\| \cdot \|\partial_j \partial_k u\|$$

$$= 2b_2 \sum_{k,h} \|\partial_j \partial_k u\|^2 = 2b_2 \|\Delta u\|^2$$

and, in a similar way,

$$\Re \left\langle \partial_h \partial_s u, M(\partial_s \Delta a_{hk})\partial_k u\right\rangle \le 2b_3 \sum_{s,h,k} \|\partial_k u\| \cdot \|\partial_s \partial_h u\|$$

$$\le b_3 \sum_{s,h,k} \left(\|\partial_k u\|^2 + \|\partial_s \partial_h u\|^2\right)$$

$$= b_3 d \|\Delta u\|^2 + b_3 d^2 \sum_k \|\partial_k u\|^2$$

$$= b_3 d \|\Delta u\|^2 - b_3 d^2 \left\langle u, \Delta u\right\rangle$$

$$\le b_3 d^2 \|\Delta u\| (\|\Delta u\| + \|u\|)$$

Therefore we find the inequality

$$2\Re \left\langle \Delta u, [\Delta, G_2]u\right\rangle \le (2b_2 + b_3 d^2) \|\Delta u\| (\|\Delta u\| + \|u\|) \qquad (13)$$

and, since $[\Delta, G_3]$ is a second order differential operator, we find also a constant b_4 such that

$$\|[\Delta, G_3]u\| \le b_4 (\|\Delta u\| + \|u\|) .$$

This, together with (12), proves (11) with $b = 2b_2 + b_3 d^2 + 2b_4$.   □

Theorem 3.3. *For all $\lambda > max\{\lambda_1, b\}$ the range of $\lambda - G_1$ is dense.*

Proof. We now show that the range of $\lambda - G_1$ contains all $f \in H^2(\mathbb{R}^d, \mathbb{C})$. For all $\nu > 0$ let $u_\nu \in \text{Dom}(\Delta^2)$ be the unique solution of $\lambda u_\nu - E u_\nu = f$. Multiplying this equation by $\overline{u}_\nu$ and integrating we find

$$\lambda \|u_\nu\|^2 - \frac{\nu}{2} \langle u_\nu, \Delta u_\nu \rangle - \langle u_\nu, G_1 u_\nu \rangle = \langle u_\nu, f \rangle.$$

Taking the real part, since the operators Δ and G_1 are dissipative so that $\Re \langle u_\nu, \Delta u_\nu \rangle \leq 0$ and $\Re \langle u_\nu, G_1 u_\nu \rangle \leq 0$, we find the inequality $\lambda \|u_\nu\|^2 \leq \|u_\nu\| \cdot \|f\|$, namely $\lambda \|u_\nu\| \leq \|f\|$, proving that the norm of u_ν is uniformly bounded by $\lambda^{-1} \|f\|$.

In a similar way, applying the Laplacian to both sides of $\lambda u_\nu - E u_\nu = f$, left multiplying by $\Delta \overline{u}_\nu$ and integrating, we find the equation

$$\lambda \|\Delta u_\nu\|^2 - \frac{\nu}{2} \langle \Delta u_\nu, \Delta^2 u_\nu \rangle - \langle \Delta u_\nu, \Delta G_1 u_\nu \rangle = \langle \Delta u_\nu, \Delta f \rangle.$$

Now $\langle \Delta u_\nu, \Delta^2 u_\nu \rangle \leq 0$, therefore, by Proposition 3.3, taking the real part of both sides, we find the inequality

$$\lambda \|\Delta u_\nu\|^2 - b \|\Delta u_\nu\|^2 - b \|\Delta u_\nu\| \cdot \|u_\nu\| \leq \|\Delta u_\nu\| \cdot \|\Delta f\|.$$

Therefore, dividing by $\|\Delta u_\nu\|$ which is non-zero unless $u_\nu = 0$, since $\|u_\nu\| \leq \lambda^{-1} \|f\|$, we obtain the a priori estimate

$$(\lambda - b) \|\Delta u_\nu\| \leq b \|u_\nu\| + \|\Delta f\| \leq \|\Delta f\| + \lambda^{-1} b \|f\|$$

showing that the norm of Δu_ν is uniformly bounded with respect to ν.

We now prove that u_ν converges in $L^2(\mathbb{R}^d; \mathbb{C})$ as ν tends to 0. To this end note that, for all $\nu, \mu > 0$, we have

$$\lambda(u_\nu - u_\mu) - G_1(u_\nu - u_\mu) = \frac{\nu}{2} \Delta u_\nu - \frac{\mu}{2} \Delta u_\mu.$$

Multiplying by $\overline{u}_\nu - \overline{u}_\mu$, integrating and using once more the dissipativity of G_1 we find the inequality

$$2\lambda \|u_\nu - u_\mu\| \leq \nu \|\Delta u_\nu\| + \mu \|\Delta u_\mu\|.$$

Since $\|\Delta u_\nu\|$ is uniformly bounded with respect to ν, it follows that $\|u_\nu - u_\mu\|$ converges to 0 as ν, μ go to 0 and the family $(u_\nu)_{\nu>0}$ converges to a $u \in L^2(\mathbb{R}^d; \mathbb{C})$.

It follows that $G_1 u_\nu = \lambda u_\nu - f$ also converges in $L^2(\mathbb{R}^d; \mathbb{C})$, therefore, since G_1 is closed, u belongs to $\text{Dom}(G_1) = \text{Dom}(\Delta)$ and $\lambda u - G_1 u = f$. Thus f belongs to the range of $\lambda - G_1$ and the proof is complete.   □

Theorem 3.4. *The operator G_1 generates a strongly continuous contraction semigroup on $L^2(\mathbb{R}^d; \mathbb{C})$.*

4. G generates a semigroup

In this section we show that the operators G and G^* generate strongly continuous contraction semigroups on $L^2(\mathbb{R}^d; \mathbb{C})$ with $d \leq 3$ and prove some useful properties.

Recall that an operator X is *relatively bounded* with respect to G if $\mathrm{Dom}(G) \subseteq \mathrm{Dom}(X)$ and there exists two constants $c_1, c_2 > 0$ such that

$$\|Xu\| \leq c_1 \|Gu\| + c_2 \|u\|$$

for all $u \in \mathrm{Dom}(G)$. In this case, letting c the infimum of all constants c_1 such that the above inequality holds, we say that X is c-relatively bounded with respect to G.

Proposition 4.1. *Let V be a function of the form $V = V_1 + V_2$ with $V_1 \in L^2(\mathbb{R}^d; \mathbb{C})$ and $V_2 \in L^\infty(\mathbb{R}^d; \mathbb{C})$, $d \leq 3$. The multiplication operator by V defined on $\mathrm{Dom}(\Delta)$ is relatively bounded with respect to Δ and G_1 with relative bound 0.*

Proof. For all $u \in \mathrm{Dom}(\Delta)$ we have

$$\|Vu\| \leq \|V_1\| \cdot \|u\|_\infty + \|V_2\|_\infty \cdot \|u\|$$

where $\|u\|_\infty$ denotes the essential supremum norm of u. Indeed, by Theorem IX.28 (a)[16] p. 55, if $u \in \mathrm{Dom}(\Delta)$ then u belongs to the domain of the multiplication operator by V_1. Moreover, for all $\varepsilon > 0$, there exists a constant $c(\varepsilon)$ (independent of u) such that

$$\|u\|_\infty \leq \varepsilon \|\Delta u\| + c(\varepsilon) \|u\|.$$

Therefore we have

$$\|Vu\| \leq \varepsilon \|V_1\| \cdot \|\Delta u\| + (c(\varepsilon) \|V_1\| + \|V_2\|_\infty) \|u\|$$

and relative boundedness with relative bound 0 with respect to Δ follows from the arbitrariness of ε.

Relative boundedness with respect to G_1 follows from Proposition 2.1 (9) and the inequalities applied with a fixed choice of the ε there, e.g. $3/4$,

$$\|Vu\| \leq (2m\varepsilon \|V_1\|) \|H_0 u\| + (c(\varepsilon) \|V_1\| + \|V_2\|_\infty) \|u\|$$
$$\leq (4m\varepsilon \|V_1\|) \|G_1 u\|$$
$$+ \left(4m\varepsilon \|V_1\| c(\sigma, \eta, m, 3/4)^{1/2} + c(\varepsilon) \|V_1\| + \|V_2\|_\infty\right) \|u\|$$

again by the arbitrarity of ε. □

Theorem 4.1. *Let V be a real-valued function of the form $V = V_1 + V_2$ with $V_1 \in L^2(\mathbb{R}^d; \mathbb{R})$ and $V_2 \in L^\infty(\mathbb{R}^d; \mathbb{R})$, $d \le 3$. The operators G and G^* defined on domain $\mathrm{Dom}(\Delta)$ as $G_0 - iH$ and $G_0 + iH$ generate strongly continuous contraction semigroups on $L^2(\mathbb{R}^d; \mathbb{C})$. The linear manifold $C_c^\infty(\mathbb{R}^d; \mathbb{C})$ is a core for both.*

Proof. For all $u \in \mathrm{Dom}(\Delta)$ we have $(G_0 + iH)u = (G_1 + iV)u$. Therefore, since the multiplication operator iV is dissipative and relatively bounded with respect to G_1 with relative bound 0, the conclusion follows from a well-known result on perturbation of contraction semigroup (see e.g. Ref. 11 Theorem 2.7 p. 128).

It is now straightforward to check that linear manifold $C_c^\infty(\mathbb{R}^d; \mathbb{C})$ is a core for G since $\mathrm{Dom}(G) = \mathrm{Dom}(\Delta)$ and smooth functions with compact support are a core for Δ.

Let $(u_n)_{n \ge 1}$ be a sequence in $C_c^\infty(\mathbb{R}^d; \mathbb{C})$ converging in norm to a u in $L^2(\mathbb{R}^d; \mathbb{C})$ such that $(Gu_n)_{n \ge 1}$ converges in $L^2(\mathbb{R}^d; \mathbb{C})$ to a vector v. Fix $\varepsilon \in]0, 1[$. Since iV is relatively bounded with respect to G_1, by Proposition 2.1 9, for all $n, m \ge 1$, we have

$$(1 - \varepsilon) \left\| H_0(u_n - u_m) \right\|^2 \le \left\| G_1(u_n - u_m) \right\|^2 + c(\sigma, \eta, \varepsilon) \left\| (u_n - u_m) \right\|^2$$

$$\le (1 + \varepsilon) \left\| G(u_n - u_m) \right\|^2 + c_2(\sigma, \eta, \varepsilon) \left\| (u_n - u_m) \right\|^2$$

where $c_2(\sigma, \eta, \varepsilon)$ is another positive constant. Therefore, $(\Delta u_n)_{n \ge 1}$ is Cauchy and converges to Δu because Δ is a closed operator. Thus $u \in \mathrm{Dom}(G_1) = \mathrm{Dom}(G)$ and $v = \lim_{n \to \infty} Gu_n = Gu$.

Proofs for G^* are the same. □

Proposition 4.2. *Let $d \le 3$ and $0 \le \theta < d/2$. The multiplication operator by $|x|^{-\theta}$ defined on $\mathrm{Dom}(\Delta)$ is relatively bounded with relative bound 0 with respect to Δ and G_1.*

In particular, for $d = 3$ the multiplication operator by a Coulomb potential $\kappa |x|^{-1}$ is relatively bounded with relative bound 0 with respect to Δ and G_1.

Proof. The multiplication operator by $|x|^{-\theta}$ can be written as the sum of two functions $V_1 \in L^2(\mathbb{R}^d; \mathbb{R})$ and $V_2 \in L^\infty(\mathbb{R}^d; \mathbb{R})$,

$$V_1(x) = \min\{|x|^{-\theta}, 1\}, \qquad V_2(x) = |x|^{-\theta} - V_1(x). \qquad □$$

Acknowledgment. LPM gratefully acknowledges financial support from CONACYT Estancias Postdoctorales 000000000133238.

References

1. L. Accardi, F. Fagnola, S. Hachicha, Generic q-Markov semigroups and speed of convergence of q-algorithms, *Infin. Dimens. Anal. Quantum Probab. Relat. Top.* **9**, 567-594 (2006).
2. L. Accardi, Y.G. Lu, I. Volovich, *Quantum Theory and its Stochastic Limit.* (Springer, Berlin, 2001).
3. R. Alicki, Quantum Dynamical Semigroups and Dissipative Collisions of Heavy Ions, *Z. Phys. A* **307**, 279-285 (1982).
4. R. Alicki, A. Frigerio, Scattering theory for quantum dynamical semigroups II. *Ann. Inst. H. Poincaré A* **38**, 187–197 (1983).
5. P. Cannarsa, V. Vespri, Generation of analytic semigroups by elliptic operators with unbounded coefficients. *SIAM J. Math. Anal.* **18**, 857–872 (1987).
6. R. Carbone, F. Fagnola, S. Hachicha, Generic quantum Markov semigroups: the Gaussian gauge invariant case *Open Syst. Inf. Dyn.* **14**, 425–444 (2007).
7. R. Carbone, E. Sasso, Hypercontractivity for a quantum Ornstein-Uhlenbeck semigroup *Probab. Theory Rel. Fields* **140**, 505-522 (2008).
8. A.M. Chebotarev, F. Fagnola, Sufficient conditions for conservativity of quantum dynamical semigroups. *J. Funct. Anal.* **153**, 382–404 (1998).
9. E.B. Davies, Quantum dynamical semigroups and the neutron diffusion equation. *Rep. Math. Phys.* **11**, 169–188 (1977).
10. L. Diosi, Quantum Master Equation of a Particle in a Gas Environment. *Europhysics Lett.* **30**, 63–68 (1995).
11. K.-J. Engel, R. Nagel, *A Short Course on Operator Semigroups.* Universitext. Springer, New York, 2006.
12. F. Fagnola, Diffusion processes in Fock space. *Quantum Probability and Related Topics* **IX**, 189–214 (1994).
13. F. Fagnola, C.M. Mora, Stochastic Schrödinger equations and applications to Ehrenfest-type theorems. Preprint 2012. `arXiv:1207.2939`
14. F. Fagnola, R. Quezada, Two-photon absorption and emission process. *Infin. Dimens. Anal. Quantum Probab. Relat. Top.* **8**, 573-591 (2005).
15. M. Gregoratti, Classical dilations à la Hudson-Parthasarathy of Markov semigroups. *Stoch. Anal. Appl.* **26**, 1025–1052 (2008).
16. M. Reed, B. Simon, *Methods of modern mathematical physics. Vol II: Fourier analysis and self-adjointness*, Academic Press, San Diego, 1975.
17. A.J. Scott, G.J. Milburn, Quantum nonlinear dynamics of continuously measured systems, *Phys. Rev. A* **63**, 42101 (2001).
18. K.P. Singh, J.M. Rost, Femtosecond photoionization of atoms under noise, *Phys. Rev. A* **76**, 063403 (2007).
19. B. Wong-Dzung, L^p-theory of degenerate-elliptic and parabolic operators of second order. *Proc. Roy. Soc. Edinburgh Sect. A* **95**, 95–113 (1983).

QUANTUM OBSERVABLES ON A COMPLETELY SIMPLE SEMIGROUP

PHILIP FEINSILVER

Department of Mathematics
Southern Illinois University
Carbondale, IL
62901 USA

Completely simple semigroups arise as the support of limiting measures of random walks on semigroups. Such a limiting measure is supported on the kernel of the semigroup. Forming tensor powers of the random walk leads to a hierarchy of the limiting kernels. Tensor squares lead to quantum observables on the kernel. Recall that zeons are bosons modulo the basis elements squaring to zero. Using zeon powers leads naturally to quantum observables which reveal the structure of the kernel. Thus asymptotic information about the random walk is related to algebraic properties of the zeon powers of the random walk.

1. Introduction

This work is based on the connection between graphs and semigroups developed by Budzban and Mukherjea[2] in the context of their study of the Road Coloring Problem [RCP]. We restrict throughout to finite transformation semigroups, that is, semigroups of functions acting on a finite set, V. In their interpretation of the RCP, a principal question is whether one can determine the rank of the kernel of a semigroup from its generators. Questions in general about the kernel of a semigroup are of interest. We will explain this terminology as we proceed.

The study of probability measures on semigroups by Mukherjea,[6] Mukherjea-Högnas[5] in particular reveals the result that the convolution powers of a measure on a finite semigroup, S, converge (in the Cesàro sense) to a measure on the kernel, K, the minimal ideal of the semigroup. The kernel is always *completely simple* (see A for precise details). In this context, a completely simple semigroup is described as follows. It is a union of disjoint isomorphic groups, called "local groups". The kernel K can be organized as a two-dimensional grid with rows labelled by *partitions* of the underlying set V and columns labelled by *range classes*, a family of equinumerous sets into whose elements the blocks of a given partition are mapped by the members of $\mathcal{K}$. The cardinality of a range class is the *rank* of the kernel. The idempotent e determined by a pair $(\mathcal{P}, \mathcal{R})$, with $\mathcal{P}$ a partition of V and $\mathcal{R}$ a range class, acting as the identity of the local group, fixes each of the elements of the range class. Each of the mutually isomorphic local groups is isomorphic to a permutation group acting on its range class. We think of the kernel as composed of *cells* each labelled by a pair $(\mathcal{P}, \mathcal{R})$ and containing the corresponding local group.

First we recall the basic theory needed about probability measures on semigroups, continuing with the related context of walks on graphs. Briefly, the adjacency matrix of the graph is rescaled to a stochastic matrix, A. A decomposition of the adjacency matrix into binary stochastic matrices, C_i, i.e., matrices representing functions acting on the vertices of the graph, is called a *coloring*. The coloring functions generate the semigroup $\mathcal{S}$ and the question is to determine features of the kernel $\mathcal{K} \subset \mathcal{S}$ from the generators C_i. Here we present novel techniques by embedding the semigroup in a hierarchy of tensor powers. We will find particular second-order tensors to provide the operators giving us "quantum observables". The constructions restrict to trace-zero tensors using zeon Fock space instead of the full tensor space.

For convenient reference, we put some notations here.

1.1. *Notations*

We treat vectors as "row vectors", i.e., $1 \times n$ matrices, with $\mathbf{e}_i$ as the corresponding standard basis vectors. The column vector corresponding to v is $v^\dagger$. For a matrix M, M^* denotes its transpose. And $\mathrm{diag}(v)$ denotes the diagonal matrix with entries v_i of the vector v.

The vector having components all equal to 1 is denoted u. And the matrix $J = u^\dagger u$ has entries all ones. We use the convention that the identity matrix, I, as well as u and J, denote matrices of the appropriate size according to the context.

2. Probability measures on finite semigroups

Since we have a discrete finite set, a probability measure is given as a function on the elements. The semigroup algebra is the algebra generated by the elements $w \in \mathcal{S}$. In general, we consider formal sums $\sum f(w)\, w$. The function μ defines a probability measure if

$$0 \le \mu(w) \le 1, \forall w \in \mathcal{S}, \text{ and } \sum \mu(w) = 1 .$$

The corresponding element of the semigroup algebra is thus $\sum \mu(w)\, w$. The product of elements in the semigroup algebra yields the convolution of the coefficient functions. Thus, for the convolution of two measures μ_1 and μ_2 we have

$$\sum_{w \in \mathcal{S}} \mu_1 * \mu_2(w)\, w = \left(\sum_{w \in \mathcal{S}} \mu_1(w)\, w \right) \left(\sum_{w' \in \mathcal{S}} \mu_2(w')\, w' \right)$$

$$= \sum_{w,w' \in \mathcal{S}} \mu_1(w) \mu_2(w')\, ww'$$

Hence the convolution powers, $\mu^{(n)}$ of a single measure $\mu = \mu^{(1)}$ satisfy

$$\sum_{w \in \mathcal{S}} \mu^{(n)}(w)\, w = \left(\sum_{w \in \mathcal{S}} \mu^{(1)}(w)\, w \right)^n$$

in the semigroup algebra.

2.1. Invariant measures on the kernel

We can consider the set of probability measures with support the kernel of a finite semigroup of matrices (see A for more details). Given such a measure μ, it is *idempotent* if $\mu * \mu = \mu$, i.e., it is idempotent with respect to convolution. An idempotent measure on a finite group must be the uniform distribution on the group, its Haar measure. In general, we have

Theorem 2.1. *[Ref. 5, Th. 2.2]*
An idempotent measure μ on a finite semigroup $\mathcal{S}$ is supported on a completely simple subsemigroup, $\mathcal{K}'$ of $\mathcal{S}$. With Rees product decomposition $\mathcal{K}' = \mathcal{X}' \times \mathcal{G}' \times \mathcal{Y}'$, μ is a direct product of the form $\alpha \times \omega \times \beta$, where α is a measure on $\mathcal{X}'$, ω is Haar measure on $\mathcal{G}'$, and β is a measure on $\mathcal{Y}'$.

In our context, we will have the measure μ supported on the kernel $\mathcal{K}$ of $\mathcal{S}$. We identify $\mathcal{X}$ with the partitions of the kernel and $\mathcal{Y}$ with the ranges. The local groups are mutually isomorphic finite groups with the Haar measure assigning $1/|G|$ to each element of the local group G. Thus, if $k \in \mathcal{K}$ has Rees product decomposition (k_1, k_2, k_3) we have

$$\mu(k) = \alpha(k_1)\beta(k_3)/|G|$$

where $|G|$ is the common cardinality of the local groups, $\alpha(k_1)$ is the α-measure of the partition of k, and $\beta(k_3)$ is the β-measure of the range of k.

Remark 2.1. Given a graph with n vertices, we identify V with the numbers $\{1, 2, \ldots, n\}$. Generally, given n, we are considering semigroups of transformations acting on the set $\{1, 2, \ldots, n\}$.

For a function on $\{1, 2, \ldots, n\}$, the notation $f = [x_1 x_2 \ldots x_n]$ means that $f(i) = x_i$, $i \leq 1 \leq n$. And the rank of f is the number of distinct values $\{x_i\}_{1 \leq i \leq n}$.

Example 2.1. Here is an example with $n = 6$. Take two functions $r = [451314]$, $b = [245631]$ and generate the semigroup $\mathcal{S}$. We find the elements of minimal rank, in this case it is 3. The structure of the kernel is summarized in a table with rows labelled by the partitions and columns labelled by the range classes. The entry is the idempotent with the given partition and range. It is the identity for the local group of matrices with the given partition and range.

	$\{1, 3, 4\}$	$\{1, 4, 5\}$	$\{2, 3, 6\}$	$\{2, 5, 6\}$
$\{\{1, 2\}, \{3, 5\}, \{4, 6\}\}$	[113434]	[115454]	[223636]	[225656]
$\{\{1, 6\}, \{2, 4\}, \{3, 5\}\}$	[143431]	[145451]	[623236]	[625256]

The kernel has 48 elements, the local groups being isomorphic to the symmetric group S_3.

Each cell consists of functions with the given partition, acting as permutations on the range class. They are given by matrices whose nonzero columns are in the positions labelled by the range class. The entries in a nonzero column are in rows labelled by the elements comprising the block of the partition mapping into the element labelling that column. We label the partitions

$$\mathcal{P}_1 = \{\{1, 2\}, \{3, 5\}, \{4, 6\}\}, \qquad \mathcal{P}_2 = \{\{1, 6\}, \{2, 4\}, \{3, 5\}\}$$

and the range classes

$$\mathcal{R}_1 = \{1, 3, 4\}, \quad \mathcal{R}_2 = \{1, 4, 5\}, \quad \mathcal{R}_3 = \{2, 3, 6\}, \quad \mathcal{R}_4 = \{2, 5, 6\}.$$

The local group with partition $\mathcal{P}_1$ and range $\mathcal{R}_3$, for example, are the idempotent [223636] noted in the above table and the functions

$$\{ [663232], [662323], [336262], [332626], [226363] \}$$

isomorphic to S_3 acting on the range class $\{2, 3, 6\}$. For the function [332626], in the matrix semigroup we have the correspondence

$$[332626] \longleftrightarrow \begin{bmatrix} 0\,0\,1\,0\,0\,0 \\ 0\,0\,1\,0\,0\,0 \\ 0\,1\,0\,0\,0\,0 \\ 0\,0\,0\,0\,0\,1 \\ 0\,1\,0\,0\,0\,0 \\ 0\,0\,0\,0\,0\,1 \end{bmatrix}.$$

Using the methods developed below, one finds the measure on the partitions to be

$$\alpha = [1/3, 2/3]$$

while that on the ranges is

$$\beta = [4/9, 2/9, 1/9, 2/9]$$

as required by invariance.

3. Graphs, semigroups, and dynamical systems

We start with a regular d-out directed graph on n vertices with adjacency matrix $\mathcal{A}$. Number the vertices once and for all and identify vertex i with i and vice versa.

Form the stochastic matrix $A = d^{-1}\mathcal{A}$. We assume that A is irreducible and aperiodic. In other words, the graph is strongly connected and the limit

$$\lim_{m \to \infty} A^m = \Omega$$

exists and is a stochastic matrix with identical rows, the invariant distribution for the corresponding Markov chain, which we denote by $\pi = [p_1, \ldots, p_n]$. The limiting matrix satisfies

$$\Omega = A\Omega = \Omega A = \Omega^2$$

so that the rows and columns are fixed by A, eigenvectors with eigenvalue 1.

Decompose $A = \frac{1}{d}C_1 + \cdots + \frac{1}{d}C_d$, into binary stochastic matrices, "colors", corresponding to the d choices moving from a given vertex to another. Each coloring matrix C_i is the matrix of a function f on the vertices as in the discussion in albegra.[2] Let $\mathcal{S} = \mathcal{S}((C_1, C_2, \ldots, C_d))$ be the semigroup generated by the matrices C_i, $1 \le i \le d$. Then we may write the decomposition of A into colors in the form

$$A = \mu^{(1)}(1)C_1 + \cdots + \mu^{(1)}(d)C_d = \sum_{w \in \mathcal{S}} \mu^{(1)}(w)\, w$$

with $\mu^{(1)}$ a probability measure on $\mathcal{S}$, thinking of the elements of $\mathcal{S}$ as words w, strings from the alphabet generated by $\{C_1, \ldots, C_d\}$. We have seen in the previous section that

$$A^m = \sum_{w \in \mathcal{S}} \mu^{(m)}(w)\, w \tag{1}$$

where $\mu^{(m)}$ is the mth convolution power of the measure $\mu^{(1)}$ on $\mathcal{S}$.

The main limit theorem is the following

Theorem 3.1. *[Ref. 5, Th. 2.7]*
Let the support of $\mu^{(1)}$ generate the semigroup $\mathcal{S}$. Then the Cesàro limit

$$\lambda(w) = \lim_{M \to \infty} \frac{1}{M} \sum_{m=1}^{M} \mu^{(m)}(w)$$

exists for each $w \in \mathcal{S}$. The measure λ is concentrated on $\mathcal{K}$, the kernel of the semigroup $\mathcal{S}$. It has the canonical decomposition

$$\lambda = \alpha \times \omega \times \beta$$

corresponding to the Rees product decomposition of $X \times G \times Y$ of $\mathcal{K}$.

Remark 3.1. See Ref. 1 for related material, including a proof of the above theorem.

We use the notation $\langle \cdot \rangle$ to denote averaging with respect to λ over random elements $K \in \mathcal{K}$. Forming $M^{-1} \sum_{m=1}^{M}$ on both sides of (1), letting $M \to \infty$ we see immediately that

$$\Omega = \langle K \rangle$$

the average element of the kernel. We wish to discover further properties of the kernel by considering the behavior of tensor powers of the semigroup elements.

From the above discussion, we have the following.

Proposition 3.1. *Let $\phi : \mathcal{S} \to \mathcal{S}'$ be a homomorphism of finite matrix semigroups. Let*

$$A_\phi = \mu^{(1)}(1)\phi(C_1) + \cdots + \mu^{(1)}(d)\phi(C_d) = \sum_{w \in \mathcal{S}} \mu^{(1)}(w)\, \phi(w)$$

then the Cesàro limit

$$\lim_{M \to \infty} \frac{1}{M} \sum_{m=1}^{M} A_\phi^m$$

exists and equals

$$\Omega_\phi = \langle \phi(K) \rangle$$

the average over the kernel K with respect to the measure $\alpha \times \omega \times \beta$ as for $\mathcal{S}$.

Proof. Checking that

$$A_\phi^m = \sum_{w \in \mathcal{S}} \mu^{(m)}(w)\, \phi(w)$$

the result follows immediately. $\square$

Ω_ϕ satisfies the relations

$$A_\phi \Omega_\phi = \Omega_\phi A_\phi = \Omega_\phi^2 = \Omega_\phi \ .$$

Proposition 3.2. *The rows of Ω_ϕ span the set of left eigenvectors of A_ϕ for eigenvalue 1. The columns of Ω_ϕ span the set of right eigenvectors of A_ϕ for eigenvalue 1.*

Proof. For a left invariant vector v, we have

$$v = vA_\phi = vA_\phi^m = v\Omega_\phi \qquad \text{(taking the Cesàro limit)}$$

Thus, v is a linear combination of the rows of Ω_ϕ. Similarly for right eigenvectors. $\square$

Now, as Ω_ϕ is an idempotent, we have $\operatorname{rank} \Omega_\phi = \operatorname{tr} \Omega_\phi$ so

Corollary 3.1. *The dimension of the eigenspace of right/left invariant vectors of A_ϕ equals $\operatorname{tr} \Omega_\phi$.*

Remark 3.2. We will only consider mappings ϕ that preserve nonnegativity. A_ϕ in general will be *substochastic*, i.e., it may have zero rows or rows that do not sum to one. As noted in B, the Abel limits exist for the A_ϕ in agreement with the Cesàro limits. Abel limits are suitable for computations, e.g. with Maple or Mathematica.

4. Tensor hierarchy

Starting with a set V of n elements, let $F(V) = \{f \colon V \to V\}$. For a field of scalars, take the rationals, $\mathbb{Q}$, as the most natural for our purposes and consider the vector space $\mathcal{V} = \mathbb{Q}^n$ with $\operatorname{End}(\mathcal{V})$ the space of $n \times n$ matrices acting as endomorphisms of $\mathcal{V}$. We have the mapping

$$F(V) \xrightarrow{\mathrm{Mt}} \operatorname{End}(\mathcal{V})$$

taking $f \in F(V)$ to $\mathrm{Mt}(f) \in \operatorname{End}(\mathcal{V})$ defined by

$$(\mathrm{Mt}(f))_{ij} = \begin{cases} 1, & \text{if } f(i) = j \\ 0, & \text{otherwise} \end{cases} \tag{2}$$

Denote by $F_\circ(V)$ the semigroup consisting of $F(V)$ with the operation of composition, where we compose maps to the right: for $i \in V$, $i(f_1 f_2) = f_2(f_1(i))$. The mapping Mt gives a representation of the semigroup $F_\circ(V)$ by endomorphisms of $\mathcal{V}$, i.e., $\mathrm{Mt}(f_1 f_2) = \mathrm{Mt}(f_1)\mathrm{Mt}(f_2)$.

For W of the form $\mathrm{Mt}(f)$ corresponding to a function $f \in F(V)$, the resulting components of $W^{\otimes k}$ are components of the map on tensors given by the kth kronecker power of W. At each level k, there is an induced map

$$\operatorname{End}(\mathcal{V}) \to \operatorname{End}(\mathcal{V}^{\otimes k}) \,, \qquad \mathrm{Mt}(f) \to \mathrm{Mt}(f)^{\otimes k}$$

satisfying

$$(\mathrm{Mt}(f_1 f_2))^{\otimes k} = (\mathrm{Mt}(f_1)\mathrm{Mt}(f_2))^{\otimes k} = \mathrm{Mt}(f_1)^{\otimes k}\mathrm{Mt}(f_2)^{\otimes k} \tag{3}$$

giving, for each k, a representation of the semigroup $F_\circ(V)$ as endomorphisms of $\mathcal{V}^{\otimes k}$.

What is the function, f_k, corresponding to $\mathrm{Mt}(f)^{\otimes k}$, i.e., such that $\mathrm{Mt}(f_k) = \mathrm{Mt}(f)^{\otimes k}$? For degree 1, we have from (2)

$$\mathbf{e}_i \mathrm{Mt}(f) = \mathbf{e}_{f(i)} \tag{4}$$

And for the induced map at degree k, taking products in $\mathcal{V}^{\otimes k}$, for a multi-set I,

$$\mathbf{e}_I \mathrm{Mt}(f)^{\otimes k} = (\mathbf{e}_{i_1}\mathrm{Mt}(f)) \otimes (\mathbf{e}_{i_2}\mathrm{Mt}(f)) \otimes \cdots \otimes (\mathbf{e}_{i_k}\mathrm{Mt}(f))$$
$$= \mathbf{e}_{f(i_1)} \otimes \mathbf{e}_{f(i_2)} \otimes \cdots \otimes \mathbf{e}_{f(i_k)}$$

We see that the degree k maps are those induced on multi-subsets of V mapping

$$\{i_1, \ldots, i_k\} \to \{f(i_1), \ldots, f(i_k)\}$$

with no further conditions, i.e., there is no restriction that the indices be distinct. This thus gives the second quantization of $\mathrm{Mt}(f)$ corresponding to the induced map, the second quantization of f, extending the domain of f from V to $\mathcal{V}^{\otimes}$, the space of all tensor powers of $\mathcal{V}$.

Our main focus in this work is on the degree two component, where we have a natural correspondence of 2-tensors with matrices.

4.1. The degree 2 component of $\mathcal{V}^{\otimes}$

Working in degree 2, we denote indices $I = (i, j)$, as usual, instead of (i_1, i_2).
For given n, X, X', etc., are vectors in $\mathcal{V}^{\otimes 2} \approx \mathbb{Q}^{n^2}$.

As a vector space, $\mathcal{V}$ is isomorphic to $\mathbb{Q}^n$. Denote by $\mathrm{End}(\mathcal{V})$ the space of matrices acting on $\mathcal{V}$.

Definition   The mapping

$$\mathrm{Mat} \colon \mathcal{V}^{\otimes 2} \to \mathrm{End}(\mathcal{V})$$

is the linear isomorphism taking the vector $X = (x_{ij})$ to the matrix $\tilde{X}$ with entries

$$\tilde{X}_{ij} = x_{ij} \, .$$

We will use the explicit notation $\mathrm{Mat}(X)$ as needed for clarity.

Equip $\mathcal{V}^{\otimes 2}$ with the inner product

$$\langle X, X' \rangle = \mathrm{tr}\, \tilde{X}^* \tilde{X}'$$

the star denoting matrix transpose.

Throughout, we use the convention wherein repeated Greek indices are automatically summed over. So we write

$$\langle X, X' \rangle = X_{\lambda\mu} X'_{\lambda\mu} \, .$$

4.2. Basic Identities

Multiplying X with u, we note

$$X u^\dagger = \mathrm{tr}\, \tilde{X} J = u X^\dagger \tag{5}$$

We observe

Proposition 4.1. *(Basic Relations) We have*
1. $\mathrm{Mat}(X A^{\otimes 2}) = A^* \tilde{X} A$.
2. $\mathrm{Mat}(A^{\otimes 2} X^\dagger) = A \tilde{X} A^*$.

Proof. We check #1 as #2 is similar. The matrix $A^{\otimes 2}$ has components

$$A^{\otimes 2}_{ab,cd} = A_{ac}A_{bd}$$

And we have

$$(XA^{\otimes 2})_{cd} = \sum_{a,b} x_{ab}A_{ac}A_{bd}$$

$$= (A^*\tilde{X}A)_{cd} \qquad \square$$

We see immediately

Proposition 4.2. *For any A and X,*

$$\tilde{X} = A^*\tilde{X}A \Leftrightarrow XA^{\otimes 2} = X$$

$$\tilde{X} = A\tilde{X}A^* \Leftrightarrow A^{\otimes 2}X^\dagger = X^\dagger$$

4.3. Trace Identities

Using equation (5), we will find some identities for traces of these quantities.

Proposition 4.3. *We have*

1. $XA^{\otimes 2}u^\dagger = \operatorname{tr}(\tilde{X}AJA^*)$.
2. $uA^{\otimes 2}X^\dagger = \operatorname{tr}(\tilde{X}A^*JA)$.
3. *If A is stochastic, then* $XA^{\otimes 2}u^\dagger = \operatorname{tr}(\tilde{X}J)$.

Proof. We have, using equation (5) and Basic Relation 1,

$$XA^{\otimes 2}u^\dagger = \operatorname{tr}\operatorname{Mat}(XA^{\otimes 2})J$$

$$= \operatorname{tr}(A^*\tilde{X}AJ)$$

and rearranging terms inside the trace yields #1. Then #3 follows since A stochastic implies $AJ = J = JA^*$. And #2 follows similarly, using the second Basic Relation in the equation $uA^{\otimes 2}X^\dagger = \operatorname{tr}\operatorname{Mat}(A^{\otimes 2}X^\dagger)J$. $\square$

Using equation (5) directly for X, we have

$$X(I - A^{\otimes 2})u^\dagger = \operatorname{tr}(\tilde{X}(J - AJA^*))$$

$$u(I - A^{\otimes 2})X^\dagger = \operatorname{tr}(\tilde{X}(J - A^*JA))$$

And the first line reduces to

$$X(I - A^{\otimes 2})u^\dagger = 0$$

for stochastic A.

4.4. *Convergence to tensor hierarchy*

Proposition 4.4. *For $\ell \geq 1$, define*

$$A_{\otimes \ell} = \frac{1}{d}C_1{}^{\otimes \ell} + \cdots + \frac{1}{d}C_d{}^{\otimes \ell} .$$

Then the Cesàro limit of the powers $A_{\otimes \ell}^m$ exists and equals

$$\Omega_{\otimes \ell} = \langle K^{\otimes \ell} \rangle$$

the average taken over the kernel $\mathcal{K}$ of the semigroup $\mathcal{S}$ generated by $\{C_1, \ldots, C_d\}$.

We imagine the family $\{\mathcal{K}^{\otimes \ell}\}_{\ell \geq 1}$, as a hierarchy of kernels corresponding to the family of semigroups generated by tensor powers $C_i{}^{\otimes \ell}$ in each level ℓ. Since each element $k \in \mathcal{K}$ is a stochastic matrix, in fact a binary stochastic matrix, as each corresponds to a function, we have $kJ = J$ at every level. We thus have a mechanism to move down the hierarchy by the relation

$$k^{\otimes \ell}(J \otimes I^{\otimes(\ell-1)}) = J \otimes k^{\otimes(\ell-1)}$$

which is composed of identical blocks $k^{\otimes(\ell-1)}$.

Note that if $\operatorname{rank}\Omega_{\otimes \ell} = 1$, the invariant distribution can be computed as an element of the nullspace of $I - A^{\otimes \ell}$. Otherwise, $\Omega_{\otimes \ell}$ can be found computationally as the Abel limit

$$\lim_{s \uparrow 1}(1 - s)(I - sA^{\otimes \ell})^{-1} .$$

5. The principal observables: M and N operators

5.1. *Graph-theoretic context*

For the remainder of this paper, we work mainly in the context where the stochastic matrix $A = \frac{1}{2}\mathcal{A}$, with $\mathcal{A}$ the adjacency matrix of a 2-out regular digraph, assumed strongly connected and aperiodic. In other words, A is an ergodic transition matrix for the Markov chain induced on the vertices, V, of the graph. A decomposes into two coloring matrices, which we call from now on R and B, for "red" and "blue", respectively. Thus

$$A = \tfrac{1}{2}(R + B)$$

where R and B are binary $n \times n$ stochastic matrices.

Example 5.1. For a working example, consider

$$A = \begin{bmatrix} 0 & 0 & 1/2 & 1/2 \\ 0 & 0 & 1/2 & 1/2 \\ 1/2 & 0 & 0 & 1/2 \\ 0 & 1/2 & 1/2 & 0 \end{bmatrix}.$$

The powers of $\overset{.}{A}$ converge to the idempotent matrix

$$\Omega = \begin{bmatrix} 1/6 & 1/6 & 1/3 & 1/3 \\ 1/6 & 1/6 & 1/3 & 1/3 \\ 1/6 & 1/6 & 1/3 & 1/3 \\ 1/6 & 1/6 & 1/3 & 1/3 \end{bmatrix}.$$

As one possible decomposition, we may take

$$R = \begin{bmatrix} 0 & 0 & 0 & 1 \\ 0 & 0 & 1 & 0 \\ 1 & 0 & 0 & 0 \\ 0 & 1 & 0 & 0 \end{bmatrix}, \qquad B = \begin{bmatrix} 0 & 0 & 1 & 0 \\ 0 & 0 & 0 & 1 \\ 0 & 0 & 0 & 1 \\ 0 & 0 & 1 & 0 \end{bmatrix}.$$

In the notation used previously, we write

$$R = [4312] \qquad \text{and} \qquad B = [3443].$$

5.2. *Level 2 of the tensor hierarchy*

Let

$$A_{\otimes 2} = \tfrac{1}{2}(R^{\otimes 2} + B^{\otimes 2})$$

Then fixed points X of $A_{\otimes 2}$ are vectors corresponding to solutions to the matrix equations

$$\tfrac{1}{2}\left(R\tilde{X}R^* + B\tilde{X}B^*\right) = \tilde{X}$$

and

$$\tfrac{1}{2}\left(R^*\tilde{X}R + B^*\tilde{X}B\right) = \tilde{X}$$

5.2.1. *M and N operators*

In Section 4.1, we defined the map from vectors to matrices $X \to \tilde{X} = \text{Mat}(X)$. Now it is convenient to have a reverse map from matrices to vectors, say

$$Y \to Y_{\text{vec}}$$

where here we use Y to denote a typical matrix to distinguish our conventional use of X as a vector.

Denote by $\mathcal{N}$ the space of solutions to

$$\tfrac{1}{2}\left(RYR^* + BYB^*\right) = Y .$$

The nonnegative solutions denote by $\mathcal{N}_+$, and the space of nonnegative solutions with trace zero by $\mathcal{N}_0$.

Similarly, $\mathcal{M}$ denotes the space of solutions to

$$\tfrac{1}{2}\left(R^*YR + B^*YB\right) = Y$$

with $\mathcal{M}_+$ nonnegative solutions, and $\mathcal{M}_0$, nonnegative trace zero solutions.

Starting from the relation

$$\Omega_{\otimes 2} = \langle K^{\otimes 2}\rangle$$

We have solutions in the space $\mathcal{M}$ in the form

$$X\Omega_{\otimes 2} = \langle K^*\tilde{X}K\rangle = \sum{}' k^*\tilde{X}k$$

and solutions in $\mathcal{N}$ in the form

$$\Omega_{\otimes 2}X = \langle K\tilde{X}K^*\rangle = \sum{}' k\tilde{X}k^*$$

where the primed summation denotes an averaged sum, here with respect to the measure λ on $\mathcal{K}$.

5.2.2. *Diagonal of $N \in \mathcal{N}$*

For any Y, observe that

$$(RYR^*)_{ij} = R_{i\lambda}Y_{\lambda\mu}R_{j\mu} = Y_{iRjR}$$

since R acts as a function on the indices. Similarly for B.

Proposition 5.1. *Let $N \in \mathcal{N}_+$. Then the diagonal of N is constant.*

Proof. Assume that $N_{11} = \max\limits_{i} N_{ii}$, is the largest diagonal entry. Start with

$$2\,N_{11} = N_{1R\,1R} + N_{1B\,1B}$$

Since N_{11} is maximal, both terms on the right-hand side agree with N_{11}. Generally, if $N_{ii} = N_{11}$ then $N_{iRiR} = N_{iBiB} = N_{11}$. Proceeding inductively yields $N_{ii} = N_{11}$ for all i since the graph is strongly connected.   □

5.2.3. *Diagonal of $M \in \mathcal{M}$*

Define the diagonal matrix $\omega = \mathrm{diag}(\pi)$, where π is the invariant distribution for A.

Proposition 5.2. *Let $M \in \mathcal{M}_+$. Then the diagonal of M is a scalar multiple of ω.*

Proof. Denote the diagonal of M by D, i.e., $M_{ii} = D_i$. Start with

$$2\,M_{ij} = (R^*MR + B^*MB)_{ij} = R_{\lambda i}M_{\lambda\mu}R_{\mu j} + B_{\lambda i}M_{\lambda\mu}B_{\mu j}$$

Let $\bar{\delta}_{ij} = 1 - \delta_{ij}$, i.e., one if $i \neq j$, zero otherwise. Take $i = j$ and split off the diagonal terms on the right to get

$$
\begin{aligned}
2\,D_i &= R_{\lambda i}D_\lambda R_{\lambda i} + B_{\lambda i}D_\lambda B_{\lambda i} + R_{\lambda i}M_{\lambda\mu}R_{\mu i}\bar{\delta}_{\lambda\mu} + B_{\lambda i}M_{\lambda\mu}B_{\mu i}\bar{\delta}_{\lambda\mu} \\
&= R_{\lambda i}D_\lambda + B_{\lambda i}D_\lambda + R_{\lambda i}M_{\lambda\mu}R_{\mu i}\bar{\delta}_{\lambda\mu} + B_{\lambda i}M_{\lambda\mu}B_{\mu i}\bar{\delta}_{\lambda\mu} \\
&= 2A_{\lambda i}D_\lambda + R_{\lambda i}M_{\lambda\mu}R_{\mu i}\bar{\delta}_{\lambda\mu} + B_{\lambda i}M_{\lambda\mu}B_{\mu i}\bar{\delta}_{\lambda\mu}
\end{aligned}
$$

using the fact that the entries of R and B are 0's and 1's. Thus,

$$D_i - (DA)_i = \tfrac{1}{2}\left(R_{\lambda i}M_{\lambda\mu}R_{\mu i}\bar{\delta}_{\lambda\mu} + B_{\lambda i}M_{\lambda\mu}B_{\mu i}\bar{\delta}_{\lambda\mu}\right)$$

Now sum both sides over i. Since A is stochastic, $\sum(DA)_i = \sum D_i$ and the left-hand side vanishes. Since the right-hand side is a sum of nonnegative terms adding to zero, the right-hand side is identically zero. Thus,

$$D = DA$$

is a left fixed point of A. By irreducibility, it must be a multiple of π. $\qquad\square$

5.3. *Specification of operators N and M*

In the following, we will use M and N to refer to the specific operators defined by

$$M = \mathrm{Mat}(u\Omega_{\otimes 2}) = \langle K^* J K \rangle$$

and

$$N = \mathrm{Mat}(\Omega_{\otimes 2}I_{\mathrm{vec}}) = \langle K K^* \rangle$$

Another special operator in $\mathcal{M}$ is defined as

$$\tilde{M} = \mathrm{Mat}(\Omega_{\mathrm{vec}}\Omega_{\otimes 2}) = \langle K^* \Omega K \rangle$$

We can pass to trace-zero operators by subtracting off the diagonals defining

$$M_0 = M - \tau\omega \tag{6}$$
$$N_0 = J - N \tag{7}$$

For now, τ is defined by the above relation. We will see that the diagonal of N is $\mathrm{diag}(u) = I$. The reason for complementing with respect to J, i.e. subtracting from 1 everywhere will become clear after Proposition 5.4.

Note that at level 1,

$$u\Omega = uu^\dagger\pi = n\pi = n\pi A \tag{8}$$
$$\Omega u^\dagger = u^\dagger = Au^\dagger \tag{9}$$

So for level two, we define the fields

$$\pi_{\otimes 2} = u\Omega_{\otimes 2}$$
$$u^\dagger_{\otimes 2} = \Omega_{\otimes 2}I_{\mathrm{vec}}$$

with analogous expressions for each level ℓ. Indeed, at level two these satisfy

$$\pi_{\otimes 2}A_{\otimes 2} = \pi_{\otimes 2}$$
$$A_{\otimes 2}u^\dagger_{\otimes 2} = u^\dagger_{\otimes 2}$$

extending relations (8) and (9). The families $\{u_{\otimes\ell}\}$ and $\{\pi_{\otimes\ell}\}$ provide extensions to scalar fields the basic all-ones vector u and invariant distribution π of level one.

5.4. *Basic level two relations*

In detailing the structure of the kernel, we need notations corresponding to the range classes. Given an idempotent $e \in \mathcal{K}$, define $\underset{\sim}{\rho}(e)$ to be the *state vector* of the range of e, namely, it is a 0-1 vector with $\underset{\sim}{\rho}(e)_i = 1$ exactly when $ie = e$. The corresponding matrix

$$\rho(e) = \mathrm{diag}(\underset{\sim}{\rho}(e))$$

Proposition 5.3. *We have the relations:*

$$\mathrm{Mat}(\Omega_{\mathrm{vec}}\Omega_{\otimes 2}) = \tilde{M} = \frac{n}{r^2}\,\langle\underset{\sim}{\rho}^\dagger\underset{\sim}{\rho}\rangle \qquad \mathrm{Mat}(\Omega_{\otimes 2}\Omega_{\mathrm{vec}}) = \Omega\Omega^* = \left(\sum p_i^2\right)J$$

$$\mathrm{Mat}(I_{\mathrm{vec}}\Omega_{\otimes 2}) = \langle K^*K\rangle = n\omega \qquad \mathrm{Mat}(\Omega_{\otimes 2}I_{\mathrm{vec}}) = N$$

$$\mathrm{Mat}(J_{\mathrm{vec}}\Omega_{\otimes 2}) = M \qquad\qquad\quad \mathrm{Mat}(\Omega_{\otimes 2}J_{\mathrm{vec}}) = J$$

Note we are writing J_{vec} for parallel formulation as an alternative for u.

Proof. We will give the proof of the first relation later, as it relies on Friedman's Theorem. Consider

$$\mathrm{Mat}(\Omega_{\otimes 2}\Omega_{\mathrm{vec}}) = \langle K\Omega K^* \rangle = \langle \Omega K^* \rangle$$
$$= \Omega\Omega^* = u^\dagger \pi \pi^\dagger u = (\pi\pi^\dagger)u^\dagger u$$

where we use the fact that elements of $\mathcal{K}$ are stochastic matrices. Next, observe that k^*k is always diagonal. Indeed

$$(k^*k)_{ij} = k_{\lambda i}k_{\lambda j}$$

indicates a summation over λ such that k maps λ to both i and j. In fact it is the size of the block of the partition that k maps into i when $i = j$. Furthermore, for any vector v, we have the tautology

$$\mathrm{diag}(v) = \mathrm{diag}(\mathrm{diag}(v)u^\dagger)$$

We have

$$\mathrm{Mat}(I_{\mathrm{vec}}\Omega_{\otimes 2}) = \langle K^*K \rangle$$

and we compute

$$\langle K^*Ku^\dagger \rangle = \langle K^*u^\dagger \rangle = \Omega^*u^\dagger = n\,\pi^\dagger .$$

hence the result. And

$$\mathrm{Mat}(\Omega_{\otimes 2}I_{\mathrm{vec}}) = \langle KK^* \rangle = N$$

by our definition. We have

$$\mathrm{Mat}(J_{\mathrm{vec}}\Omega_{\otimes 2}) = \langle K^*JK \rangle$$

the definition of M. And finally,

$$\mathrm{Mat}(\Omega_{\otimes 2}J_{\mathrm{vec}}) = \langle KJK^* \rangle = J$$

immediately. □

For some trace calculations, note that since $\mathrm{tr}\,\omega = 1$, equation (6) effectively says

$$\tau = \mathrm{tr}\,M$$

Now, we note that

$$\mathrm{tr}\,NJ = \mathrm{tr}\,\langle KK^*J \rangle = \mathrm{tr}\,\langle K^*JK \rangle = \mathrm{tr}\,M = \tau$$

as well.

5.4.1. *Interpretation of τ*

In describing the kernel, we have two basic constants, the rank r and τ. We want to interpret τ as describing features of the kernel. Start with

Proposition 5.4. *Consider an idempotent $e \in \mathcal{K}$, determined by a partition and range $(\mathcal{P}, \mathcal{R})$. For all k in the local group G_e for which e acts as the identity,*

$$kk^* = ee^*$$

is a 0 -1 matrix such $(kk^*)_{ij} = 1$ if i and j are in the same block of $\mathcal{P}$.

Proof. Writing out $k_{i\lambda}k_{j\lambda}$ shows that this sum counts 1 precisely when both i and j map to the same range element, i.e., they are in the same block of $\mathcal{P}$. $\qquad \square$

We have an interpretation of the entries N_{ij} as the probability that vertices i and j appear in the same block of a partition. Thus, the entry $(J - N)_{ij}$ gives the probability that i and j are *split*, i.e., they never appear together in the same block of a partition.

Remark 5.1. Note that ee^* is the same for all cells in a given row of the kernel. And is independent of k in a given cell. Thus,

$$N = \langle KK^* \rangle = \langle EE^* \rangle$$

where it is sufficient in this last average to consider only the idempotents in any single column of the kernel.

The column sum $(uee^*)_i$ is the size of the block of $\mathcal{P}$ containing i. Now $\operatorname{tr} Jee^* = \operatorname{tr} Jkk^*$ sums all the ones in kk^*, and for all local k,

$\operatorname{tr} k^*Jk = \operatorname{tr} Jkk^* = $ equals the sum of the squares of the blocksizes of $\mathcal{P}$.

It follows that

Proposition 5.5. $\tau = \operatorname{tr} M$ *is the average sum of squares of the blocksizes over the partitions $\mathcal{P}$ of $\mathcal{K}$.*

We continue with some relations useful for deriving and describing properties of the kernel.

5.5. *Useful relations between an idempotent and its range matrix*

We make some useful observations regarding an idempotent e and its range matrix $\rho(e)$.

Proposition 5.6. *For any idempotent e,*

    1. $\rho(e)e = \rho(e)\,, \qquad e^*\rho(e) = \rho(e)\,, \qquad e\rho(e) = e.$

    2. $ee^*\rho(e) = e$.

Proof. For #1, the first and last relations follow since e fixes its range. Transposing the first yields the second. And #2 follows directly. □

For $k \in \mathcal{K}$, let $\underset{\sim}{\rho}(k)$ denote the state vector of the range of k and $\rho(k)$ the corresponding diagonal matrix.

Corollary 5.1.

    1. For any $k \in \mathcal{K}$, $k\rho(k) = k$.

    2. $e^*\underset{\sim}{\rho}(e)^\dagger = \underset{\sim}{\rho}(e)^\dagger$.

    3. $e\underset{\sim}{\rho}(e)^\dagger = u^\dagger.$

Proof. For #1, write $k = ke$ with $\rho(k) = \rho(e)$. Then

$$k\rho(k) = ke\rho(e) = ke = k$$

as in the Proposition. And #2 and #3 follow from the Proposition via $\mathrm{diag}(v)u^\dagger = v^\dagger$. □

6. Projections

We can compute the row and column projections in the kernel using our operators M and N.

The idempotents having the same range $\mathcal{R}$, say, form a *left-zero semigroup*. That is, if e_1 and e_2 have the same range, then $e_1e_2 = e_1$, since e_1 determines the partition and the range is fixed by both idempotents. Correspondingly, the idempotents having the same partition $\mathcal{P}$, say, form a *right-zero semigroup*. That is, if e_1 and e_2 have the same partition, then $e_1e_2 = e_2$, since the partition for both is the same and an element in the range of e_2 is mapped by e_1 into an element in the same block of the common partition, which in turn is mapped back to the original element by e_2.

Remark 6.1. Note that in this terminology Proposition 5.6 says that the pair $\{e, \rho(e)\}$ form a left-zero semigroup.

Notation 6.1. It is convenient to fix an enumeration of the partitions $\{\mathcal{P}_1, \mathcal{P}_2, \ldots\}$ and ranges $\{\mathcal{R}_1, \mathcal{R}_2, \ldots\}$. We set ρ_j to be the matrix $\mathrm{diag}(\underline{\rho}_j)$ where $\underline{\rho}_j$ is the state vector for $\mathcal{R}_j$.

6.1. *Column projections*

Denote e_{ij} the idempotents for range $\mathcal{R}_j$. Define the column projection to be the average

$$P_j = \alpha_\lambda e_{\lambda j} \ .$$

The measure α is the component of λ on the partitions.

Proposition 6.1.

1. $P_j^2 = P_j$ *is an idempotent.*

2. $P_j = N \rho_j.$

Proof. For #1, with u a vector of all ones,

$$\begin{aligned} P_j^2 &= \alpha_\lambda \alpha_\mu e_{\lambda j} e_{\mu j} \\ &= \alpha_\lambda \alpha_\mu u_\mu e_{\lambda j} \\ &= \alpha_\lambda e_{\lambda j} = P_j \end{aligned}$$

using the fact that the idempotents with a common range form a left-zero semigroup and that the α's are a probability distribution.

For #2, recall the relation $e^* \rho(e) = \rho(e)$ from Proposition 5.6, and we have

$$N \rho_j = \alpha_\lambda e_{\lambda j} e_{\lambda j}^* \rho_j = \alpha_\lambda e_{\lambda j} \rho_j = P_j$$

with $e_{ij} \rho_j = e_{ij}$ as each e_{ij} fixes the common range. $\qquad \square$

6.2. *Row projections*

Analogously, define $Q_i = \beta_\mu e_{i\mu}$, the average idempotent with common partition $\mathcal{P}_i$. As in #1 of Prop. 6.1, $Q_i^2 = Q_i$, where now the idempotents form a right-zero semigroup.

We need the fact [to be proved later] that the average of the range matrices is r times $\omega = \operatorname{diag}(\pi)$. We state this in the form

$$\omega = r^{-1}\beta_\mu\rho_\mu \tag{10}$$

where β is the part of the measure λ on the range classes.

Proposition 6.2.

$$Q_i = r\, e_i e_i^* \omega$$

where e_i is any idempotent with partition $\mathcal{P}_i$.

Proof. First, recall that $e_{ij}e_{ij}^* = e_i e_i^*$ for any idempotent e_{ij} with partition $\mathcal{P}_i$. And $e_i e_i^* \rho(e_i) = e_i$. Using equation (10) we have

$$r\, e_i e_i^* \omega = e_i e_i^* \beta_\mu \rho_\mu$$
$$= \beta_\mu e_{i\mu} e_{i\mu}^* \rho_\mu$$
$$= \beta_\mu e_{i\mu} = Q_i$$

as required. □

6.3. *Average idempotent*

We now compute the average idempotent $\langle E \rangle$.

Theorem 6.1. *The average idempotent satisfies*

$$\langle E \rangle = rN\,\omega$$

Proof. From Proposition 6.2, we need to average Q_i over the partitions. We get

$$\langle E \rangle = \alpha_\lambda Q_\lambda = r\,\alpha_\lambda e_{\lambda 1} e_{\lambda 1}^* \omega = r\,N\omega$$

using the idempotents with common range $\mathcal{R}_1$. □

7. Equipartitioning

In the current framework, as in Ref. 2, we express Friedman's Theorem this way:

Theorem 7.1. *[Refs. 2,4]*
For $k \in \mathcal{K}$, let $\underset{\sim}{\rho}(k)$ denote the 0-1 vector with support the range of k. Then

$$\pi k = \frac{1}{r}\underset{\sim}{\rho}(k)$$

where r is the rank of the kernel.

We will refer to this theorem as BM/F.

Corollary 7.1. *For any $k \in \mathcal{K}$, we have*

$$\pi k k^* = \frac{1}{r} u$$

Proof. By Corollary 5.1, we have

$$k\rho(k) = k \qquad \text{which implies} \qquad k\underset{\sim}{\rho}(k)^\dagger = u^\dagger$$

Now transpose. $\square$

This corollary can be rephrased as stating that each block of any partition has π-probability equal to $1/r$. Here we will give an independent proof with a functional analytic flavor.

7.1. *A and* Δ

Working with two colors R and B, it is convenient to introduce the operator Δ

$$\Delta - \tfrac{1}{2}(R - B)$$

Then we check that:

$$A_{\otimes 2} = \tfrac{1}{2}(R^{\otimes 2} + B^{\otimes 2}) = A^{\otimes 2} + \Delta^{\otimes 2}$$

Hence, elements $Y \in \mathcal{N}$ satisfy

$$AYA^* + \Delta Y \Delta^* = Y$$

and those in $\mathcal{M}$ satisfy

$$A^* Y A + \Delta^* Y \Delta = Y$$

In particular, N and M satisfy these equations accordingly:

$$ANA^* + \Delta N \Delta^* = N \tag{11}$$
$$A^* M A + \Delta^* M \Delta = M \tag{12}$$

7.2. *Friedman's Theorem d'après Budzban-Mukherjea*

First, some notation

Notation 7.1. Let $\mathcal{E}(\mathcal{K})$ denote the set of all idempotents of the kernel $\mathcal{K}$.

Start with

Proposition 7.1. *For every $k \in \mathcal{K}$, $\pi \Delta k = 0$. Equivalently,* $\pi R k = \pi B k = \pi k$.

Proof. Multiplying eq. (11) on the left by π and on the right by $\pi^\dagger$, we have, using $\pi A = \pi$, $A^* \pi^\dagger = \pi^\dagger$,

$$\pi N \pi^\dagger + \pi \Delta N \Delta^* \pi^\dagger = \pi N \pi^\dagger$$

Since $N = \langle K K^* \rangle$, we have

$$\pi \Delta N \Delta^* \pi^\dagger = 0$$
$$= \langle \pi \Delta K K^* \Delta^* \pi^\dagger \rangle = \langle \|\pi \Delta K\|^2 \rangle$$

All terms are nonnegative so that $\pi \Delta k = 0$ for every k. From the definition of Δ the alternative formulation results after averaging $\pi R k = \pi B k$ to $\pi A k = \pi k$. $\qquad \square$

And directly from the definition of N

Corollary 7.2. *We have $\pi \Delta N = 0$.*

We extend to words in the semigroup generated by R and B.

Proposition 7.2. *For every $w \in \mathcal{S}$, and $k \in \mathcal{K}$, $\pi w k = \pi k$.*

Proof. This follows by induction on l, the length of w. Proposition 7.1 is the result for $l = 1$. Let w_{l+1} be a word of length $l + 1$, then $w_{l+1} =$ one of $R w_l$, $B w_l$, where w_l has length l. Since $w_l k \in \mathcal{K}$ for $k \in \mathcal{K}$, we have, by Proposition 7.1,

$$\pi w_{l+1} k = \pi R w_l k = \pi B w_l k$$
$$= \pi w_l k \qquad \text{(averaging)}$$
$$= \pi k \qquad \text{(by induction)} \qquad \square$$

Proposition 7.3. *For any $e \in \mathcal{E}(\mathcal{K})$, $w \in \mathcal{S}$,*

$$\pi w e = (1/r)\, \underset{\sim}{\rho}(e) \qquad and \qquad \pi w e e^* = (1/r)\, u.$$

Proof. From Corollary 7.2, $\pi \Delta N = 0$. Equivalently, $N\Delta^* \pi^\dagger = 0$. Hence, multiplying equation (11) on the right by $\pi^\dagger$, using $A^* \pi^\dagger = \pi^\dagger$, we find

$$AN\pi^\dagger = N\pi^\dagger$$

So $N\pi^\dagger$ is fixed by A, hence is a constant vector. Transposing back, say $\pi N = au$ for some constant a. Now let $e \in \mathcal{E}(\mathcal{K})$ and take the e_i with the same range as e in the averaging defining N as $\langle EE^* \rangle$. Then from Proposition 7.2 we have $\pi e e_i = \pi e_i$. Multiplying by e_i^* and averaging yields $\pi e N = \pi N$, hence $\pi e N e^* = \pi N e^*$. Since the idempotents of a column are a left-zero semigroup, $ee_i = e$, $e_i^* e^* = e^*$. Thus,

$$eNe^* = e\,\alpha_\mu e_\mu e_\mu^* e^* = ee^*$$

Thus on the one hand $\pi e N e^* = \pi N e^*$ and on the other $\pi e N e^* = \pi e e^*$. Or $\pi e e^* = \pi N e^* = aue^* = au$ for any idempotent e. Thus, by Prop. 5.6, #2, $\pi e = a\rho(e)$ and $\pi e u^\dagger = 1 = ar$ yields $a = 1/r$, where r is the rank of the kernel. So, for any $w \in \mathcal{S}$, $\pi w e = \pi e = (1/r)\rho(e)$ and Cor. 5.1, #3 yields $\pi w e e^* = (1/r)u$, as required. $\square$

Now taking $w \in \mathcal{K}$ yields $\pi k = (1/r)\rho(k)$, the theorem of Budzban-Mukherjea/Friedman.

Averaging the second relation in the above Proposition yields

Corollary 7.3. *For any $w \in \mathcal{S}$, $\pi w N = \frac{1}{r} u$.*

7.2.1. *More on projections*

Referring to the column projections P_j of Schrödinger algebra6, we have

$$\pi P_j = \pi N \rho_j = \tfrac{1}{r} u \rho_j = \tfrac{1}{r} \rho_j$$

And for row projections Q_j, via Proposition 7.3,

$$\pi Q_j = r\pi e_i e_i^* \omega = u\omega = \pi$$

Since Q_j is a stochastic matrix, $Q_j u^\dagger = u^\dagger$, we have the result

Proposition 7.4. *Each of the row projections Q_j commutes with Ω.*

7.2.2. *Local groups*

Let G_{ij} be the local group with partition $\mathcal{P}_i$ and range $\mathcal{R}_j$. For convenience, when averaging over G_{ij}, we use the abbreviated form $\displaystyle\sum{}' = \frac{1}{|G_{ij}|} \sum .$

An important fact is

Lemma 7.1. *The average group element is* $u^\dagger \underset{\sim}{\rho}/r$, *specifically,*

$$\sideset{}{'}\sum_{G_{ij}} k = r^{-1} u^\dagger \underset{\sim}{\rho}_j$$

Proof. From the structure of the kernel, for a fixed e, $e\mathcal{K}e$ is the local group having e as local identity. So take $e = e_{ij}$, the identity for G_{ij}. Since $\Omega = \langle K \rangle$, we have, via Prop. 7.3,

$$\langle e K e \rangle = e \Omega e = e u^\dagger \pi e = u^\dagger \pi e = r^{-1} u^\dagger \underset{\sim}{\rho}(e) \qquad \square$$

Now we see that to average over the kernel, we need only average the above relation over the ranges. This yields

$$\Omega = r^{-1} u^\dagger \beta_\mu \underset{\sim}{\rho}_\mu = r^{-1} u^\dagger u \beta_\mu \rho_\mu = r^{-1} J \beta_\mu \rho_\mu \tag{13}$$

since $\underset{\sim}{\rho}(e) = u\rho(e)$. Comparing diagonals, we thus derive equation (10)

$$\omega = r^{-1} \beta_\mu \rho_\mu$$

used in computing the row projections, finding as well

$$\pi = r^{-1} \beta_\mu \underset{\sim}{\rho}_\mu$$

We are now in a position to prove the first relation of Proposition 5.3:

Theorem 7.2.

$$\operatorname{Mat}(\Omega_{\mathrm{vec}} \Omega_{\otimes 2}) = \frac{n}{r^2} \langle \underset{\sim}{\rho}^\dagger \underset{\sim}{\rho} \rangle = \frac{n}{r^2} \langle \rho J \rho \rangle$$

Proof.

$$\langle K^* \Omega K \rangle = \langle K^* u^\dagger \pi K \rangle = r^{-1} \langle K^* u^\dagger \underset{\sim}{\rho}(K) \rangle$$

$$= r^{-1} \sideset{}{'}\sum_{i,j} \sideset{}{'}\sum_{G_{ij}} k^* u^\dagger \underset{\sim}{\rho}_j$$

summing over the cells of the kernel, the primes indicating averaging. By Lemma 7.1, we have

$$r^{-1} \sideset{}{'}\sum_{i,j} \sideset{}{'}\sum_{G_{ij}} k^* u^\dagger \underset{\sim}{\rho}_j = r^{-2} \sideset{}{'}\sum_{i,j} \underset{\sim}{\rho}_j{}^\dagger u u^\dagger \underset{\sim}{\rho}_j = nr^{-2} \sideset{}{'}\sum_{i,j} \underset{\sim}{\rho}_j{}^\dagger \underset{\sim}{\rho}_j$$

$$= nr^{-2} \sideset{}{'}\sum_{i,j} \rho_j u^\dagger u \rho_j = nr^{-2} \langle \rho J \rho \rangle \qquad \square$$

8. Properties of M, N, and Ω

Here we look at some relations among the main operators of interest, including $\tilde{M}$.

Start with

Proposition 8.1.

1. $\Omega N = r^{-1} J$.

2. $M\Omega = n\Omega^* \Omega$.

Proof. $\Omega N = u^\dagger \pi N = r^{-1} u^\dagger u = r^{-1} J$ as required. And, noting that $J\Omega = n\Omega$,

$$M\Omega = \langle K^* J K \Omega \rangle = \langle K^* J \Omega \rangle = n\langle K^* \rangle \Omega = \Omega^* \Omega$$

via $k\Omega = \Omega$, for all $k \in \mathcal{K}$.   □

And consequently,

Corollary 8.1. NM *commutes with* Ω.

Proof. First,

$$NM\Omega = nN\Omega^*\Omega = n(\Omega N)^*\Omega = (n/r)J\Omega = (n^2/r)\Omega$$

And second,

$$\Omega NM = r^{-1} JM = r^{-1}\langle JK^* JK \rangle = r^{-1}\langle (KJ)^* JK \rangle$$
$$= r^{-1}\langle J^2 K \rangle = r^{-1} n J\Omega = (n^2/r)\Omega$$

as required.   □

Observe in the proof the relation

$$JM = n^2 \Omega$$

The case of doubly stochastic A turns out to be particularly interesting.

Recall the basic fact that two symmetric matrices commute if and only if their product is symmetric.

Theorem 8.1. *If A is doubly stochastic, then $\{M, N, J\}$ generate a commutative algebra.*

Proof. If A is doubly stochastic, then $u = n\pi$ and $n\Omega = J$. Thus $JM = n^2\Omega = nJ$ is symmetric. And $JN = n\Omega N = (n/r)J$ is symmetric. So J commutes with M and N. We check that M and N commute:

$$
\begin{aligned}
MN &= \langle K^* u^\dagger u K \rangle N \\
&= n^2 \langle K^* \pi^\dagger \pi K N \rangle \\
&= (n^2/r) \langle K^* \pi^\dagger u \rangle \\
&= (n^2/r) \langle \Omega K \rangle^* \\
&= (n^2/r) (\Omega^*)^2 = (n/r) J.
\end{aligned}
$$

which is symmetric as well. □

Another feature involves $\tilde{M}$.

Theorem 8.2.

$$
N\tilde{M} = \frac{n}{r}\Omega
$$

Proof. We have

$$
N\tilde{M} = nr^{-2}N\langle \rho J \rho \rangle
$$

Note that for any P_j, $P_j J = J$. Fixing a range, we drop subscripts for convenience. And with $P = P_j$ the corresponding column projection,

$$
N\rho J\rho = PJ\rho = J\rho
$$

By equation (13), averaging yields $r\Omega$. Hence the result. □

9. Zeon hierarchy

Now we pass from the tensor hierarchy to the zeon hierarchy by looking at representations of the semigroup $F_\circ(V)$ acting on zeon Fock space. We begin with the basic definitions and constructions. Then we proceed with statements parallelling those of Schrödinger algebra4, including proofs when they illustrate important differences between the two systems.

Remark 9.1. The introductory material in this section is largely taken from Ref. 3.

Consider the exterior algebra generated by a chosen basis $\{\mathbf{e}_i\} \subset V$, with relations $\mathbf{e}_i \wedge \mathbf{e}_j = -\mathbf{e}_j \wedge \mathbf{e}_i$. Denoting multi-indices by roman capital letters $I = (i_1, i_2, \ldots, i_k)$, J, K, etc., at level k, a basis for $V^{\wedge k}$ is given by

$$
\mathbf{e}_I = \mathbf{e}_{i_1} \wedge \cdots \wedge \mathbf{e}_{i_k}
$$

with I running through all k-subsets of $\{1, 2, \ldots, n\}$, i.e., k-tuples with distinct components. For $\mathrm{Mt}(f)$, $f \in F(V)$, define the matrix

$$(\mathrm{Mt}(f)^{\vee k})_{\mathrm{IJ}} = |(\mathrm{Mt}(f)^{\wedge k})_{\mathrm{IJ}}|$$

taking absolute values entry-wise. It is important to observe that we are not taking the fully symmetric representation of $\mathrm{End}(\mathcal{V})$, which would come by looking at the action on boson Fock space, spanned by symmetric tensors. However, note that the fully symmetric representation is given by maps induced by the action of $\mathrm{Mt}(f)$ on the algebra generated by commuting variables $\{\mathbf{e}_i\}$. We take this viewpoint as the starting point of the construction of the zeon Fock space, $\mathcal{Z}$, to be defined presently.

Definition 9.1. A *zeon algebra* is a commutative, associative algebra generated by elements $\mathbf{e}_i$ such that $\mathbf{e}_i^2 = 0$, $i \geq 1$.

For a standard zeon algebra, $\mathcal{Z}$, the elements $\mathbf{e}_i$ are finite in number, n, and are the basis of an n-dimensional vector space, $\mathcal{V} \approx \mathbb{Q}^n$. We assume no further relations among the generators $\mathbf{e}_i$. Then the kth *zeon tensor power* of $\mathcal{V}$, denoted $\mathcal{V}^{\vee k}$, is the degree k component of the graded algebra $\mathcal{Z}$, with basis

$$\mathbf{e}_{\mathrm{I}} = \mathbf{e}_{i_1} \cdots \mathbf{e}_{i_k}$$

analogously to the exterior power except now the variables commute. The assumptions on the $\mathbf{e}_i$ imply that $\mathcal{V}^{\vee k}$ is isomorphic to the subspace of symmetric tensors spanned by elementary tensors with no repeated factors. *As vector spaces,*

$$\mathcal{V}^{\vee k} \approx \mathcal{V}^{\wedge k}$$

The *zeon Fock space* is $\mathcal{Z}$ presented as a graded algebra

$$\mathcal{Z} = \mathbb{Q} \oplus \left(\bigoplus_{k \geq 1} \mathcal{V}^{\vee k} \right)$$

Since $\mathcal{V}$ is finite-dimensional, k runs from 1 to $n = \dim \mathcal{V}$.

A linear operator $W \in \mathrm{End}(\mathcal{V})$ extends to the operator $W^{\vee k} \in \mathrm{End}(\mathcal{V}^{\vee k})$. The *second quantization* of W is the induced map on $\mathcal{Z}$.

For the exterior algebra, the IJ^{th} component of $W^{\wedge k}$ is the determinant of the corresponding submatrix of W, with rows indexed by I and columns by J. Having dropped the signs, the IJ^{th} component of $W^{\vee k}$ is the permanent of the corresponding submatrix of W.

For W of the form $\mathrm{Mt}(f)$ corresponding to a function $f \in F(V)$, the resulting components of $W^{\vee k}$ are exactly the absolute values of the entries of $W^{\wedge k}$, as we wanted. At each level k, there is an induced map

$$\mathrm{End}(\mathcal{V}) \to \mathrm{End}(\mathcal{V}^{\vee k}) , \qquad \mathrm{Mt}(f) \to \mathrm{Mt}(f)^{\vee k}$$

satisfying

$$(\mathrm{Mt}(f_1 f_2))^{\vee k} = (\mathrm{Mt}(f_1)\mathrm{Mt}(f_2))^{\vee k} = \mathrm{Mt}(f_1)^{\vee k}\mathrm{Mt}(f_2)^{\vee k} \qquad (14)$$

giving, for each k, a representation of the semigroup $F_{\mathrm{o}}(V)$ as endomorphisms of $\mathcal{V}^{\vee k}$. However, for general W_1, W_2, the homomorphism property, (14), no longer holds, i.e., $(W_1 W_2)^{\vee k}$ does not necessarily equal $W_1^{\vee k} W_2^{\vee k}$. It is not hard to see that a sufficient condition is that W_1 have at most one non-zero entry per column or that W_2 have at most one non-zero entry per row. For example, if one of them is diagonal, as well as the case where both correspond to functions.

What is the function, f_k, corresponding to $\mathrm{Mt}(f)^{\vee k}$, i.e., such that $M(f_k) = \mathrm{Mt}(f)^{\vee k}$? For degree 1, we have from (2)

$$\mathbf{e}_i \mathrm{Mt}(f) = \mathbf{e}_{f(i)} \qquad (15)$$

And for the induced map at degree k, taking products in $\mathcal{Z}$,

$$\mathbf{e}_{\mathrm{I}} \mathrm{Mt}(f)^{\vee k} = (\mathbf{e}_{i_1} \mathrm{Mt}(f)) (\mathbf{e}_{i_2} \mathrm{Mt}(f)) \cdots (\mathbf{e}_{i_k} \mathrm{Mt}(f))$$

$$= \mathbf{e}_{f(i_1)} \mathbf{e}_{f(i_2)} \cdots \mathbf{e}_{f(i_k)}$$

We see that the degree k maps are those induced on k-subsets of V mapping

$$\{i_1, \ldots, i_k\} \to \{f(i_1), \ldots, f(i_k)\}$$

with the property that the image in the zeon algebra is zero if $f(i_l) = f(i_m)$ for any pair i_l, i_m. Thus the second quantization of $\mathrm{Mt}(f)$ corresponds to the induced map, the second quantization of f, extending the domain of f from V to the power set 2^V.

Now we proceed to the degree 2 component in the zeon case.

9.1. The degree 2 component of $\mathcal{Z}$

Working in degree 2, we denote indices $\mathrm{I} = (i, j)$, as usual, instead of (i_1, i_2). For given n, X, X', etc., are vectors in $\mathcal{V}^{\vee 2} \approx \mathbb{Q}^{\binom{n}{2}}$.
As a vector space, $\mathcal{V}$ is isomorphic to $\mathbb{Q}^n$.

Denote by $\mathrm{Sym}(\mathcal{V})$ the space of symmetric matrices acting on $\mathcal{V}$.

Definition   The mapping

$$\mathrm{Mat}\colon \mathcal{V}^{\vee 2} \to \mathrm{Sym}(\mathcal{V})$$

is the linear embedding taking the vector $X = (x_{ij})$ to the symmetric matrix $\hat{X}$ with components

$$\hat{X}_{ij} = \begin{cases} x_{ij}, & \text{for } i < j \\ 0, & \text{for } i = j \end{cases}$$

and the property $\hat{X}_{ji} = \hat{X}_{ij}$ fills out the matrix.

We will use the explicit notation $\mathrm{Mat}(X)$ as needed for clarity.

Equip $\mathcal{V}^{\vee 2}$ with the inner product

$$\langle X, X' \rangle = \frac{1}{2} \mathrm{tr}\, \hat{X}\hat{X}' = X_{\lambda\mu} X'_{\lambda\mu}$$

9.2. Basic Identities

Multiplying X with u, we observe that

$$X u^{\dagger} = \frac{1}{2} \mathrm{tr}\, \hat{X} J = u X^{\dagger} \tag{16}$$

Observe also that if D is diagonal, then $\mathrm{tr}\, D = \mathrm{tr}\, DJ$.

Proposition 9.1. *(Basic Relations) We have*

1. $\mathrm{Mat}(X A^{\vee 2}) = A^* \hat{X} A - D^+$, *where* D^+ *is a diagonal matrix satisfying* $\mathrm{tr}\, D^+ = \mathrm{tr}\, A^* \hat{X} A$.

2. $\mathrm{Mat}(A^{\vee 2} X^{\dagger}) = A \hat{X} A^* - D^-$, *where* D^- *is a diagonal matrix satisfying* $\mathrm{tr}\, D^- = \mathrm{tr}\, A \hat{X} A^*$.

3. If A *and* X *have nonnegative entries, then* D^+ *and* D^- *have nonnegative entries. In particular, in that case, vanishing trace for* $D^{\pm}$ *implies vanishing of the corresponding matrix.*

Proof. The components of $X A^{\vee 2}$ are

$$(X A^{\vee 2})_{ij} = \theta_{ij} \theta_{\lambda\mu} (x_{\lambda\mu} A_{\lambda i} A_{\mu j} + x_{\lambda\mu} A_{\mu i} A_{\lambda j})$$
$$= \theta_{ij} (A^* \hat{X} A)_{ij}$$

with the *theta symbol* for pairs of single indices

$$\theta_{ij} = \begin{cases} 1, & \text{if } i < j \\ 0, & \text{otherwise} \end{cases}$$

Note that the diagonal terms of $\hat{X}$ vanish anyway. And $A^*\hat{X}A$ will be symmetric if $\hat{X}$ is. Since the left-hand side has zero diagonal entries, we can remove the theta symbol and compensate by subtracting off the diagonal, call it D^+. Taking traces yields #1. And #2 follows similarly. $\qquad\square$

Remark 9.2. Observe that D^+ and D^- may be explicitly given by

$$D_{ii}^+ = 2\,x_{\lambda\mu}A_{\lambda i}A_{\mu i}$$
$$D_{ii}^- = 2\,x_{\lambda\mu}A_{i\lambda}A_{i\mu}$$

where for D^+ the A elements are taken within a given column, while for D^-, the A elements are in a given row.

Connections between zeons and nonnegativity that we develop here give some indication that their natural place is indeed in probability theory and quantum probability theory.

Parallelling Proposition 4.2, we see the role of nonnegativity appearing.

Proposition 9.2. *Let X and A be nonnegative. Then*

$$\hat{X} = A^*\hat{X}A \Rightarrow XA^{\vee 2} = X$$
$$\hat{X} = A\hat{X}A^* \Rightarrow A^{\vee 2}X^\dagger = X^\dagger$$

Proof. We have

$$D^+ = A^*\hat{X}A - \mathrm{Mat}(XA^{\vee 2})$$

If $\hat{X} = A^*\hat{X}A$, then since $\hat{X}$ has vanishing trace, $\operatorname{tr} A^*\hat{X}A = 0$. So $\operatorname{tr} D^+ = 0$, hence $D^+ = 0$, and $\hat{X} = A^*\hat{X}A = \mathrm{Mat}(XA^{\vee 2})$. The second implication follows similarly. $\qquad\square$

9.3. *Trace Identities*

We continue with trace identities for zeons. The proofs follow from the above relations much as they do in the tensor case.

Proposition 9.3. *We have*
  1. $XA^{\vee 2}u^\dagger = \frac{1}{2}\operatorname{tr}(\hat{X}A(J - I)A^*)$.
  2. $uA^{\vee 2}X^\dagger = \frac{1}{2}\operatorname{tr}(\hat{X}A^*(J - I)A)$.
  3. If A is stochastic, then $XA^{\vee 2}u^\dagger = \frac{1}{2}\operatorname{tr}(\hat{X}(J - AA^*))$.

Using equation (16) directly for X, we have

$$X(I - A^{\vee 2})u^\dagger = \tfrac{1}{2} \operatorname{tr}(\hat{X}(J - AJA^* + AA^*)) \qquad (17)$$

$$u(I - A^{\vee 2})X^\dagger = \tfrac{1}{2} \operatorname{tr}(\hat{X}(J - A^*JA + A^*A)) \qquad (18)$$

For zeons a new feature appears. For stochastic A, equation (17) yields

Lemma 9.1. *"integration-by-parts for zeons"*

$$X(I - A^{\vee 2})u^\dagger = \tfrac{1}{2} \operatorname{tr} A^* \hat{X} A \qquad (19)$$

9.4. Zeon hierarchy. M and N operators

Consider the semigroup generated by the matrices $C_i^{\vee \ell}$, corresponding to the action on ℓ-sets of vertices. The map $w \to w^{\vee \ell}$ is a homomorphism of matrix semigroups. We now have our main working tool in this context.

Proposition 9.4. *For $1 \le \ell \le n$, define*

$$A_\ell = \frac{1}{d} C_1^{\vee \ell} + \cdots + \frac{1}{d} C_d^{\vee \ell}.$$

Then the Cesàro limit of the powers A_ℓ^m exists and equals

$$\Omega_\ell = \langle K^{\vee \ell} \rangle$$

the average taken over the kernel $\mathcal{K}$ of the semigroup $\mathcal{S}$ generated by $\{C_1, \ldots, C_d\}$.

With r the rank of the kernel, we see that Ω_ℓ vanishes for $\ell > r$.

Here we have A_ℓ in general *substochastic*, i.e., it may have zero rows or rows that do not sum to one. As noted in the B, the Abel limits exist for the A_ℓ. They agree with the Cesàro limits.

9.5. M and N operators via zeons

Here we will show how the operators M and N appear via zeons. The main tool are the trace identities. Note that we automatically get N_0 while the diagonal of M is cancelled out in the inner product with $\hat{X}$.

Proposition 9.5. *For any X,*

$$X\Omega_2 u^\dagger = Xu_2^\dagger = \langle X, u_2 \rangle = \tfrac{1}{2} \operatorname{tr} \hat{X} \hat{u}_2 = \tfrac{1}{2} \operatorname{tr} \hat{X} N_0$$

where $N_0 = \langle J - KK^* \rangle$, taken over the kernel.
And

$$u\Omega_2 X^\dagger = \pi_2 X^\dagger = \langle \pi_2, X \rangle = \tfrac{1}{2} \operatorname{tr} \hat{\pi}_2 \hat{X} = \tfrac{1}{2} \operatorname{tr} \hat{X} M$$

where $M = \langle K^* JK \rangle$, taken over the kernel.

Proof. We have $\Omega_2 = \langle K^{\vee 2} \rangle$. Using Proposition 9.3 for stochastic matrices,

$$X\Omega_2 u^\dagger = \langle X K^{\vee 2} u^\dagger \rangle = \tfrac{1}{2} \operatorname{tr} \langle \hat{X}(J - KK^*) \rangle = \tfrac{1}{2} \operatorname{tr} \hat{X} N_0$$

Similarly,

$$u\Omega_2 X^\dagger = \langle u K^{\vee 2} X^\dagger \rangle = \tfrac{1}{2} \operatorname{tr} \langle \hat{X} K^*(J - I)K) \rangle = \tfrac{1}{2} \operatorname{tr} \langle \hat{X} K^* JK \rangle$$

where, since K^*K is diagonal, $\operatorname{tr} \hat{X} K^* K = 0$ as the diagonal of $\hat{X}$ vanishes. $\square$

From these relations we see that

$$u_2^\dagger = \Omega_2 u^\dagger \quad \Rightarrow \quad N_0 = \langle J - KK^* \rangle = \operatorname{Mat}(u_2)$$

and

$$\pi_2 = u\Omega_2 \quad \Rightarrow \quad M_0 = \langle K^* JK \rangle - \tau\omega = \operatorname{Mat}(\pi_2)$$

Cf., Proposition 5.3. Here M and N only require applying Ω_2 to u.

9.6. *The spaces* $\mathcal{M}_0$ *and* $\mathcal{N}_0$

Now we have

$$A_2 = \tfrac{1}{2}(R^{\vee 2} + B^{\vee 2}) = A^{\vee 2} + \Delta^{\vee 2} .$$

Define the mappings $F, G \colon \operatorname{Sym}(\mathcal{V}) \to \operatorname{Sym}(\mathcal{V})$ by

$$F(Y) = AYA^* + \Delta Y\Delta^* \qquad \text{and} \qquad G(Y) = A^* YA + \Delta^* Y\Delta .$$

consistent with

$$F(Y) = \tfrac{1}{2}(RYR^* + BYB^*) \qquad \text{and} \qquad G(Y) = \tfrac{1}{2}(R^* YR + B^* YB) .$$

First, some observations

Proposition 9.6. *For any vector v, $G(\operatorname{diag}(v)) = \operatorname{diag}(vA)$.*

Proof. Let $Y = G(\operatorname{diag}(v))$. Then $2Y_{ij} = R_{\lambda i} v_\lambda R_{\lambda j} + B_{\lambda i} v_\lambda B_{\lambda j}$. A term like R_{li} means that R maps l to i. Since R is a function, l can only map to both i and j if $i = j$. So Y is diagonal. If $i = j$, then we get, using $R_{ij}^2 = R_{ij}$, $B_{ij}^2 = B_{ij}$,

$$2Y_{ii} = R_{\lambda i} v_\lambda + B_{\lambda i} v_\lambda = 2 v_\lambda A_{\lambda i}$$

as required. $\square$

For example, taking $v = u$, we see that

$$G(I) = I \text{ if and only if } A \text{ is doubly stochastic.}$$

First, observe

Proposition 9.7. $F(J) = J$ and $G(\omega) = \omega$. That is, $J \in \mathcal{N}_+$ and $\omega \in \mathcal{M}_+$.

Proof. We have $F(J) = AJA^* + \Delta J\Delta^*$. With $AJ = J = JA^*$ and $\Delta J = 0$, this reduces to J. Second, from the previous Proposition, we have $G(\omega) = G(\text{diag}(\pi))) = \text{diag}(\pi A) = \text{diag}(\pi) = \omega$. □

Now recall Propositions 5.1 and 5.2. We see that an element of $\mathcal{M}_+$ differs by a multiple of ω from an element of $\mathcal{M}_0$. For an element $N \in \mathcal{N}_+$ with entries bounded by 1, with diagonal entries all equal to one, we see that $J - N$ gives a corresponding element in $\mathcal{N}_0$. We have seen the special operators M, N, M_0 and N_0 as interesting examples.

Note that equation (19) applied to R and B and then averaged yields

$$X(I - A_2)u^\dagger = \tfrac{1}{2}\operatorname{tr} G(\hat{X}) \tag{20}$$

The next theorem shows the general relation between fixed points of A_2 and fixed points of F and G. Note that we are not restricting *a priori* to nonnegative solutions except in case 2 of the theorem. This allows us to associate trace-zero M and N operators with averages over zeon powers of the kernel elements.

Theorem 9.1.
1. $F(\hat{X}) = \hat{X}$ if and only if $A_2 X^\dagger = X^\dagger$.
2. For *nonnegative solutions* X, $G(\hat{X}) = \hat{X}$ if and only if $X A_2 = X$.

Proof. First, from Proposition 9.1, applying the Basic Relations for R and B and averaging, we get

$$\operatorname{Mat}(XA_2) = G(\hat{X}) - \tfrac{1}{2}\left(D_R^+ + D_B^+\right) \tag{21}$$

$$\operatorname{Mat}(A_2 X^\dagger) = F(\hat{X}) - \tfrac{1}{2}\left(D_R^- + D_B^-\right) \tag{22}$$

where the subscripts on the D's indicate the diagonals corresponding to R and B respectively. Recalling the remark following Proposition 9.1, note that for D_R^-, say, we have terms like $R_{i\lambda}x_{\lambda\mu}R_{i\mu}$. Since R is a function, we must have $\lambda = \mu$, but then $x_{\lambda\mu} = 0$. In other words, D_R^- and D_B^- vanish so that, in fact, $\operatorname{Mat}(A_2 X^\dagger) = F(\hat{X})$. And #1 follows immediately.

For #2, if $G(\hat{X}) = \hat{X}$, then the trace of the diagonal terms in equation (21) vanish, so under the assumption of nonnegativity, they must vanish. And replacing $G(\hat{X})$ by $\hat{X}$, we have $X A_2 = X$.

Assume $X A_2 = X$. Taking traces in equation (21), the trace of the left-hand side vanishes and the trace of the diagonal terms equals $\operatorname{tr} G(\hat{X})$. Now, use the integration-by-parts formula for A_2, equation (20). The left-hand side vanishes, hence $\operatorname{tr} G(\hat{X}) = 0$, implying the vanishing of the terms D_R^+ and D_B^+ as well. Then equation (21) reads $\operatorname{Mat}(X) = G(\hat{X})$ as required.

$\square$

9.7. *Quantum observables via zeons in summary*

We have Ω_2 as the Cesàro limit of the powers A_2^m. Then with u here denoting the all-ones vector of dimension $\binom{n}{2}$, we have

$$u_2^\dagger = \Omega_2 u^\dagger \qquad \text{and} \qquad \pi_2 = u \Omega_2$$

1. $N_0 = \langle J - KK^* \rangle = \operatorname{Mat}(u_2)$ satisfies $A N_0 A^* + \Delta N_0 \Delta^* = N_0$. It is a nonnegative solution with $\operatorname{tr} N_0 = 0$. $N = \langle KK^* \rangle$ and $J = u^\dagger u$ are also nonnegative solutions, with trace equal to n.

2. $M_0 = M - \tau \omega$. From Proposition 5.3, we have $\langle K^* K \rangle = n\omega$. This allows us to express

$$M_0 = \langle K^* J K \rangle - \tau/n \langle K^* K \rangle .$$

$M_0 = \operatorname{Mat}(\pi_2)$ satisfies $A^* M_0 A + \Delta^* M_0 \Delta = M_0$. It is a nonnegative solution with $\operatorname{tr} M_0 = 0$. $M = \langle K^* J K \rangle$ and $\omega = \operatorname{diag}(\pi)$ are also nonnegative solutions, with trace equal to τ and 1 respectively.

3. We have $\tilde{M} = \langle K^* \Omega K \rangle = nr^{-2} \langle \underset{\sim}{\rho}^\dagger \underset{\sim}{\rho} \rangle$. The diagonal of $\underset{\sim}{\rho}^\dagger \underset{\sim}{\rho}$ is ρ so the diagonal of $\tilde{M}$ is $r\omega$. Thus, $\tilde{M}_0 = \tilde{M} - r\omega$ has zero diagonal. Hence $\tilde{M}_0$ and M_0 are nonnegative, symmetric, with vanishing trace, elements of $\mathcal{M}_0$, corresponding to solutions to $X A_2 = X$. Notice that $\operatorname{tr} \Omega_2 = 1$ would imply that the space of such solutions equals 1 so in that case $\tilde{M}_0$ and M_0 are proportional. Thus, M_0 could be computed using knowledge of the range classes.

9.8. *Levels in the zeon hierarchy*

At level 1, we have left and right eigenvectors of A: $\pi_1 = \pi$ and $u_1^+ = u^+$, respectively. All of the components of u are equal to 1. And $\Omega = \Omega_1 = u^+ \pi$.

At every level ℓ, $1 \le \ell \le r$, we have an Abel limit

$$\Omega_\ell = \lim_{s \uparrow 1} (1 - s)(I - s A_\ell)^{-1}$$

When π_ℓ and u_ℓ are unique, we have $\Omega_\ell = u_\ell^+ \pi_\ell$.

At level r, as at level 1, we have $\Omega_r = u_r^+ \pi_r$.

9.8.1. *Interpretation*

Start at level r. Then the nonzero entries of π_r are in the columns corresponding to the ranges of the kernel, $\mathcal{K}$. The nonzero entries of u_r^+ are in rows corresponding to cross-sections of the partition classes of $\mathcal{K}$. I.e., for each partition class, find all possible cross-sections weighted by one, take the total, and you get a lower bound on the corresponding entries of u_r^+. In particular, for a right group, all cross-sections appear with the same weight, which can be scaled to 1.

At each level, if unique, π_ℓ and u_ℓ give you information about the recurrent class and corresponding cross-sections.

A sequence of pairs (π_ℓ, u_ℓ) can be constructed inductively from (π_r, u_r) as consecutive marginals. Thus, for $\ell < r$, let

$$\pi_\ell(\mathrm{I}) = \sum_{\mathrm{J} \supset \mathrm{I}} \pi_{\ell+1}(\mathrm{J})$$

Note that in the sum each J differs from I by a single vertex. Starting from level r, we inductively move down the hierarchy and determine a fixed point of A_ℓ, π_ℓ, for each level ℓ.

For u's, proceed as follows:

$$u_\ell(\mathrm{I}) = \sum_i p_i \, u_{\ell+1}(\mathrm{I} \cup \{i\})$$

where the weights p_i are the components of the invariant measure π.

With this construction, one recovers at level one, a multiple of $\pi_1 = \pi$ and a constant vector u_1, a multiple of u.

Note that the rank r is the maximum level ℓ such that $\Omega_\ell \neq 0$.

For our closing observation, we remark that by introducing an absorbing state at each level, R_ℓ, B_ℓ, and A_ℓ, can be extended to stochastic matrices as at level one.

Remark 9.3. For further information as to the structure of the zeon hierarchy see Ref. 1.

10. Conclusion

At this point the interpretation of the eigenvalues and eigenvectors of the operators M and N is unavailable. It would be quite interesting to discover their meaning. For further work, it would be nice to have similar constructions for at least some classes of infinite semigroups, e.g. compact semigroups. Another avenue to explore would be to develop such a theory for semigroups of operators on Hilbert space.

Acknowledgments

The author is indebted to G. Budzban for introducing him to this subject.

This article is motivated by the QP32 Conference in Levico, Italy, May-June, 2011. The author wishes to thank the organizers for a truly enjoyable meeting.

Appendix A. Semigroups and kernels

A finite semigroup, in general, is a finite set, $\mathcal{S}$, which is closed under a binary associative operation. For any subset T of $\mathcal{S}$, we will write $\mathcal{E}(T)$ to refer to the set of idempotents in T.

Theorem A.1. *[Ref. 5, Th. 1.1]*
Let $\mathcal{S}$ be a finite semigroup. Then $\mathcal{S}$ contains a minimal ideal $\mathcal{K}$ called the kernel which is a disjoint union of isomorphic groups. In fact, $\mathcal{K}$ is isomorphic to $\mathcal{X} \times \mathcal{G} \times \mathcal{Y}$ where, given $e \in \mathcal{E}(\mathcal{S})$, then $e\mathcal{K}e$ is a group and

$$\mathcal{X} = \mathcal{E}(\mathcal{K}e), \qquad \mathcal{G} = e\mathcal{K}e, \qquad \mathcal{Y} = \mathcal{E}(e\mathcal{K})$$

and if (x_1, g_1, y_1), $(x_2, g_2, y_2) \in \mathcal{X} \times \mathcal{G} \times \mathcal{Y}$ then the multiplication rule has the form

$$(x_1, g_1, y_1)(x_2, g_2, y_2) = (x_1, g_1 \phi(y_1, x_2) g_2, y_2)$$

where $\phi: \mathcal{Y} \times \mathcal{X} \to \mathcal{G}$ is the sandwich function.

The product structure $\mathcal{X} \times \mathcal{G} \times \mathcal{Y}$ is called a Rees product and any semigroup that has a Rees product is *completely simple*. The kernel of a finite semigroup is always completely simple.

Appendix A.1. *Kernel of a matrix semigroup*

An extremely useful characterization of the kernel for a semigroup of matrices is known.

Theorem A.2. *[Ref. 5, Props. 1.8, 1.9]*
Let $\mathcal{S}$ be a finite semigroup of matrices. Then the kernel, $\mathcal{K}$, of $\mathcal{S}$ is the set of matrices with minimal rank.

Suppose $\mathcal{S} = \mathcal{S}((C_1, C_2, \ldots, C_d))$ is a semigroup generated by binary stochastic matrices C_i. Then $\mathcal{S}$ is a finite semigroup with kernel $\mathcal{K} = \mathcal{X} \times \mathcal{G} \times \mathcal{Y}$, the Rees product structure of Theorem A.1. Let k, k' be elements of $\mathcal{K}$. Then with respect to the Rees product structure $k = (k_1, k_2, k_3)$ and $k' = (k_1', k_2', k_3')$. We form the ideals $\mathcal{K}k$, $k'\mathcal{K}$, and their intersection $k'\mathcal{K}k$:

(1) $\mathcal{K}k = \mathcal{X} \times \mathcal{G} \times \{k_3\}$ is a minimal left ideal in $\mathcal{K}$ whose elements all have the same range, or nonzero columns as k. We call this type of semigroup a *left group*.

(2) $k'\mathcal{K} = \{k_1'\} \times \mathcal{G} \times \mathcal{Y}$ is a minimal right ideal in $\mathcal{K}$ whose elements all have the same partition of the vertices as k'. A block $\mathcal{B}_j$ in the partition can be assigned to each nonzero column of k' by

$$\mathcal{B}_j = \{i : k_{i,j}' = 1\}$$

A semigroup with the structure $k'\mathcal{K}$ is a *right group*.

(3) $k'\mathcal{K}k$, the intersection of $k'\mathcal{K}$ and $\mathcal{K}k$, is a maximal group in $\mathcal{K}$ (an H-class in the language of semigroups). It is best thought of as the set of functions specified by the partition of k' and the range of k, acting as a group of permutations on the range of k. The idempotent of $k'\mathcal{K}k$ is the function which is the identity when restricted to the range of k.

Appendix B. Abel limits

We are working with stochastic and substochastic matrices. We give the details here on the existence of Abel limits. Let P be a substochastic matrix. The entries p_{ij} satisfy $0 \leq p_{ij} \leq 1$ with $\sum_j p_{ij} \leq 1$. Inductively, P^n is substochastic for every integer $n > 0$, with P^n stochastic if P is.

Proposition B.1. *The Abel limit*

$$\Omega = \lim_{s \uparrow 1} (1 - s)(I - sP)^{-1}$$

exists and satisfies

$$\Omega = \Omega^2 = P\Omega = \Omega P$$

Proof. We take $0 < s < 1$ throughout.

For each i, j, the matrix elements $\langle e_i, P^n e_j \rangle$ are uniformly bounded by 1. Denote

$$Q(s) = (1 - s)(I - sP)^{-1}$$

So we have

$$\langle e_i, Q(s)e_j \rangle = (1-s)\sum_{n=0}^{\infty} s^n \langle e_i, P^n e_j \rangle \leq (1-s)\sum_{n=0}^{\infty} s^n = 1$$

so the matrix elements of $Q(s)$ are bounded by 1 uniformly in s as well. Now take any sequence $\{s_j\}$, $s_j \uparrow 1$. Take a further subsequence $\{s_{j11}\}$ along which the 11 matrix elements converge. Continuing with a diagonalization procedure going successively through the matrix elements, we have a subsequence $\{s'\}$ along which all of the matrix elements converge, i.e., along which $Q(s')$ converges. Call the limit Ω.

Writing $(I - sP)^{-1} - I = sP(I - sP)^{-1}$, multiplying by $1 - s$ and letting $s \uparrow 1$ along s' yields

$$\Omega = P\Omega = \Omega P$$

writing the P on the other side for this last equality. Now, $s\Omega = sP\Omega$ implies $(1 - s)\Omega = (I - sP)\Omega$. Similarly, $(1 - s)\Omega = \Omega(I - sP)$, so

$$Q(s)\Omega = \Omega Q(s) = \Omega \tag{B.1}$$

Taking limits along s' yields $\Omega^2 = \Omega$.

For the limit to exist, we check that Ω is the only limit point of $Q(s)$. From (B.1), if Ω_1 is any limit point of $Q(s)$, $\Omega_1 \Omega = \Omega \Omega_1 = \Omega$. Interchanging rôles of Ω_1 and Ω yields $\Omega_1 \Omega = \Omega \Omega_1 = \Omega_1$ as well, i.e., $\Omega = \Omega_1$. $\square$

If P is stochastic, it has u^+ as a nontrivial fixed point. In general we have

Corollary B.1. *Let* $\Omega = \lim_{s\uparrow 1}(1 - s)(I - sP)^{-1}$ *be the Abel limit of the powers of* P.

P has a nontrivial fixed point if and only if $\Omega \neq 0$.

Proof. If $\Omega \neq 0$, then $P\Omega = \Omega$ shows that any nonzero column of Ω is a nontrivial fixed point. On the other hand, if $Pv = v \neq 0$, then, as in the above proof, $Q(s)v = v$ and hence $\Omega v = v$ shows that $\Omega \neq 0$. $\square$

Note that since Ω is a projection, its rank is the dimension of the space of fixed points.

References

1. Greg Budzban and Philip Feinsilver. A hierarchical structure of transformation semigroups with applications to probability limit measures. *J. Difference Equ. Appl.*, 18(8):1405–1434, 2012.

2. Greg Budzban and Arunava Mukherjea. A semigroup approach to the road coloring problem. In *Probability on algebraic structures (Gainesville, FL, 1999)*, volume 261 of *Contemp. Math.*, pages 195–207. Amer. Math. Soc., Providence, RI, 2000.

3. Ph. Feinsilver, Zeon algebra, Fock space, and Markov chains, *Communications on stochastic analysis* **2**(2008), 263–275.

4. Joel Friedman. On the road coloring problem. *Proc. Amer. Math. Soc.*, 110(4):1133–1135, 1990.

5. Göran Högnäs and Arunava Mukherjea. *Probability measures on semigroups.* Probability and its Applications (New York). Springer, New York, second edition, 2011.

6. Arunava Mukherjea. Recurrent random walks in nonnegative matrices: attractors of certain iterated function systems. *Probab. Theory Related Fields*, 91(3-4):297–306, 1992.

A MATHEMATICAL TREATMENT FOR THE
CONTEXTUAL DEPENDENT BIO-SYSTEMS

TOSHIHIDE HARA* and MASANORI OHYA

*Department of Information Sciences, Tokyo University of Science,
2641 Yamazaki, Noda City, Chiba, Japan
* E-mail: hara@is.noda.tus.ac.jp*

In bio-systems, there exist several phenomena breaking the laws of total probability such as the lactose-glucose interference in E. coli growth. We call such phenomena the contextual dependent adaptive systems. Recently we introduced a new mathematical framework to treat the probability in those systems. In this paper, we discuss the essence of this mathematical frame with a simple example "a state change of tongue for sweetness".

Keywords: Adaptive dynamics; lifting; contextual dependent adaptive systems; laws of total probability.

1. Introduction

There exist several phenomena breaking the law of total probability such as quantum interference in cognitive science, the lactose-glucose interference in E. coli growth. These phenomena will require us a change of usual probability law. Recently we have studied such "contextual dependent adaptive systems" with a new mathematical framework.[1,3–6,9] This framework is constructed based on the concepts of the lifting[2] and the adaptive dynamics.[7,8]

In this paper, we show that the mathematical framework enables us to understand phenomena with an example "a state change of tongue for sweetness".

2. New views of joint probability in contextual dependent systems

Let us discuss how to use the concept of lifting to explain phenomena breaking the usual probability law.

Let $\mathcal{A}, \mathcal{B}$ be C*-algebras describing the systems for a study, more specifically, let $\mathcal{A}, \mathcal{B}$ be the sets of all observables in Hilbert spaces $\mathcal{H}$, $\mathcal{K}$;

$\mathcal{A} = \mathcal{O}(\mathcal{H})$, $\mathcal{B} = \mathcal{O}(\mathcal{K})$. Let $\mathcal{E}^*$ be a lifting from $\mathcal{S}(\mathcal{H})$ to $\mathcal{S}(\mathcal{H} \otimes \mathcal{K})$, so that its dual map $\mathcal{E}$ is a mapping from $\mathcal{A} \otimes \mathcal{B}$ to $\mathcal{A}$. There are several liftings for various different cases to be considered: (1) If $\mathcal{K}$ is $\mathbb{C}$, then the lifting $\mathcal{E}^*$ is nothing but a channel from $\mathcal{S}(\mathcal{H})$ to $\mathcal{S}(\mathcal{H})$. (2) If $\mathcal{H}$ is $\mathbb{C}$, then the lifting $\mathcal{E}^*$ is a channel from $\mathcal{S}(\mathcal{H})$ to $\mathcal{S}(\mathcal{K})$. Further $\mathcal{K}$ or $\mathcal{H}$ can be decomposed as $\mathcal{K} = \otimes_i \mathcal{K}_i$ (resp. $\oplus_i \mathcal{K}_i$), and so for $\mathcal{H}$, so that $\mathcal{B}$ can be $\otimes_i \mathcal{B}_i$ (resp. $\oplus_i \mathcal{B}_i$) and so for $\mathcal{A}$. See ref. 2 for details.

The adaptive dynamics is considered that the dynamics of a state or an observable after an instant (say the time t_0) attached to a system of interest is affected by the existence of some other observable and state at that instant. Let $\rho \in \mathcal{S}(\mathcal{H})$ and $A \in \mathcal{A}$ be a state and an observable before t_0, and let $\sigma \in \mathcal{S}(\mathcal{H} \otimes \mathcal{K})$ and $Q \in \mathcal{A} \otimes \mathcal{B}$ be a state and an observable to give an effect to the state ρ and the observable A. In many cases, the effect to the state is dual to that to the observable, so that we will discuss the effect to the state only. This effect is described by a lifting $\mathcal{E}^*_{\sigma Q}$, so that the state ρ becomes $\mathcal{E}^*_{\sigma Q}\rho$ first, then it will be $\text{tr}_\mathcal{K} \mathcal{E}^*_{\sigma Q}\rho \equiv \rho_{\sigma Q}$. The adaptive dynamics is the whole process such as

$$Adaptive\ Dynamics: \ \rho \Rightarrow \mathcal{E}^*_{\sigma Q}\rho \Rightarrow \rho_{\sigma Q} = tr_\mathcal{K} \mathcal{E}^*_{\sigma Q}\rho$$

That is, what we need is how to construct the lifting $\mathcal{E}^*_{\sigma Q}$ for each problem to be studied by choosing σ and Q properly.

Here let us discuss how to use the lifting $\mathcal{E}^*_{\sigma Q}$ above to understand the breaking of total probability law. The expectation value of another observable $B \in \mathcal{A}$ or $\mathcal{A} \otimes \mathcal{B}$ in the adaptive state $\rho_{\sigma Q}$ is

$$\text{tr}\rho_{\sigma Q} B = \text{tr}_\mathcal{H} \text{tr}_\mathcal{K} B \mathcal{E}^*_{\sigma Q}\rho.$$

Now suppose that there are two event systems $A = \{a_k \in \mathbb{R}, F_k \in \mathcal{A}\}$ and $B = \{b_j \in \mathbb{R}, E_j \in \mathcal{A}\}$, where we do not assume F_k, E_j are projections, but they satisfy the conditions $\sum_k F_k = I$, $\sum_j E_j = I$ as POVM (positive operator valued measure) corresponding to the partition of a probability space in classical system. Then the "joint-like" probability obtaining a_k and b_j might be given by the formula

$$P(a_k, b_j) = tr E_j \,\square\, F_k \mathcal{E}^*_{\sigma Q}\rho, \tag{1}$$

where $\square$ is a certain operation (relation) between A and B, more generally one can take a certain operator function $f(E_j, F_k)$ instead of $E_j \,\square\, F_k$. If σ, Q are independent from any F_k, E_j and the operation $\square$ is the usual tensor product $\otimes$ so that A and B can be considered in two independent systems, then the above "joint-like" probability becomes the joint probability. However if not such a case, e.g., Q is related to A and B, the situation will be more subtle. Therefore the problem is how to set the operation $\boxdot$ and how to construct the lifting $\mathcal{E}^*_{\sigma Q}$ in order to describe the particular problems associated to systems of interest.

In the next section, we discuss how to apply the formula (1) to a problem breaking the usual probability law.

3. State change of tongue for sweetness

Let us consider a simple and intuitive example. When one takes sugar S and chocolate C and he is asked whether it is sweet (1) or not so (2). Then the simple classical probability law:

$$P\left(C = 1\right) = P\left(C = 1|S = 1\right) P\left(S = 1\right) + P\left(C = 1|S = 2\right) P\left(S = 2\right)$$

may not be satisfied, because we usually feel the following probabilities:

$$P\left(C = 1\right) \approx 1,$$
$$P\left(S = 1\right) \approx 1,$$
$$P\left(S = 2\right) \approx 0,$$
$$P\left(C = 1|S = 2\right) \approx 1,$$
$$P\left(C = 1|S = 1\right) < \frac{1}{2}.$$

Thus the LHS $P\left(C = 1\right)$ will be very close to 1 but the RHS will be less than $\frac{1}{2}$.

Let us discuss this breaking by means of adaptive dynamics. Let $e_1 \equiv \begin{pmatrix} 1 \\ 0 \end{pmatrix}$ and $e_2 \equiv \begin{pmatrix} 0 \\ 1 \end{pmatrix}$ be the orthogonal vectors in Hilbert space $\mathbb{C}^2$ describing the tongue states, sweet and non-sweet, respectively. Here we start from the neutral pure state ρ_0 because we consider two sweet things, then the initial state of tongue is such as

$$\rho = \rho_0 \equiv |x_0\rangle \langle x_0|,$$

where $x_0 = \frac{1}{\sqrt{2}} (e_1 + e_2)$.

When one takes "sugar", the operator corresponding to taking "sugar" will be given as

$$S = \begin{pmatrix} \lambda_1 & 0 \\ 0 & \lambda_2 \end{pmatrix},$$

where $|\lambda_1|^2 + |\lambda_2|^2 = 1$ and $\sigma_S = S^*S$. Taking sugar, he will taste that it is sweet with the probability $|\lambda_1|^2$ and non-sweet with the probability $|\lambda_2|^2$, so $|\lambda_1|^2$ should be much higher than $|\lambda_2|^2$ for a usual sugar. This comes from the following change of the neutral initial tongue (i.e., non-adaptive) state:

$$\rho_0 \to \rho_S = \Lambda_S^*(\rho_0) \equiv \frac{S\rho S^*}{\mathrm{tr} S\rho S^*},$$

which is the state when he takes the sugar.

The subtle point of the present problem is that the state of tongue is neither ρ_S nor ρ_0 at the instant just after taking sugar. Note here that if we kill the subjectivity (personal character?) of one's tongue, then the state ρ_S can be understood as

$$E_1\rho_S E_1 + E_2\rho_S E_2,$$

where $E_1 \equiv |e_1\rangle\langle e_2|, E_2 \equiv |e_2\rangle\langle e_2|$. For some time duration, the tongue becomes dull to sweetness, so the tongue state can be written by means of a certain "exchanging" operator $X = \begin{pmatrix} 0 & 1 \\ 1 & 0 \end{pmatrix}$ such that

$$\rho_S \underset{\text{dull to sweetness}}{\dashrightarrow} \rho_S^a = X\rho_S X,$$

where "a" means the adaptive change. Then similarly as sugar, when one takes a chocolate, the state will be $\rho_{S\to C}^a$ given by

$$\rho_{S\to C}^a = \Lambda_C^*(\rho_S^a) \equiv \frac{C\rho_S^a C^*}{\mathrm{tr} C\rho_S^a C^*},$$

where $|\mu_1|^2 + |\mu_2|^2 = 1$.

Here we introduce the following nonlinear demolition lifting:

$$\varepsilon_{SXC}^*(\rho_0)\left(= \mathcal{E}_{\sigma Q}^*(\rho_0)\right) \equiv \rho_S \otimes \rho_{S\to C}^a$$
$$= \Lambda_S^*(\rho_0) \otimes \Lambda_C^*(X\Lambda_S^*(\rho_0)X),$$

which implies the joint probabilities $P(S = j, C = k)$ $(j, k = 1, 2)$ as

$$P(S = j, C = k) = \mathrm{tr}(E_j \otimes E_k)\varepsilon_{SXC}^*(\rho_0).$$

The probability that one tastes sweetness of the chocolate after tasting sugar is

$$P(C = 1|S = 1)P(S = 1) + P(C = 1|S = 2)P(S = 2)$$

$$= P\left(C = 1, S = 1\right) + P\left(C = 1, S = 2\right)$$
$$= \mathrm{tr}\left(E_1 \otimes E_1\right) \varepsilon^*_{SXC}\left(\rho_0\right) + \mathrm{tr}\left(E_1 \otimes E_2\right) \varepsilon^*_{SXC}\left(\rho_0\right)$$
$$= \frac{|\lambda_2|^2 |\mu_1|^2}{|\lambda_2|^2 |\mu_1|^2 + |\lambda_1|^2 |\mu_2|^2}.$$

Note that this probability is much less than

$$P\left(C = 1\right) = \mathrm{tr} E_1 \Lambda^*_C\left(\rho_0\right) = |\mu_1|^2,$$

which is the probability of sweetness tasted by the neutral tongue ρ_0. In this sense, the low of total probability;

$$P\left(C = 1\right) = P\left(C = 1, S = 1\right) + P\left(C = 1, S = 2\right)$$

is not satisfied in a contextual dependent system.

4. Conclusion

The above consideration implies that our mathematical framework can be used to understand the contextual phenomena. In bio sciences, "Design of experiments" claims that the order of several experiments should be randomized in order to omit the unexpected effects coming from the order of experiments. Although such a randomization is useful for minimizing the unexpected error, there has not existed why we need such a randomization. Our framework might be useful to explain the reason.

References

1. L. Accardi, A. Khrennikov, and M. Ohya. The problem of quantum-like representation in economy cognitive science, and genetics. *QP-PQ: Quantum Probability and White Noise Analysis (Quantum Bio-Informatics II)*, 24:1–8, 2008.
2. L. Accardi and M. Ohya. Compound channels, transition expectations, and liftings. *Applied mathematics & optimization*, 39(1):33–59, 1999.
3. M. Asano, I. Basieva, A. Khrennikov, M. Ohya, and I. Yamato. A general quantum information model for the contextual dependent systems breaking the classical probability law. *Arxiv preprint arXiv:1105.4769*, 2011.
4. M. Asano, M. Ohya, and A. Khrennikov. Quantum-like model for decision making process in two players game. *Foundations of Physics*, 41(3):538–548, 2011.
5. M. Asano, M. Ohya, Y. Tanaka, A. Khrennikov, and I. Basieva. Quantum-like representation of bayesian updating. In *AIP Conference Proceedings*, volume 1327, page 57, 2011.
6. I. Basieva, A. Khrennikov, M. Ohya, and I. Yamato. Quantum-like interference effect in gene expression: glucose-lactose destructive interference. *Systems and Synthetic Biology*, 5:59–68, 2011.

7. M. Ohya. On compound state and mutual information in quantum information theory. *IEEE Transactions on Information Theory*, 29(5):770–774, 1983.
8. M. Ohya. Adaptive dynamics and its applications to chaos and NPC problem. *Quantum Bio-Informatics*, 21:181–216, 2008.
9. M. Ohya and I. Volovich. *Mathematical Foundations of Quantum Information and Computation and Its Applications to Nano-and Bio-systems*. Springer Verlag, 2011.

ON THE SPECTRAL GAP OF THE N-PHOTON
ABSORPTION-EMISSION PROCESS

RAÚL HERMIDA* and ROBERTO QUEZADA**

Universidad Autónoma Metropolitana, Iztapalapa Campus
Av. San Rafael Atlixco 186, Col. Vicentina
09340 Iztapalapa D.F., Mexico City.
* E-mail: hermidaoc@gmail.com
** E-mail: roqb@xanum.uam.mx

We give an estimate for the off-diagonal (quantum) gap for the n-photon absorption-emission process.

Keywords: n-photon absorption and emission, Dirichlet form, invariant subspaces, spectral gap.

1. Introduction

Multi-photon processes (i.e., simultaneous absorption or emission of two, three or more photons) where predicted theoretically by Göpper-Mayer in 1931. Since the experimental observation of the two-photon absorption emission in the 1960's, many experimental efforts have focused on the multi-photon processes to create states with typical quantum features. Although the more studied multi-photon process is that of two-photon absorption and/or emission; recently, studies of higher-order multi-photon processes, in particular three- and four-photon absorption or emission have received considerable attention, see Ref.[4,5,12,13,15] and the references therein. Apart from academic interest, investigation of multi-photon absorption emission processes is motivated by potential applications for fluorescence imaging Ref.,[11] emission of coherent light (lasing) Ref.,[10] optical data storage Ref.,[14] among others. For instance, three-photon absorption has been also used to produce stimulated emission, as well as to produce spatially confined excitation useful for three dimensional data storage and imaging, see Ref.[4,9,10] and the references therein.

The present work aims at contributing to the study of spectral properties of the n-photon absorption and emission process. In particular, we give an estimate for the off-diagonal (quantum) gap of this process. We apply the method of invariant subspaces introduced by Carbone and Fagnola in Ref.[2] to reduce the computation of the off-diagonal gap to the computation of an infinite number of minima of the Dirichlet form restricted to the subspaces of the family.

In Section 2 we introduce the generator of the n-photon absorption-emission process. Section 3 is devoted to compute the invariant states and to establish the approach to equilibrium property of this semigroup. The Dirichlet form as well as its main properties are studied in Section 4 and in Section 5 we give a suitable decomposition in invariant subspaces of $L_2(\mathsf{h})$, the space of Hilbert-Schmidt operators. Finally, in Section 6, we compute an estimate for the off-diagonal spectral gap of the n-photon absorption-emission process for all $n \geq 2$.

2. The generator

Let h be the Hilbert space $\ell_2(\mathbb{C})$, of all square summable complex sequences, and let a, $a^\dagger$, $N = a^\dagger a$ be the annihilation, creation and number operators, respectively. Recall that operators a, $a^\dagger$ are defined on the subspace

$$\mathcal{D}\left(N^{1/2}\right) = \left\{ u \in \mathsf{h} : \sum_{k \geq 0} k|u_k|^2 < \infty \right\},$$

by means of

$$ae_k = \sqrt{k}e_{k-1} \quad \text{and} \quad a^\dagger e_{k-1} = \sqrt{k}e_k, \quad \text{for} \quad k \geq 1,$$

where $(e_k)_{k \geq 0}$ is the canonical orthonormal basis of h.

For any fixed positive integer n we denote by G the operator defined on $\mathcal{D}(N^n)$ by means of

$$G = -\frac{\lambda^2}{2}a^n a^{\dagger n} - \frac{\mu^2}{2}a^{\dagger n}a^n - i\omega a^{\dagger n}a^n$$

with $0 \leq \lambda < \mu$, $\omega \in \mathbb{R}$.

Clearly G generates a strongly continuous semigroup of contractions $(P_t)_{t \geq 0}$ with

$$P_t = e^{-\frac{t}{2}(\lambda^2(N+1)\cdots(N+n)+(\mu^2+2i\omega)N(N-1)\cdots(N-n+1))}.$$

For every $x \in \mathcal{B}(\mathsf{h})$, the von Neumann algebra of all bounded operators on h, the Gorini-Kossakowski-Sudarshan and Lindblad (GKSL) formal generator is a sesquilinear form defined by

$$\mathcal{L}(x)[u, v] = \langle Gu, xv \rangle + \sum_{\ell=1}^{2} \langle L_\ell u, xL_\ell v \rangle + \langle u, xGv \rangle, \tag{1}$$

for $u, v \in \mathcal{D}(N^n)$, where $L_1 = \mu a^n$, $L_2 = \lambda a^{\dagger^n}$. It is easy to verify that all conditions for existence and conservativity (i.e. identity preservation) of the minimal semigroup associated with G, L_1, L_2 ((H-min) in Ref.[6]) hold true and this quantum Markov semigroup (qms) $\mathcal{T} = (\mathcal{T}_t)_{t \geq 0}$ satisfies the so called Lindblad equation

$$\langle v, \mathcal{T}_t(x)u \rangle = \langle v, P_t^* x P_t u \rangle + \sum_{\ell=1}^{2} \int_0^t \langle L_\ell P_{t-s} v, \mathcal{T}_s(x) L_\ell P_{t-s} u \rangle \, ds, \quad (2)$$

for all $u, v \in$ h. We denote by $\mathcal{L}$ the *true* generator of the minimal qms $\mathcal{T} = (\mathcal{T}_t)_{t \geq 1}$, i.e.,

$$\frac{d}{dt} \mathcal{T}_t(x) = \mathcal{L}\big(\mathcal{T}_t(x)\big), \quad \text{for all } x \in \mathrm{Dom}(\mathcal{L}), \quad \text{where}$$
$$\mathrm{Dom}(\mathcal{L}) = \{x \in \mathcal{B}(h) : \mathcal{L}(x) \in \mathcal{B}(h)\} \quad \text{and} \quad \mathcal{L}(x) = \mathcal{L}(x) \; \forall x \in \mathrm{Dom}(\mathcal{L}). \quad (3)$$

From now on, for every integer $x \geq 0$, we shall use the notation: $[x]_n = \prod_{r=0}^{n-1}(x-r)$, for $x \geq n$ and $[x]_n = 0$, for $0 \leq x \leq n-1$. The following properties of $[x]_n$ will be useful.

Proposition 2.1.

$$[x+n]_n = \prod_{r=1}^{n}(x+r), \quad [x+1]_n = \frac{x+1}{x+1-n}[x]_n$$
$$[x]_{n+1} = (x-n)[x]_n, \quad [x]_{n-1} = \frac{[x]_n}{x+1-n}. \quad (4)$$

Proof. All properties follow from the observation that

$$[x]_n = \prod_{r=0}^{n-1}(x-r) = x(x-1)\cdots(x-(n-1)) = \frac{x!}{(x-n)!}. \qquad \square$$

A direct computation shows that

$$a^n e_k = \prod_{r=0}^{n-1} \sqrt{k-r} \; e_{k-n} = \sqrt{\prod_{r=0}^{n-1}(k-r)} \; e_{k-n} = \sqrt{[k]_n} \; e_{k-n}.$$

In a similar way we get, $a^{\dagger n} e_k = \sqrt{[k+n]_n} \; e_{k+n}$.
Hence if $x = \sum_{j,k} x_{jk} |e_j\rangle \langle e_k|$, we get an explicit expression for $\mathcal{L}$:

$$\mathcal{L}(x) = \sum_{j,k \geq 0} |e_j\rangle \langle e_k| \left\{ \mu^2 x_{j-n,k-n}[j]_n^{1/2}[k]_n^{1/2} \right.$$

$$+ \lambda^2 x_{j+n,k+n}[j+n]_n^{1/2}[k+n]_n^{1/2} - \frac{1}{2} x_{jk} \left[\mu^2 \left([j]_n + [k]_n \right) \right. \tag{5}$$

$$\left. \left. + \lambda^2 \left([j+n]_n + [k+n]_n \right) - 2i\omega \left([j]_n - [k]_n \right) \right] \right\}.$$

3. Invariant states and convergence to equilibrium: the case $0 < \lambda < \mu$.

The pre-dual semigroup $(\mathcal{T}_{*t})_{t \geq 0}$ is defined by means of the duality relation

$$tr\left(\rho \mathcal{T}_t(x) \right) = tr\left(\mathcal{T}_{*t}(\rho)x \right),$$

for all $x \in \mathcal{B}(\mathsf{h})$ and $\rho \in L_1(\mathsf{h})$, the space of finite trace operators on h.

Definition 3.1. The state ρ is invariant for a QMS $(\mathcal{T}_t)_{t \geq 0}$ if for every $x \in \mathcal{B}(\mathsf{h})$ and any $t \geq 0$ we have that $tr(\rho \mathcal{T}_t(x)) = tr(\rho x)$. Equivalently, $\mathcal{T}_{*t}(\rho) = \rho$, $\forall t \geq 0$.

Notice that if ρ is an invariant state in the domain of the pre-dual generator $\mathcal{L}_*$, then

$$0 = \frac{d}{dt} \mathcal{T}_{*t}(\rho) \mid_{t=0} = \mathcal{L}_*(\mathcal{T}_{*t}(\rho)) \mid_{t=0} = \mathcal{L}_*(\rho).$$

Therefore ρ is invariant for $(\mathcal{T}_t)_{t \geq 0}$ if and only if $\mathcal{L}_*(\rho) = 0$.

By direct computation, from (5) one can show that the formal generator of the pre-dual semigroup has the form,

$$\mathcal{L}_*(x) = \frac{1}{2} \sum_{j,k \geq 0} \left\{ x_{jk} \left[\mu^2 \left([j]_n + [k]_n \right) + \lambda^2 \left([j+n]_n + [k+n]_n \right) \right. \right.$$

$$\left. + 2i\omega \left([j]_n - [k]_n \right) \right] - 2\lambda^2 x_{j-n,k-n}[j]_n^{1/2}[k]_n^{1/2} \tag{6}$$

$$\left. - 2\mu^2 x_{j+n,k+n}[j+n]_n^{1/2}[k+n]_n^{1/2} \right\} |e_j\rangle \langle e_k|.$$

The following formal computations aims at finding diagonal detailed balance invariant states. If $\rho = \sum_{j \geq 0} \rho_j |e_j\rangle \langle e_j|$, we have that

$$0 = \mathcal{L}_*(\rho)$$

$$= -\sum_{j \geq 0} \left(\lambda^2 \rho_{j-n}[j]_n + \mu^2[j+n]_n \rho_{j+n} - \rho_j[j]_n \mu^2 - [j+n]_n \lambda^2 \rho_j \right) |e_j\rangle \langle e_j|,$$

if the following detailed balance conditions hold:

$$\lambda^2[j]_n \rho_{j-n} = \mu^2[j]_n \rho_j, \quad \text{and} \quad \mu^2[j+n]_n \rho_{j+n} = \lambda^2[j+n]_n \rho_j. \tag{7}$$

In the first one we assume that $j \geq n$. We know that $[j]_n = 0$ for $j < n$, hence we have the equivalent conditions:

$$\rho_j = \left(\frac{\lambda}{\mu}\right)^2 \rho_{j-n} = \nu^2 \rho_{j-n}, \quad \text{and} \quad \rho_{j+n} = \left(\frac{\lambda}{\mu}\right)^2 \rho_j = \nu^2 \rho_j \tag{8}$$

where $\nu = \frac{\lambda}{\mu} < 1$. By direct computation we get the following states satisfying the infinitesimal detailed balance conditions (8)

$$\rho_{(r)} = (1 - \nu^2) \sum_{k \geq 0} \nu^{2k} |e_{kn+r}\rangle \langle e_{kn+r}|, \quad \text{for} \quad 0 \leq r \leq n-1. \tag{9}$$

The following result was proved in Ref.[8]

Proposition 3.1.

(i) The states $\rho_{(r)}$, $0 \leq r \leq n - 1$, are invariant for the n-photon absorption-emission semigroup.

(ii) Each invariant state is a convex linear combination of the set $(\rho_{(r)})_{0 \leq r < n}$, i.e., it has the form

$$\rho = \sum_{r=0}^{n-1} \alpha_r \rho_{(r)}, \quad 0 < \alpha_r \leq 1, \quad \sum_{r=0}^{n-1} \alpha_r = 1. \tag{10}$$

In particular it is diagonal.

By $\{X_1, X_2, \cdots\}'$ we denote the generalized commutator of the (possibly unbounded) operators $X_1, X_2, \cdots$. This is the sub-algebra of $\mathcal{B}(\mathsf{h})$ of all operators y such that $yX_k \subseteq X_k y$ (i.e. $\text{Dom}(X_k) \subseteq \text{Dom}(X_k y)$ and $yX_k u = X_k yu$ for all $u \in \text{Dom}(X_k)$) for all $k \geq 1$.

Let p_r be the support projection of the invariant state $\rho_{(r)}$. By Theorem 3.1 and Proposition 3.1 of Ref.,[8] each support projection is harmonic, i.e., $\mathcal{T}_t(p_r) = p_r$, $\forall t \geq 0$.

Theorem 3.1. *For any normal initial state σ, there exists an invariant state σ_∞ such that,*

$$\sigma_\infty = \lim_{t \to \infty} \mathcal{T}_{*t}(\sigma),$$

in the weak topology of $L_1(\mathsf{h})$. Moreover, $\sigma_\infty = \sum_{0 \leq r \leq n-1} tr(\sigma p_r) \rho_{(r)}$.

Proof. The proof of the limit follows from an application of the relation

$$\{a^n, a^{\dagger n}\}' = \{a^n, a^{\dagger n}, a^n a^{\dagger n}\}',$$

for every $n \geq 1$; whose proof is immediate (see Lemma 7.1 in Ref.[7]), a result of Frigerio-Verri and a result of Fagnola-Rebolledo (see Theorems 4.3 and 4.4 in the former reference). The explicit form of σ_∞ follows from the harmonicity (i.e., $\mathcal{T}_t$-invariance) of the support projections p_r. Indeed, since σ_∞ is diagonal by Theorem 3.1 it has the form $\sigma_\infty = \sum_{0 \leq r \leq n-1} \alpha_r \rho_{(r)}$, and writing $\sigma = \sum_{0 \leq r \leq n-1} p_r \sigma p_r + \sigma_{off}$, where σ_{off} denotes the off-diagonal part of σ, we have that

$$\alpha_r = tr(p_r \sigma_\infty) = \lim_{t \to \infty} tr\Big(p_r \mathcal{T}_{*t}\Big(\sum_{0 \leq s \leq (n-1)} p_s \sigma p_s + \sigma_{off} \Big)\Big) = tr(\sigma p_r). \qquad \square$$

4. The Dirichlet form

Following Ref.[3] we replace the von Neumann algebra $\mathcal{B}(h)$ by the Hilbert space $L_2(h)$ of Hilbert-Schmidt operators provided with the inner product $\langle y, x \rangle = tr(y^* x)$, so that one can use the Hilbert structure. For every faithful invariant state ρ and $\theta \in [0, 1]$, the embedding of $\mathcal{B}(h)$ into $L_2(h)$ is defined by means of

$$\iota : \mathcal{B}(h) \to L_2(h), \qquad \iota(x) = \rho^{\theta/2} x \rho^{\frac{1-\theta}{2}}.$$

Now we define $T_t(\iota(x)) = \iota(\mathcal{T}_t(x))$ for $t \geq 0$ and $x \in \mathcal{B}(h)$. The operators T_t can be extended to the whole $L_2(h)$ and they define a unique strongly continuous semigroup of contractions $T = (T_t)_{t \geq 0}$ on $L_2(h)$ (see Carbone,[1] Theorem 2.0.3). Moreover, if L is the generator of T, then $\iota(D(\mathcal{L}))$ is contained in the domain of L and

$$L\big(\rho^{\theta/2} x \rho^{\frac{1-\theta}{2}}\big) = \rho^{\theta/2} \mathcal{L}(x) \rho^{\frac{1-\theta}{2}}$$

for all $x \in D(\mathcal{L})$.

The Dirichlet form, defined for $x \in \text{Dom}(L)$, is the quadratic form $\mathcal{E}$ associated with L, i.e., it is defined by means of

$$\mathcal{E}(x) = -\Re\langle x, L(x) \rangle.$$

For $y = \sum_{k,l \geq 0} x_{kl} |e_k\rangle \langle e_l| \in \iota(D(\mathcal{L}))$, let

$$x = \iota^{-1}(y) = \sum_{k,l \geq 0} x_{kl} \left|\rho^{-\frac{\theta}{2}} e_k\right\rangle \left\langle \rho^{\frac{\theta-1}{2}} e_l\right|$$

$$= \sum_{k,l \geq 0} \rho_k^{-\frac{\theta}{2}} \rho_l^{\frac{\theta-1}{2}} x_{kl} |e_k\rangle \langle e_l|. \tag{11}$$

Then we have that

$$
\mathcal{L}(x) = \sum_{j,l \geq 0} |e_j\rangle \langle e_l| \left\{ \mu^2 \rho_{j-n}^{-\frac{\theta}{2}} \rho_{l-n}^{\frac{\theta-1}{2}} x_{j-n,\,l-n} [j]_n^{1/2} [l]_n^{1/2} \right.
$$

$$
+ \lambda^2 \rho_{j+n}^{-\frac{\theta}{2}} \rho_{l+n}^{\frac{\theta-1}{2}} x_{j+n,\,l+n} [j+n]_n^{1/2} [l+n]_n^{1/2}
$$

$$
- \frac{1}{2} \rho_j^{-\frac{\theta}{2}} \rho_l^{\frac{\theta-1}{2}} x_{j,l} \left[\mu^2 \left([j]_n + [l]_n \right) + \lambda^2 \left([j+n]_n + [l+n]_n \right) \right.
$$

$$
\left. \left. - 2i\omega \left([j]_n - [l]_n \right) \right] \right\}.
$$

It follows from (8) that

$$
\rho_{j-n}^{-\theta/2} = \nu^\theta \rho_j^{-\theta/2}, \qquad \rho_{l-n}^{\frac{\theta-1}{2}} = \nu^{-(\theta-1)} \rho_l^{\frac{\theta-1}{2}},
$$

$$
\rho_{j+n}^{-\theta/2} = \nu^{-\theta} \rho_j^{-\theta/2} \quad \text{and} \quad \rho_{l+n}^{\frac{\theta-1}{2}} = \nu^{-(\theta-1)} \rho_l^{\frac{\theta-1}{2}}. \tag{12}
$$

Hence,

$$
\rho^{\theta/2} \mathcal{L}(x) \rho^{\frac{1-\theta}{2}} = \sum_{j,l \geq 0} \rho^{\frac{\theta}{2}} |e_j\rangle \langle e_l| \rho^{\frac{1-\theta}{2}} \rho_j^{-\frac{\theta}{2}} \rho_l^{\frac{\theta-1}{2}} \left\{ \mu^2 \nu x_{j-n,\,l-n} [j]_n^{1/2} [l]_n^{1/2} \right.
$$

$$
+ \lambda^2 \nu^{-1} x_{j+n,\,l+n} [j+n]_n^{1/2} [l+n]_n^{1/2}
$$

$$
- \frac{1}{2} x_{j,l} \left[\mu^2 \left([j]_n + [l]_n \right) + \lambda^2 \left([j+n]_n + [l+n]_n \right) \right.
$$

$$
\left. \left. - 2i\omega \left([j]_n - [l]_n \right) \right] \right\}, \tag{13}
$$

consequently

$$
L(y) = \sum_{j,k \geq 0} |e_j\rangle \langle e_k| \left\{ \lambda\mu x_{j-n,\,k-n} [j]_n^{1/2} [k]_n^{1/2} \right.
$$

$$
+ \lambda\mu x_{j+n,\,k+n} [j+n]_n^{1/2} [k+n]_n^{1/2}
$$

$$
- \frac{1}{2} x_{j,k} \left[\mu^2 \left([j]_n + [k]_n \right) + \lambda^2 \left([j+n]_n + [k+n]_n \right) \right.
$$

$$
\left. \left. - 2i\omega \left([j]_n - [k]_n \right) \right] \right\}. \tag{14}
$$

Proposition 4.1. *For every* $x \in L_2(\mathsf{h})$ *of the form* $x = \sum_{\ell,m \geq 0} x_{\ell m}$ $|e_\ell\rangle \langle e_m|$ *the Dirichlet form can be written as*

$$
\mathcal{E}(x) = \frac{1}{2} \sum_{j,k \geq 0} \left| \mu[j+n]_n^{1/2} x_{j+n,\,k+n} - \lambda[k+n]_n^{1/2} x_{j,k} \right|^2
$$

$$
+ \frac{1}{2} \sum_{j,k \geq 0} \left| \mu[k+n]_n^{1/2} x_{j+n,\,k+n} - \lambda[j+n]_n^{1/2} x_{j,k} \right|^2
$$

$$
+ \sum_{j \geq 0} \sum_{k=0}^{n-1} \frac{1}{2} \mu^2 [j+n]_n \left(|x_{k,\,j+n}|^2 + |x_{j+n,\,k}|^2 \right). \tag{15}
$$

Proof. If $x = \sum_{\ell,m \geq 0} x_{\ell m} |e_\ell\rangle \langle e_m|$, then $x^* = \sum_{\ell,m \geq 0} \overline{x_{\ell m}} |e_m\rangle \langle e_\ell|$. After some computations one can show using (14), that the explicit expression for the Dirichlet form is

$$
\begin{aligned}
\mathcal{E}(x) = {}& - \Re\langle x, \, L(x)\rangle = -\Re(tr(x^*L(x))) \\
= {}& \sum_{j,k \geq 0} \overline{x_{jk}} \Big\{ \lambda\mu x_{j-n,\,k-n}[j]_n^{1/2}[k]_n^{1/2} + \lambda\mu x_{j+n,\,k+n}[j+n]_n^{1/2}[k+n]_n^{1/2} \\
& - \frac{1}{2} x_{j,\,k} \Big[\mu^2 \left([j]_n + [k]_n\right) + \lambda^2 \left([j+n]_n + [k+n]_n\right) \\
& - 2i\omega \left([j]_n - [k]_n\right) \Big] \Big\}.
\end{aligned}
$$

We can write,

$$
\begin{aligned}
\mathcal{E}(x) = {}& - \Re tr(x^*L(x)) \\
= {}& - \Re\Big[\sum_{j,k \geq 0} \overline{x_{jk}} \Big\{ \lambda\mu x_{j-n,\,k-n}[j]_n^{1/2}[k]_n^{1/2} \\
& + \lambda\mu x_{j+n,\,k+n}[j+n]_n^{1/2}[k+n]_n^{1/2} - \frac{1}{2} x_{j,\,k} \Big[\mu^2 \left([j]_n + [k]_n\right) \\
& + \lambda^2 \left([j+n]_n + [k+n]_n\right) - 2i\omega \left([j]_n - [k]_n\right) \Big] \Big\} \Big] \qquad (16) \\
= {}& \sum_{j,k \geq 0} \Big\{ - \lambda\mu\Re(\overline{x_{jk}}x_{j-n,k-n}))[j]_n^{1/2}[k]_n^{1/2} \\
& - \lambda\mu\Re(\overline{x_{jk}}x_{j+n,k+n}))[j+n]_n^{1/2}[k+n]_n^{1/2} \\
& + \frac{1}{2}|x_{jk}|^2 \Big[\mu^2 \left([j]_n + [k]_n\right) + \lambda^2 \left([j+n]_n + [k+n]_n\right) \Big] \Big\}.
\end{aligned}
$$

Using that $[j]_n, [k]_n = 0 \quad \forall j, k = 0, 1, \ldots, n-1$ and with the changes of variable $j' = j - n$, $k' = k - n$ the first summand in the former identity (16) takes the new form

$$
\begin{aligned}
\sum_{j,k \geq 0} & -\lambda\mu\Re(\overline{x_{jk}}x_{j-n,k-n}))[j]_n^{1/2}[k]_n^{1/2} \\
& = -\frac{1}{2} \sum_{j,k \geq n} 2\lambda\mu\Re(\overline{x_{jk}}x_{j-n,k-n}))[j]_n^{1/2}[k]_n^{1/2} \\
& = -\frac{1}{2} \sum_{j',k' \geq 0} 2\lambda\mu\Re(\overline{x_{j'+nk'+n}}x_{j',k'}))[j'+n]_n^{1/2}[k'+n]_n^{1/2}.
\end{aligned}
$$

For the third summand in (16), using the same change of variable we get

$$\sum_{j,k\geq 0} \frac{1}{2}|x_{jk}|^2 \left[\mu^2 \left([j]_n + [k]_n\right) + \lambda^2 \left([j+n]_n + [k+n]_n\right) \right]$$

$$= \sum_{j,k\geq 0} \frac{1}{2}|x_{jk}|^2\mu^2 \left([j]_n + [k]_n\right) + \sum_{j,k\geq 0} \frac{1}{2}|x_{jk}|^2\lambda^2 \left([j+n]_n + [k+n]_n\right)$$

$$= \sum_{j,k\geq n} \frac{1}{2}|x_{jk}|^2\mu^2 \left([j]_n + [k]_n\right) + \sum_{j\geq n}\sum_{k=0}^{n-1} \frac{1}{2}|x_{jk}|^2\mu^2[j]_n$$

$$+ \sum_{k\geq n}\sum_{j=0}^{n-1} \frac{1}{2}|x_{jk}|^2\mu^2[k]_n + \sum_{j,k\geq 0} \frac{1}{2}|x_{jk}|^2\lambda^2 \left([j+n]_n + [k+n]_n\right)$$

$$= \sum_{j',k'\geq 0} \frac{1}{2}|x_{j'+n,\,k'+n}|^2\mu^2 \left([j'+n]_n + [k'+n]_n\right)$$

$$+ \sum_{j\geq 0}\sum_{k=0}^{n-1} \frac{1}{2}\mu^2[j+n]_n \left(|x_{k,\,j+n}|^2 + |x_{j+n,\,k}|^2\right)$$

$$+ \sum_{j,k\geq 0} \frac{1}{2}|x_{jk}|^2\lambda^2 \left([j+n]_n + [k+n]_n\right).$$

So that, from (16) we get the result. □

Proposition 4.2. *If $\mathcal{E}(x) = 0$, then x is diagonal.*

Proof. If $\mathcal{E}(x) = 0$, then each one of the summands in (15) must be zero, from the first and second summands we get $\forall j, k \geq 0$

$$\mu[j+n]_n^{1/2}x_{j+n,\,k+n} = \lambda[k+n]_n^{1/2}x_{j,\,k}$$

$$\mu[k+n]_n^{1/2}x_{j+n,\,k+n} = \lambda[j+n]_n^{1/2}x_{j,\,k}$$

therefore

$$x_{j+n,\,k+n} = \frac{\lambda}{\mu}\frac{[k+n]_n^{1/2}}{[j+n]_n^{1/2}}x_{j,k} = \frac{\lambda}{\mu}\frac{[j+n]_n^{1/2}}{[k+n]_n^{1/2}}x_{j,k}. \qquad (17)$$

But $x_{j,k} \neq 0$ implies that

$$\frac{[k+n]_n^{1/2}}{[j+n]_n^{1/2}} = \frac{[j+n]_n^{1/2}}{[k+n]_n^{1/2}} \Rightarrow [k+n]_n = [j+n]_n \Rightarrow j = k,$$

hence x is diagonal. □

Remark 4.1. It follows from (17) that the elements $x \in KerL$ satisfy $x_{n+r} = \nu x_r$. Hence for the elements $x \in KerL$ we have that $x_{nj+r} = \nu^j x_r$.

Let $W := span\{\rho_{(r)}^{1/2} : 0 \leq r \leq n-1\}$

Theorem 4.1. *We have, $KerL = \iota(Ker\mathcal{L}) = W$.*

Proof. The support projection of $\rho_{(r)}$ is given by

$$p_r = \sum_{k \geq 0} |e_{kn+r}\rangle \langle e_{kn+r}|.$$

Moreover we have from (10) that $\rho_{kn+r}^{1/2} = \alpha_r^{1/2} \left(1 - \nu^2\right)^{1/2} \nu^k$. Then

$$\iota(p_r) = \rho^{\frac{\theta}{2}} \sum_{k \geq 0} |e_{kn+r}\rangle \langle e_{kn+r}| \rho^{\frac{1-\theta}{2}}$$

$$= \sum_{k \geq 0} \left|\rho^{\frac{\theta}{2}} e_{kn+r}\right\rangle \left\langle \rho^{\frac{1-\theta}{2}} e_{kn+r}\right|$$

$$= \sum_{k \geq 0} \rho_{kn+r}^{1/2} |e_{kn+r}\rangle \langle e_{kn+r}| = \sum_{k \geq 0} \alpha_r^{1/2} \left(1 - \nu^2\right)^{1/2} \nu^k |e_{kn+r}\rangle \langle e_{kn+r}|$$

$$= \alpha_r^{1/2} \sum_{k \geq 0} \left(1 - \nu^2\right)^{1/2} \nu^k |e_{kn+r}\rangle \langle e_{kn+r}| = \alpha_r^{1/2} \rho_r^{1/2}.$$

On the other side, since $\mathcal{L}(p_r) = 0$ (the support projections p_r are harmonic),

$$L(\rho_r^{1/2}) = \rho^{\frac{\theta}{2}} \mathcal{L}(\rho_r^{1/2}) \rho^{\frac{1-\theta}{2}} = \rho^{\frac{\theta}{2}} \alpha_r^{-1/2} \mathcal{L}(p_r) \rho^{\frac{1-\theta}{2}} = 0,$$

i.e., $W \subset KerL$.

Now, if $x \in KerL$ using the result in Remark 4.1 we get

$$x = \sum_{j \geq 0} x_j |e_j\rangle \langle e_j| = \sum_{j \geq 0} \sum_{r=0}^{n-1} x_{nj+r} |e_{nj+r}\rangle \langle e_{nj+r}|$$

$$= \sum_{j \geq 0} \sum_{r=0}^{n-1} \nu^j x_r |e_{nj+r}\rangle \langle e_{nj+r}| = \sum_{r=0}^{n-1} x_r \sum_{j \geq 0} \nu^j |e_{nj+r}\rangle \langle e_{nj+r}|$$

$$= \sum_{r=0}^{n-1} x_r \left(1 - \nu^2\right)^{-1/2} \rho_r^{1/2} \in W,$$

therefore $KerL \subset W$. This finishes the proof. □

5. Invariant subspaces

For every $r \in \{0, 1, \ldots, n-1\}$ and $m \in \mathbb{Z}$ let us define the subspaces of $L_2(\mathsf{h})$,

$$\mathcal{H}_m^r = span\left\{|e_{kn+r}\rangle \langle e_{kn+r+m}| : k \in \mathbb{N} \cup \{0\} \text{ and } kn + r + m \geq 0\right\}.$$

Proposition 5.1. *The subspaces $\mathcal{H}_m^r$ have the following properties*

(i) If $x \in \mathcal{H}_m^r$, then $x^* \in \mathcal{H}_{-m}^q$ where x^* is the adjoint operator of x and $q \equiv (m+r) \bmod n$, $0 \le q \le n-1$.

(ii) The subspaces $\mathcal{H}_m^r$ are mutually orthogonal on $L_2(\mathsf{h})$ and

$$L_2(\mathsf{h}) = \bigoplus_{\substack{m \in \mathbb{Z} \\ r \in \{0,1,\ldots,n-1\}}} \mathcal{H}_m^r$$

(iii) Each $\mathcal{H}_m^r$ is invariant for L and, consequently, also for the semigroup T.

(iv) Each $\mathcal{H}_m^r$ is isometrically isomorphic to the space $\ell_2(\mathbb{N})$ of all square summable sequences.

Proof. Let $x \in \mathcal{H}_m^r$, then $x = \sum_k x_k \left| e_{kn+r} \right\rangle \left\langle e_{kn+r+m} \right|$ and

$$x^* = \sum_k \overline{x_k} \left| e_{kn+r+m} \right\rangle \left\langle e_{kn+r} \right| = \sum_k \overline{x_k} \left| e_{(k+p)n+q} \right\rangle \left\langle e_{(k+p)n+q-m} \right|$$

where $r + m = pn + q$.
If $j = k + p$ and we write $y_j = \overline{x_{j-p}}$ we obtain:

$$x^* = \sum_j y_j \left| e_{jn+q} \right\rangle \left\langle e_{jn+q-m} \right| \in \mathcal{H}_{-m}^q.$$

This proves (i).
To prove (ii), it suffices to consider the orthonormal basis.
For every k, let $\left| e_{kn+r} \right\rangle \left\langle e_{kn+r+m} \right| \in \mathcal{H}_m^r$ and $\left| e_{kn+s} \right\rangle \left\langle e_{kn+s+m'} \right| \in \mathcal{H}_{m'}^s$.
We have that

$$
\begin{aligned}
&\left\langle \left| e_{kn+r} \right\rangle \left\langle e_{kn+r+m} \right|, \left| e_{kn+s} \right\rangle \left\langle e_{kn+s+m'} \right| \right\rangle_{L_2(\mathsf{h})} \\
&= Tr \left(\left| e_{kn+s} \right\rangle \left\langle e_{kn+s+m'} \right|^* \left| e_{kn+r} \right\rangle \left\langle e_{kn+r+m} \right| \right) \\
&= Tr \left(\left| e_{kn+s+m'} \right\rangle \left\langle e_{kn+s} \right| \left| e_{kn+r} \right\rangle \left\langle e_{kn+r+m} \right| \right) \\
&= Tr \left(\left| e_{kn+s+m'} \right\rangle \left\langle e_{kn+r+m} \right| \right) \delta_{kn+s,kn+r} = 0,
\end{aligned}
\tag{18}
$$

for $s \ne r$. If $r = s$ we have that $Tr \left(\left| e_i \right\rangle \left\langle e_j \right| \right) = \delta_{i,j}$, with $i = kn + s + m'$ and $j = kn + r + m$, which proves the orthogonality.

For every basic element of $L_2(\mathsf{h})$, of the form $\left| e_i \right\rangle \left\langle e_j \right|$, by the division algorithm there exist $k, r \in \mathbb{Z}$, $0 \le r \le n-1$ such that $i = kn + r$. Taking $m = j - i$ we obtain that $\left| e_i \right\rangle \left\langle e_j \right| = \left| e_{kn+r} \right\rangle \left\langle e_{kn+r+m} \right| \in \mathcal{H}_m^r$.

Let us prove (iii). From (14), one can see that for x of the form (11), the shifts in the indices of the matrix elements of x are of length n, i.e., the (j,k)-matrix element of $L(x)$ depends on the matrix elements $(j-n, k-n)$,

(j, k) and $(j + n, k + n)$ of x. So that, any basic element in $\mathcal{H}_m^r$ of the form $|e_{kn+r}\rangle \langle e_{kn+r+m}|$ is sent by L into an operator with the only non-zero matrix elements in positions $|e_{(k-1)n+r}\rangle \langle e_{(k-1)n+r+m}|$, $|e_{kn+r}\rangle \langle e_{kn+r+m}|$ and $|e_{(k+1)n+r}\rangle \langle e_{(k+1)n+r+m}|$, so that $L(x) \in \mathcal{H}_m^r$. This proves (iii).

(iv) follows from the fact that every separable hilbert space $\mathcal{H}_m^r$ is isometrically isomorphic to $\ell_2(\mathbb{N})$. □

6. The spectral gap for $0 < \lambda < \mu$

For $m \neq 0$, $r \in \{0, 1, \ldots, n-1\}$ let us define

$$A_m^r = \inf\{\mathcal{E}(x) : \|x\| = 1, x \in \mathcal{H}_m^r\}$$

and

$$A_0^r = \inf\{\mathcal{E}(x) : \|x\| = 1, x \in \mathcal{H}_0^r, \ x \perp \rho_r^{1/2}\}.$$

Lemma 6.1. *Let us fix $m \in \mathbb{Z}$, $r \in \{0, 1, \ldots, n-1\}$ and $x \in L_2(\mathsf{h})$. By $\mathcal{H}_m^r(x)$ we denote the orthogonal projection of x onto $\mathcal{H}_m^r$. We have the identities*

$$\inf_{x \perp KerL} \frac{\mathcal{E}(\mathcal{H}_m^r(x))}{\|\mathcal{H}_m^r(x)\|^2} = \inf_{\substack{x \perp KerL \\ x \in \mathcal{H}_m^r}} \frac{\mathcal{E}(x)}{\|x\|^2} = \inf_{\substack{x \perp KerL \\ x \in \mathcal{H}_m^r \\ \|x\|=1}} \mathcal{E}(x) \tag{19}$$

$$= A_m^r.$$

Proof. The first identity is just a change of variable, indeed, $x \in \mathcal{H}_m^r$, implies $x = \mathcal{H}_m^r(x)$. In the last identity we have that for $m \neq 0$, $x \in \mathcal{H}_m^r$ implies $x \perp KerL$, since all elements in $KerL$ are diagonal, hence the identity holds. Finally, if $m = 0$, then $x = \mathcal{H}_0^r(x)$ and it is orthogonal with every $\rho_s^{1/2}$ for $r \neq s$, then orthogonality condition reduces to $x \perp \rho_r^{1/2}$. □

Definition 6.1. The spectral gap of the n-photon absorption-emission process is defined as

$$\mathrm{Gap}(L) := \inf_{x \perp KerL} \frac{\mathcal{E}(x)}{\|x\|^2}.$$

The following results was proved by R. Carbone in Ref.[1]

Proposition 6.1. *Let $(T_t)_{t \geq 0}$ be a strongly continuous semigroup of contractions on a Hilbert space h with infinitesimal generator L and invariant vector v, then the spectral gap of L is the maximal positive number ε satisfying that for every $x \in \mathsf{h}$ and $t \geq 0$*

$$\|T_t x - \langle v, x \rangle v\| \leq e^{-\varepsilon t}\|x - \langle v, x \rangle v\|.$$

The following proposition is a consequence of the definition of the gap together with Lemma 6.1.

Proposition 6.2. *For the n-photon absorption-emission process we have that*

$$Gap(L) = \inf_{m \in \mathbb{Z}} \inf_{r \in \{0,1,\dots,n-1\}} A_m^r. \tag{20}$$

6.1. *An estimate for the off-diagonal gap*

In this section we shall compute an estimate of the off-diagonal gap, i.e., we will find a lower bound for the infimum (20) for $m \neq 0$. Notice that from (16) one can see that the Dirichlet form $\mathcal{E}$ is invariant under the take of adjoint, i.e., $\mathcal{E}(x^*) = \mathcal{E}(x)$. Hence by (i) in Proposition 5.1, it suffices to consider $m > 0$. Elementary computations yields the following.

Lemma 6.2. *For every $k \geq 1$, $0 \leq r \leq n$, $m \geq 1$ and $x_k, x_{k-1} \in \mathbb{C}$, the following inequality holds:*

$$4\lambda\mu[kn + r]_n^{1/2}[kn + r + m]_n^{1/2}|x_k|\,|x_{k-1}|$$
$$\leq \mu^2\Big([kn + r]_n + [kn + r + m]_n\Big)|x_k|^2$$
$$+ 4\lambda^2 \frac{[kn + r]_n[kn + r + m]_n}{[kn + r]_n + [kn + r + m]_n}|x_{k-1}|^2.$$

Let $m > 0$ and let us take $x \in \mathcal{H}_m^r$. Hence, for some sequence $(x_k)_{k \geq 0} \in \ell_2(\mathbb{N})$, we can write $x = \sum_{k \geq 0} x_k |e_{kn+r}\rangle\langle e_{kn+r+m}|$. Therefore from (15) we get

$$\mathcal{E}(x) = \frac{1}{2}\sum_{k \geq 1} |\mu[kn + r]_n^{1/2}x_k - \lambda[kn + r + m]_n^{1/2}x_{k-1}|^2$$
$$+ \frac{1}{2}\sum_{k \geq 1} |\mu[kn + r + m]_n^{1/2}x_k - \lambda[kn + r]_n^{1/2}x_{k-1}|^2 + \frac{1}{2}|x_0|\mu^2[r + m]_n$$
$$= \frac{1}{2}\sum_{k \geq 1} \Big(([kn + r]_n + [kn + r + m]_n)(\mu^2|x_k|^2 + \lambda^2|x_{k-1}|^2) \tag{21}$$
$$- 4\mu\lambda[kn + r]_n^{\frac{1}{2}}[kn + r + m]_n^{\frac{1}{2}}\Re(x_k\overline{x}_{k-1})\Big) + \frac{1}{2}|x_0|^2[r + m]_n\mu^2$$
$$\geq \frac{1}{2}\sum_{k \geq 1} \Big(([kn + r]_n + [kn + r + m]_n)(\mu^2|x_k|^2 + \lambda^2|x_{k-1}|^2)$$
$$- 4\mu\lambda[kn + r]_n^{\frac{1}{2}}[kn + r + m]_n^{\frac{1}{2}}|x_k||x_{k-1}|\Big) + \frac{1}{2}|x_0|^2\mu^2[r + m]_n$$

$$\geq \frac{1}{2} \sum_{k \geq 1} \Big(([kn+r]_n + [kn+r+m]_n)(\mu^2 |x_k|^2 + \lambda^2 |x_{k-1}|^2)$$

$$- \mu^2 ([kn+r]_n + [kn+r+m]_n)|x_k|^2$$

$$- 4\lambda^2 \frac{[kn+r+m]_n[kn+r]_n}{[kn+r+m]_n + [kn+r]_n}|x_{k-1}|^2 \Big)$$

$$+ \frac{1}{2}\mu^2 |x_0|^2 [r+m]_n$$

$$= \frac{1}{2}|x_0|^2 \mu^2 [r+m]_n + \frac{\lambda^2}{2} \sum_{k \geq 1} \frac{([kn+r+m]_n - [kn+r]_n)^2}{[kn+r+m]_n + [kn+r]_n}|x_{k-1}|^2.$$

Lemma 6.3. *For $n \geq 2$, $k \geq 1$, $0 \leq r \leq n$ and $m \geq 1$ the following inequality holds*

$$\frac{([kn+r+m]_n - [kn+r]_n)^2}{[kn+r+m]_n + [kn+r]_n} \geq \frac{n^2 n!}{n+2} \tag{22}$$

with equality for $m = 1$, $r = 0$ and all $k \geq 1$ if $n = 2$; and equality for $m = 1$, $r = 0$, $k = 1$ if $n \geq 3$.

Proof. Let $T(x,m) = \frac{([x+m]_n - [x]_n)^2}{[x+m]_n + [x]_n}$, for $x \geq n$ and $m \geq 1$. Then

$$T(x, m+1) - T(x,m)$$

$$= \frac{([x+m+1]_n - [x]_n)^2}{[x+m+1]_n + [x]_n} - \frac{([x+m]_n - [x]_n)^2}{[x+m]_n + [x]_n}$$

$$= [x+m+1]_n + [x]_n - \frac{4[x+m+1]_n[x]_n}{[x+m+1]_n + [x]_n}$$

$$- \left([x+m]_n + [x]_n - \frac{4[x+m]_n[x]_n}{[x+m]_n + [x]_n} \right)$$

$$= [x+m+1]_n - [x+m]_n$$

$$+ 4[x]_n \left(\frac{[x+m]_n}{[x+m]_n + [x]_n} - \frac{[x+m+1]_n}{[x+m+1]_n + [x]_n} \right)$$

$$= [x+m+1]_n - [x+m]_n$$

$$+ 4[x]_n \frac{[x+m]_n([x+m+1]_n + [x]_n) - [x+m+1]_n([x+m]_n + [x]_n)}{([x+m]_n + [x]_n)([x+m+1]_n + [x]_n)}$$

$$= [x+m+1]_n - [x+m]_n$$

$$- 4[x]_n^2 \frac{[x+m+1]_n - [x+m]_n}{([x+m]_n + [x]_n)([x+m+1]_n + [x]_n)}$$

$$\geq [x+m+1]_n - [x+m]_n - 4[x]_n^2 \frac{[x+m+1]_n - [x+m]_n}{([x]_n + [x]_n)([x]_n + [x]_n)} = 0,$$

since $[x]_n \leq [x+m]_m \leq [x+m+1]_n$. Hence the function T is non-decreasing in the variable $m \geq 1$ for every $x \geq n$.

If $n = 2$, we have that for $m = 1$ and $r = 0$

$$\frac{([2k+1]_2 - [2k]_2)^2}{[2k+1] + [2k]} = \frac{(\frac{2k+1}{2k-1} - 1)^2 [2k]_2^2}{(\frac{2k+1}{2k-1} + 1)[2k]_2} = \frac{2^2}{4k(2k-1)}[2k]_2 = 2, \quad (23)$$

which proves that equality holds in (22) for all $k \geq 1$.

Let us take $n \geq 3$, $m = 1$ and show that the function $T(x) := T(x, 1)$ is non-decreasing as a function of $x \geq n$. Using properties (4) we get

$$[x+1]_n - [x]_n = \left(\frac{x+1}{x+1-n} - 1 \right) [x]_n = \frac{n}{x+1-n} [x]_n$$

and $[x+1]_n + [x]_n = \frac{2x+2-n}{x+1-n} [x]_n$, hence

$$\frac{([x+1]_n - [x]_n)^2}{[x+1]_n + [x]_n} = \frac{\frac{n^2}{(x+1-n)^2} [x]_n^2}{\frac{2x+2-n}{x+1-n} [x]_n} = \frac{n^2}{(x+1-n)(2x+2-n)} [x]_n$$

$$= \frac{n^2}{2x+2-n} [x]_{n-1}.$$

Therefore $T(x) = \frac{n^2}{2x+2-n} [x]_{n-1} = \frac{n^2 \prod_{j=0}^{n-2}(x-j)}{2x+2-n}$. Function T is non-decreasing if $\ln T$ is and we have that

$$\ln T(x) = 2 \ln n + \sum_{j=0}^{n-2} \ln(x-j) - \ln\{2x+2-n\}.$$

Taking derivative we get $\frac{d \ln T}{dx} = \sum_{j=0}^{n-2} \frac{1}{x-j} - \frac{2}{2x+2-n}$. Since $x \geq n$ we have $2x \geq n + x$ then $2x - n + 2 > 2x - n > x$; therefore $\frac{1}{x} > \frac{1}{2x+2-n}$ and this implies $\frac{-2}{2x+2-n} > -\frac{2}{x}$. From $x - j \leq x$ one obtains $\frac{1}{x-j} \geq \frac{1}{x}$. Therefore we have the inequality

$$\frac{d \ln T}{dx} \geq \sum_{j=0}^{n-2} \frac{1}{x} - \frac{2}{x} = \frac{n-1}{x} - \frac{2}{x} = \frac{n-3}{x} \geq 0,$$

if $n \geq 3$. We conclude that T is non-decresing for $n \geq 3$.

Now if $x = kn+r$, with $k \geq 1$ and $0 \leq r \leq n-1$, we see that x increases if k or r increases, so that the minimum of T is attained with $k = 1$ and $r = 0$, i.e., $x = n$. Therefore the minimum of T is

$$T(n) = \frac{n^2}{n+2} [n]_{n-1} = \frac{n^2 n!}{(n+2)(n-(n-1))!} = \frac{n^2 n!}{n+2}.$$

This proves the lemma. $\qquad \square$

Theorem 6.1. *For the n-photon absorption-emission process with $n \geq 2$, we have*

$$Gap_{off}(L) = \inf_{0 \neq m \in \mathbb{Z}} \inf_{r \in \{0,1,\ldots,n-1\}} A_m^r \geq \frac{\lambda^2}{2} \frac{n^2 n!}{n+2}, \qquad (24)$$

with equality if $n = 2$.

Proof. From (21) and the result of the above lemma we have that

$$\mathcal{E}(x) \geq \frac{1}{2}\lambda^2 \sum_{k \geq 1} \frac{n^2 n!}{n+2}|x_{k-1}|^2 = \frac{\lambda^2}{2} \frac{n^2 n!}{n+2}\|x\|^2, \qquad (25)$$

for all $x \in \mathcal{H}_m^r$, $m \geq 1$ and $0 \leq r \leq n$. This proves the Theorem.   □

Remark 6.1.

1.- In the case $n = 2$ we obtain from the right-hand side of (25) the value λ^2, that coincides with the value of the off-diagonal gap computed in Ref.[3]

2.- For $|e_0\rangle\langle e_1| \in \mathcal{H}_1^0$ we have that $\mathcal{E}(|e_0\rangle\langle e_1|) = (\frac{n+2}{n})^2 \frac{\lambda^2}{2} \frac{n^2 n!}{n+2}$, hence

$$\frac{\lambda^2}{2} \frac{n^2 n!}{n+2} \leq Gap_{off}(L) \leq (\frac{n+2}{n})^2 \frac{\lambda^2}{2} \frac{n^2 n!}{n+2}.$$

Since $\frac{n+2}{n} \to 1$ as $n \to \infty$, our estimate is sharp for big values of n.

Acknowledgement: RQ is grateful to the organizers of the *32nd Conference on Quantum Probability and Related Topics* in Levico Terme, Italy, for warm hospitality during the conference.

References

1. Carbone R., Exponential Ergodicity of Some Quantum Markov Semigroups, Ph.D Thesis, Università Degli Studi di Milano, 1-93, 2000.
2. Carbone R., and Fagnola F., Exponential L2-convergence of quantum Markov semigroups on B(h), Mat. Zametki 68, 523, 2000; translation in Math. Notes 68, 452, 2000.
3. Carbone R., Fagnola F., García J.C. and Quezada R., Spectral properties of the two-photon absorption and emission process, Journal of Math. Phys. 49, 032106, 1-18, 2008.
4. Correa D.S., De Boni L., Balogh D.T., and Mendonca C.R., Three- and four-photon excitation of poly(2- methoxy-5-(2-ethylhexyloxy)-1,4-phenylenevinylene)(MEH-PPV), Adv. Mater. 19, 2653-, 2007.

5. Delysse S., Filloux P., Dumarcher V., Fiorini C., and Nunzi J.M., Multi-photon absorption in organic dye solutions, Opt. Mater. 9, 347, 1998.
6. Fagnola F., Quantum Markov semigroups and quantum Markov flows, Proyecciones 18, 1-144, 1999.
7. Fagnola F. and Quezada R. Two Photon absorption and emission process. Infinite dimensional Analysis, Quantum probability and Related topics Vol. 8, No. 4 (2005) 573-591. World Scientific Publishing Company.
8. García J. C., Pantaleón-Martínez L., and Quezada, R. A sufficient condition for all invariant states of a QMS to be diagonal, in Proceedings of the 30th Conference on Quantum Probability and Related Topics, QP-PQ Quantum Probability and White Noise Analysis Vol. XXVII, World Scientific, 148-162, 2010.
9. Guang S.S., Ken-Tye Y., Jing Z., Hai-Yan Q., and Paras N. P., Emission, Optical Limiting and Stabilization of CdTe/CdS/ZnS Quantum Tripods System, IEEE Journal of Quantum Electronics 46, , 931-936, 2010.
10. He G.S., Markowicz P.P., Lin T.C., and Prasad P.N., Observation of stimulated emission by direct three-photon excitation, Nature (London) 415, 767-770, 2001.
11. Larson D.R., Zipfel W.R., Williams R.M., Clark S.W., Bruchez M.P., Wise F.W., and Webb W.W., Watersoluble quantum dots for multi-photon fluorescence imaging in vivo, Science 300, 1434, 2003.
12. Lu C.G., Cui Y.P., Huang W., Yun B.F., Wang Z.Y., Hu G.H., Cui J., Lu Z.F., and Qian Y., Vibrational resonance enhanced broadband multi-photon absorption in a triphenylamine derivatives, Appl. Phys. Lett. 91, 121111, 2007.
13. Matsuda H., Fujimoto Y., Ito S., Nagasawa Y., Miyasaka H., Asahi T., and Masuhara H., Development of near-infrared 35 fs laser microscope and its application to the detection of three- and four-photon fluorescence of organic microcrystals, J. Phys. Chem. B 110, 1091, 2006.
14. Parthenopoulos D.A. and Rentzepis P.M., Three-dimensional optical storage memory, Science 245, 843, 1989.
15. Yoshino F., Polyakov S., Liu M., and Stegeman G., Observation of three-photon enhanced four-photon absorption, Phys. Rev. Lett. 91, 063902, 2003.

SOME RECENT TOPICS ON WHITE NOISE THEORY

TAKEYUKI HIDA

Professor Emeritus Nagoya University and Meijo University
Hirabari-Minami, 2-903, Tenpaku-ku, Nagoya, 468-0020, Japan
E-mail: takeyuki@math.nagoya-u.ac.jp

SI SI

Faculty of Information Science and Technology, Aichi Prefectural University
Nagakute, Aichi-ken, 480-1198, Japan
E-mail: sisi@ist.aichi-pu.ac.jp

White noise theory has made significant progress in these days. Here are chosen some topics that are developing.

We shall show that one parameter groups and semi-groups that are isomorphic to subgroups of the infinite dimensional rotation group are playing new significant roles in white noise analysis. The associated Lie algebras are quite helpful to explain their probabilistic properties.

Two methods of introducing spaces of generalized white noise functionals, namely $(L^2)^-$ and $(S)^*$. We are going to introduce a third method by using renormalization technique. The space that we are concerned with is built up on the generalized linear spaces with the base of continuously many members like $\{\dot{B}(t), t \in R^1\}$. each of which has infinite length.

Keywords: White noise, infinite dimensional rotation group, half whisker

1. Infinite Dimensional Rotation Group

Take a nuclear space E which is dense in the space $L^2(R^1)$ and take the infinite dimensional rotation group $O(E)$. There are two classes of subgroups of $O(E)$; namely Class I and Class II.

The Class I is determined by choosing a complete orthonormal system $\{\xi_k\}$ such that ξ_k belongs to E for every k. Take finitely many, say n, ξ_k's which span an n-dimensional subspace E_n of E. We can define a subgroup G_n of $O(E)$ such that any $g \in G_n$ acts on $E_n^\perp$ as the identity and the restriction of g to E_n is a rotation, i.e. $g|_{E_n} \in SO(n)$.

If we take an increasing family $E_n, n \geq 1$, then, we can define a projective limit of G_n's; let it be denoted by G_∞. This is a typical example of subgroup belonging to the Class I.

Also, the windmill subgroup belongs to the Class I.

The members in Class I are defined by using a complete orthonormal base, so that the Class I depends on the choice of the base $\{\xi_n\}$. We may say they are *digital*.

To the contrary, our interest is focused on the Class II, that is, *analogue*. More precisely, a member $g \in O(E)$ belongs to the Class II if it is defined in a manner that there exists a diffeomorphism f of R^1 such that

$$(g\xi)(u) = \xi(f(u))\sqrt{|f'(u)|}.$$

Choosing a one-parameter subgroup $g_t, t \in R^1$ of Class II provides a good method of research in the white noise analysis. A one-parameter subgroup of $O(E)$ is called a *whisker*. Lie algebra involving the infinitesimal generators of the whiskers tells us the way of their actions on white noise functionals.

We can further extend this approach to the level of the study of continuous one-parameter semi-groups such as $g_t, t \geq 0$, where $g_t \in O(E)$ for every t. We are now going to this direction.

Compare the following two cases :

Digital   : Projective limit of spheres is reminded.

Analogue : Space E^* : Base involves continuously many vectors. See Ref. 11.

2. Class II Subgroups of $O(E)$

This section is devoted to a brief review of the known results and find some hints to find new good subgroups.

Each member of this class, say $\{g_t, t : real\}$, should be defined by a system of parameterized diffeomorphisms $\{\psi_t(u)$ of of $\bar{R} = R \cup \infty$: one point compactification. Namely,

$$\xi(u) \mapsto (g\xi(\psi_t(u))\sqrt{|\psi'_t(u)|}, \tag{1}$$

where $\psi'_t(u) = \frac{d}{du}\psi_t(u)$.

We are interested in a subgroup which can be made to be a local Lie group embedded in $O(E)$. In what follows the basic nuclear space is specified to D_o (see Ref. 4).

More practically, we restrict our attention to the case where $g_t, t \in R$ forms a one-parameter group such that g_t is continuous in t. Assume further that there exists the (infinitesimal) generator $A = \frac{d}{dt} g_t|_{t=o}$ of the whisker g_t.

There exists, by the assumptions of the group property and continuity, a family $\{\psi_t(u), t \in R$ such that $\psi_t(u)$ is measurable in (t, u) and satisfies the group properties

$$\psi_t \cdot \psi_t = \psi_{t+s}$$
$$\psi_0(u) = u.$$

Such a one-parameter group is called (*analytic*) *whisker*.

Let g_t^* be the adjoint operator to g_t. Then, the system $\{g_t^*\}$ again forms a one-parameter group of μ-measure (the white noise measure) preserving transformations g_t^*. The system is a flow on the white noise space $E^*(\mu), \mu$ being the white noise measure.

By J. Aezél,[1] we have an expression for $\psi_t(u)$:

$$\psi_t(u) = f(f^{-1}(u) + t) \tag{2}$$

where f is continuous and strictly monotone. Its (infinitesimal) generator α, if f is differentiable, can be expressed in the form

$$\alpha = a(u)\frac{d}{du} + \frac{1}{2}a'(u), \tag{3}$$

where

$$a(u) = f'(f^{-1}(u)). \tag{4}$$

See e.g. Refs. 4,5.

We have already established the results that there exists a three dimensional subgroup of class II with significant probabilistic meanings. The group consists of three one-parameter subgroups, the generators of which are expressed by $a(u) = 1, a(u) = u, a(u) = u^2$, respectively. Namely, we show a list:

$$s = \frac{d}{du},$$

$$\tau = u\frac{d}{du} + \frac{1}{2},$$

$$\kappa = u^2\frac{d}{du} + u$$

One of the interesting interpretations may be given in such a way that they, putting together, describe the projective invariance of Brownian motion. If we come to higher dimensional parameter case, then we are given the *conformal invariance*.

Those generators form a base of a three dimensional Lie algebra isomorphic to $sl(2, R)$. This fact can easily be seen by

$$[\tau, s] = -s$$

$$[\tau, \kappa] = \kappa$$

$$[\kappa, s] = 2\tau$$

There is a remark that the shift with generator s is sitting as a key member of the generators.

Also, one can take τ to be another key generator. The τ describes the Ornstein-Uhlenbeck Brownian motion. Its key role can be seen again in the discussion on half whiskers.

We are now in search of *new* whiskers that show some significant probabilistic properties as above three whiskers under somewhat general setup. There a whisker may be changed to a half-whisker.

3. Half whiskers

First we recall the notes,[10] p. 60, section O_∞ Ref. 1, where a new whisker with generator

$$\alpha^p = u^p \frac{d}{du} + \frac{p}{2} u^{p-1} \tag{5}$$

is suggested to be investigated, where p is not necessarily an integer. (The power p was written as α in Ref. 10, but to avoid notational confusion, we write p instead of α.)

Since fractional power p is involved, we tacitly assume that u is positive, We shall, therefore, take a white noise with time-parameter $[0, \infty)$. The basic nuclear space E is chosen to be D_{00}, involving functions on $[0, \infty)$, being isomorphic to D_0, eventually isomorphic to $C^\infty(S^1)$.

We are now ready to state an answer.

As was remarked in the last section, the power $p = 1$ is the key number and, in fact, it is exceptional. In this case the variable u can run through the whole R^1, that is, corresponds to a whisker with generator τ. In what follows we escape from the case $p = 1$.

We remind the relationship between f and $a(u)$ that appear in the expressions of $\psi_t(u)$ and α, respectively. The related formulas are the same as the case where u runs through R^1.

Assuming differentiability of f we have the formula (4). For $a(u) = u^p$, the corresponding $f(u)$ is determined. Namely,

$$u^p = f'(f^{-1}(u)).$$

An additional requirement for f is that the domain of f should be the entire $[0, \infty)$. Hence, we have

$$f(u) = c_p u^{\frac{1}{1-p}}, \tag{6}$$

where $c_p = (1 - p)^{1/(1-p)}$. Summing up, we define $\psi_t(u)$ for $u > 0$,

$$\psi_t(u) = f^{-1}(f(u) + t)$$

with the special choice of f.

We, therefore, have

$$f^{-1}(u) = (1 - p)^{-1} u^{1-p}. \tag{7}$$

We are ready to define a transformation g_t^p acting on D_{00} by

$$(g_t\xi)(u) = \xi(c_p(\frac{u^{1-p}}{1-p} + t)^{1/(1-p)}\sqrt{\frac{c_p}{1-p}(\frac{u^{1-p}}{1-p} + t)^{p/(1-p)}u^{-p}}. \tag{8}$$

Note that f is always positive and maps $(0, \infty)$ onto itself, in the ordinary order in the case $p < 1$, and in the reciprocal order in the case $p > 1$.

The exceptional case $p = 1$ is refered to the literature.[5] It has been well defined.

Then, we claim, still assuming $p \neq 1$, the following

Theorem 3.1.

i) g_t^p is a member of $O(D_{00})$ for every $t > 0$.

ii) The collection $\{g_t^p, t \geq 0\}$ forms a continuous semi-group with the product $g_t^p \cdot g_s^p = g_{t+s}^p$ for $t, s \geq 0$.

iii) The generator α^p of g_t^p is α^p given by (5).

Proof. Assertion i) comes from the structure of D_{00}.

Assertions ii) and iii) can be proved by the actual computations.

Definition 3.1. A continuous semi-group $g_t, t \geq 0$, each member of which comes from $\psi_t(u)$ is called a *half whisker.*

Theorem 3.2. *The collection of half whiskers $g_t^p, t \geq 0, p \in R$, generates a local Lie semi-group G_L:*

$$G_L = generated \ by \ \{g_{t_1}^{p_1} \cdots g_{t_n}^{p_n}\}.$$

4. Lie algebra

The collection $\{\alpha^p; p \in R^1\}$ generates a vector space $\mathbf{g}_L$. There is introduced the Lie product $[\cdot, \cdot]$. Note that the exceptional power $p = 1$ is included. The whiskers introduced before are also considered as half whiskers by letting the parameter t restrict to $[0, \infty)$. In this sense, we identify some of them.

$$\alpha^0 = s,$$

$$\alpha^1 = \tau,$$

$$\alpha^2 = \kappa.$$

With these understanding, we have

Theorem 4.1. *The space $\mathbf{g}_L$ is a Lie algebra parameterized by $p \in R^1$. It is associated with the local Lie semi-group G_L.*

Proof. We have

$$[\alpha^p, \alpha^q] = (q - p)u^{p+q-1}\frac{d}{du} + \frac{1}{2}(q - p)u^{(p+q-2)}. \tag{9}$$

The result is $(q - p)\alpha^{p+q-1}$. This proves the theorem.

In fact, we have an infinite dimensional Lie algebra, the base of which consists of one-parameter generators of half whiskers.

As for the algebra $\mathbf{g}_L$ we may consider a non-trivial cocycle, but we do not go further like the central extension. For more detail see Ref. 13.

Remark. The authors are grateful to Professor I. Volovich who suggested to remind Virasoro type Lie algebras in connection with the group $O(E)$. Another suggestions towards this direction can be seen in Refs. 7,8 and 13.

5. Renormalization

We recall two notions in white noise analysis:

i) Idea of the **Reduction** in white noise analysis is that we take a white noise $\{\dot{B}(t), t \in R^1\}$ to be the variable system of random functions. Indeed, $\{\dot{B}(t)\}$ is a system of idealized elemental random variables.

ii) The space of *generalized* white noise functionals. Namely $(L^2)^-$ and $(S)^*$. Details are found in Ref. 6, so they are omitted. However, two facts are specially reminded. One is that exponential functions such that $\exp[c < \dot{B}, \xi >], \xi \in E$, are test functionals, and the other is that each $\dot{B}(t)$ has its identity in the space $\mathcal{H}_1^{(-1)}$ of generalized linear functionals of white noise.

The meaning of the renormalization of white noise functionals is known, but idea of renormalizing white noise functionals has not been so clear. We shall propose a general method.

There are two directions to establish a class of white noise functionals. One is the introducing a space of generalized functionals. Another way is, starting with polynomials in $\dot{B}(t)$'s; they are the most basic functions when variables are given. Assume that we are in the second position.

A polynomial in $\dot{B}(t)$'s is as follows.

Let $p(t_1, t_2, \cdots, t_N)$ be a C- polynomial. Then, $p(\dot{B}(t_1), \cdots, \dot{B}(t_n))$ is a $\dot{B}$-polynomial. The degree of p is the degree of the $\dot{B}$-polynomial. The same for homogeneous case.

The collection of all $\dot{B}$-polynomials is denoted by **A**. It is a graded algebra, and it admits a decomposition:

$$\mathbf{A} = \bigoplus_n \mathbf{A}_n,$$

where $\mathbf{A}_n$ involves all homogeneous $\dot{B}$-polynomials of degree n. The sum is just an algebraic sum.

As is mentioned before The space $\mathcal{H}_1$ involves $\mathbf{A}_1$ which is total. However, in $\mathbf{A}_n, n \geq 2$, There are many important members that are not generalized white noise functionals, and they are important functionals, like

$\dot{B}(t)^n$. Thus, we are suggested to make them be modified so as to be generalized white noise functionals. This means the they should be modified so as to define continuous linear functionals on the space of test bfunctionals. In particular, continuous linear functionals of exponential functionals. In other words, the T- or S-functionals are well defined.

The actual method is as follows, say, the S-transform is used.

Approximate $\dot{B}(t)^p$ by $(\frac{\Delta B}{\Delta})^p$, the Δ containing t. Obviously, the ratio does not tends to $\dot{B}(t)^p$ in the topology of test function space. Note that

$$S[(\frac{\Delta B}{\Delta})^p](\xi) = (\xi, \chi_\Delta/\Delta)^p + O(\frac{1}{\Delta}),$$

where $O(\cdot)$ indicates the the term of order one or more. If this term is deleted, then the limit exists and is equal to $\xi(t)^p$, which can be a S-transform of a generalized white noise functional, more precisely in $\mathcal{H}_p$.

Thus, to get a generalized white noise functional, denoted by : $\dot{B}(t)^p$:, the amount of renormalization is of $O(\frac{1}{dt})$. Exact amount can be determined by the formula of the S- transform that comes from the Laplace transform of the Hermite polynomial with parameter. See Ref. 6.

The renormalization, in this line, of the product of $\dot{B}(t_j)$'s with different t_j's can be done multiplicatively. As for the sum of them their renormalization can be done in an additive manner. Summing up

Theorem 5.1. *The S-transform provides the method of renormalization of the members in the algebra* **A**. *The same for the T- transform.*

Finally we note that the S- or T- transform gives us a method of renormalization of exponentials of *quadratic* white noise functionals. It is the multiplicative renormalization, where the modified Fredholm determinant tells us the structure of renormalization.

Theorem 5.2. *The collection of renormalized polynomials in* $\dot{B}(t)$ *'s is dense in* $(L^2)^-$.

We have thus established a third method of getting a space of generalized white noise functionals.

Remark We have some hope to see some connection with the theory of spin (see Ref. 11).

Appendix

Hermite polynomials with parameter:

$$H_n(x; \sigma^2) = \frac{(-\sigma^2)^n}{n!} e^{(x^2/2\sigma^2)} \frac{d^n}{dx^n} e^{(-x^2/2\sigma^2)}$$

It is expressed in the form

$$\frac{\sigma^n}{2^{(n/2)}} \sum_0^{[n/2]} \frac{(-1)^k}{k!} \frac{(\sqrt{2}x/\sigma)^{(n-2k)}}{(n-2k)!}$$

The leading term is $x^n/n!$, then follows terms of "positive" order of σ.

References

1. J. Aczél, Vorlesungen über Funktionalgleichungen und ihre SAnwendungen, Birkhäuser, 1960.
2. I.M. Gel'fand. Generalized random processes. (in Russian), Doklady Acad. Nauk SSRR, 100 (1955), 853-856.
3. I.M. Gel'fand, M.I. Graev and N.Ya. Vilenkin, Generalized functions. vol. 5, 1962 (Russian original), Academic Press, 1966.
4. T. Hida, Stationary stochastic processes, Princeton Univ. Press, 1970.
5. T. Hida, Brownian motion. Iwanami Pub. Co. 1975, in Japanese; english transl. Springer-Verlag, 1980.
6. T. Hida and Si Si, Lectures on white noise functionals. World Sci. Pub. Co. 2008.
7. A.A. Kirillov, Kähler structures on K-orbits of the group of diffepmorphisms of a circle. Functional Analysis and its Applications. vol.21, no.2. 42-45 (1986; english. 1987).
8. A.A. Kirillov and D.V. Yur'ev, Kähler geometry of the infinite dimensional homogeneous space $M = Diff_+(S^1)/Rot(S^1)$. Functional analysis and its Applications vol.21, no.4, (1987; english ed. 284-294).
9. Si Si, Introduction to Hida distributions . World Sci. Pub Co. 2012.
10. T. Shimizu and Si Si, Professor Takeyuki Hida's mathematical notes. informal publication 2004.
11. S. Tomonaga, Theory of Spin. In Japanese Misuzu Pub. Co. 1974, new ed.2008. English translation Univ. of Chicago Press, 1997.
12. F. Smithies, Integral equations. Cambridge Univ. Press. 1958.
13. B. Khesin and R, Wendt, The geometry of Infinite-Dimensional groups. Springer, 2009.

QUANTUM LÉVY AREA AS QUANTUM MARTINGALE LIMIT

ROBIN L. HUDSON

Mathematics Department, Loughborough University,
Loughborough, Leicestershire LE11 3TU, Great Britain
E-mail: R.Hudson@lboro.ac.uk

We replace the independent one-dimensional Brownian motions in the definition of classical Lévy area by the mutually non-commuting "momentum" and "position" Brownian motions of non-Fock quantum stochastic calculus, which are independent in the sense of factorisation of joint characteristic functions. The corresponding quantum Lévy area can then be constructed in a way similar to Lévy's original construction as a quantum martingale limit.

1. Introduction

Lévy's stochastic area,[10,11] as well as being of great intrinsic interest and possessing numerous connections with classical mathematical analysis and mathematical physics,[7] has recently come to prominence in the theory of rough paths[5,12,13].

Here we present a quantum or noncommutative version of Lévy area in which the two independent one-dimensional Brownian motions used to define it are replaced by the pair of quantum Brownian motions (P, Q), forming the quantum Wiener process of[2]. Each of P and Q consists of a process of mutually commuting self-adjoint operators which can be simultaneously diagonalised as multiplication by a classical Brownian motion of appropriate variance in the Hilbert space of complex-valued functions square-integrable with respect to Wiener measure, But P and Q together satisfy the commutation relation

$$P(s)Q(t) - Q(t)P(s) = -i \min \{s, t\} I, \qquad (1)$$

where I is the identity operator. Thus the processes P and Q cannot be simultaneously diagonalised by one and the same diagonalising unitary transformation. From the point of view of quantum physics it is operationally meaningless to speak of, for example, the joint probability distribution of

$P(t)$ and $Q(t)$, and thus to ask if they are stochastically independent, since non-commuting observables cannot be simultaneously measured and there is no operational definition of their joint distribution. Despite this, $P(t)$ and $Q(t)$ can be regarded as independent in a strictly mathematical sense, in so far as for arbitrary real numbers x and y, $xP(t) + yQ(t)$ is a self-adjoint operator having the property that

$$\mathbb{E}\left[e^{i(xP(t)+yQ(t))}\right] = \mathbb{E}\left[e^{ixP(t)}\right] \mathbb{E}\left[e^{iyQ(t)}\right]$$

with expectations determined by the state for which P and Q are individually Brownian motions. Thus $P(t)$ and $Q(t)$ satisfy the condition, equivalent to independence in classical probability, that their joint quasi-characteristic function[3] $f(x,y) = \mathbb{E}\left[e^{i(xP(t)+yQ(t))}\right]$ factorises into the product of their individual characteristic functions. More generally, for arbitrary $s_1, s_2, ..., s_m$ and $t_1, t_2, ..., t_n$, the m-tuple $(P(s_1), P(s_2), ..., P(s_m))$ and the n-tuple $(Q(t_1), Q(t_2), ..., Q(t_n))$ are independent in the sense that for arbitrary real $x_1, x_2, ..., x_m$ and $y_1, y_2, ..., y_n$,

$$\mathbb{E}\left[e^{i\left(\Sigma_{j=1}^{m} x_j P(s_j) + \Sigma_{j=1}^{m} y_k Q(t_k)\right)}\right] = \mathbb{E}\left[e^{i\Sigma_{j=1}^{m} x_j P(s_j)}\right] \mathbb{E}\left[e^{i\Sigma_{j=1}^{m} y_k Q(t_k)}\right].$$

In this sense we may regard P and Q as independent Brownian motions and use them to replace the two independent classical Brownian motions which define Lévy area.

In free probability, a noncommutative analog of Lévy area for so-called free Brownian motion has been studied in[1] and[14]. Here it is expected that the less extreme form of noncommutativity and the more direct physical motivation of (1) will result in properties that are also analytically interesting.

We assume that a variance parameter $\sigma^2 > 0$ has been fixed once and for all, so that a one-dimensional Brownian motion X has the property that each $X(t)$ is $N(0, \sigma^2 t)$, that is, it is normally distributed with mean zero and variance $\sigma^2 t$.

2. Lévy area as a martingale limit

Lévy's original 1939 construction[11] of so called Lévy area, the signed area L_a^b lying between the chord joining two successive points $(X(a), Y(a))$ and $(X(b), Y(b))$ on the orbit of a planar Brownian motion (X, Y) and the orbit itself, preceded the invention of Itô calculus[9] and was obtained in effect as a martingale limit. Consider the filtration $\mathcal{F}_1([a, b[) \subset \mathcal{F}_2([a, b[) \subset \mathcal{F}_3([a, b[) \subset \cdots$ generated by the increments of X and Y over the subintervals comprising the successive dyadic dissections of the time interval

$[a, b[$, so that, denoting by $\mathcal{F}(Z_1, Z_2, ..., Z_m)$ the sigma-field generated by the random variables $Z_1, Z_2, ..., Z_m$,

$$\mathcal{F}_1\left([a, b[\right) = \mathcal{F}(X(b) - X(a), Y(b) - Y(a)),$$

$$\mathcal{F}_2\left([a, b[\right) = \mathcal{F}\left(X(\frac{a+b}{2}) - X(a), X(b) - X(\frac{a+b}{2}),\right.$$

$$\left. Y(\frac{a+b}{2}) - Y(a), Y(b) - Y(\frac{a+b}{2})\right),$$

$$...,$$

$$\mathcal{F}_{n+1}\left([a, b[\right) = \mathcal{F}\big(X(a + 2^{-n}j)(b - a)) - X(a + 2^{-n}(j-1)(b-a)),$$

$$Y(a + 2^{-n}j)(b - a)) - Y(a + 2^{-n}(j-1)(b-a)),$$

$$j = 1, 2, ..., 2^n\big),$$

$$....$$

The martingale $(M_n(a, b))_{n=1,2,...}$ consists of the successive polygonal approximations to the area formed by the endpoints of these dissections, so that, by definition,

$$M_1(a, b) = 0$$

and, using the elementary Cartesian formula

$$\frac{1}{2} \det \begin{bmatrix} 1 & x_1 & y_1 \\ 1 & x_2 & y_2 \\ 1 & x_3 & y_3 \end{bmatrix}$$

for the signed area of the triangle with vertices (x_1, y_1), (x_2, y_2) and (x_3, y_3),

$$M_2(a, b) = \frac{1}{2} \det \begin{bmatrix} 1 & X(a) & Y(a) \\ 1 & X(\frac{a+b}{2}) & Y(\frac{a+b}{2}) \\ 1 & X(b) & Y(b) \end{bmatrix}, \tag{2}$$

while for $n > 2$

$$M_{n+1}(a, b) = M_n(a, b) + \sum_{j=1}^{2^n} M_2(a + 2^{-n}(j-1)(b-a), a + 2^{-n}j(b-a)). \tag{3}$$

That $(M_n(a, b))_{n \in \mathbb{N}}$ is indeed a martingale adapted to the filtration $(\mathcal{F}_n\left([a, b[\right))_{n \in \mathbb{N}}$ is proved in Theorem 2 below. First we need the following preparatory Theorem.

Theorem 2.1. *Let X_1, Y_1, X_2 and Y_2 be mutually independent normally distributed random variables of mean zero such that X_1 and X_2 have the* *same variance* σ_1 *and* Y_1 *and* Y_2 *have the same variance* σ_2. *Then the conditional expectation*

$$\mathbb{E}\left[\frac{1}{2}(X_1 Y_2 - Y_1 X_2)\,|\mathcal{F}(X_1 + X_2, Y_1 + Y_2)\right] = 0$$

Proof. Expressing products as differences of squares and vice versa, we have

$$\frac{1}{2}(X_1 Y_2 - X_2 Y_1)$$

$$= \frac{1}{8}\left\{(X_1 + Y_2)^2 - (X_1 - Y_2)^2 - (X_2 + Y_1)^2 + (X_2 - Y_1)^2\right\}$$

$$= \frac{1}{8}\left(\left\{(X_1 + Y_2)^2 - (X_2 + Y_1)^2\right\} - \left\{(X_1 - Y_2)^2 - (X_2 - Y_1)^2\right\}\right)$$

$$= \frac{1}{8}\left\{(X_1 + Y_2 + X_2 + Y_1)(X_1 + Y_2 - X_2 - Y_1)\right.$$

$$\left. - (X_1 - Y_2 + X_2 - Y_1)(X_1 - Y_2 - X_2 + Y_1)\right\}$$

$$= \frac{1}{8}\left(\{(X_1 + X_2) + (Y_1 + Y_2)\}\{(X_1 - X_2) - (Y_1 - Y_2)\}\right.$$

$$\left. - \{(X_1 + X_2) - (Y_1 + Y_2)\}\{(X_1 - X_2) - (Y_1 - Y_2)\}\right).$$

Here $X_1 - X_2$ and $Y_1 - Y_2$ are independent of both $X_1 + X_2$ and $Y_1 + Y_2$, and hence of the σ-field which they generate. It follows that the conditional expectation

$$\mathbb{E}\left[\frac{1}{2}(X_1 Y_2 - X_2 Y_1)\,|\mathcal{F}(X_1 + X_2, Y_1 + Y_2)\right]$$

$$= \frac{1}{8}\mathbb{E}\big[\{(X_1 + X_2) + (Y_1 + Y_2)\}$$

$$\times \{(X_1 - X_2) - (Y_1 - Y_2)\}\,|\mathcal{F}(X_1 + X_2, Y_1 + Y_2)\big]$$

$$- \frac{1}{8}\mathbb{E}\big[\{(X_1 + X_2) - (Y_1 + Y_2)\}$$

$$\times \{(X_1 - X_2) + (Y_1 - Y_2)\}\,|\mathcal{F}(X_1 + X_2, Y_1 + Y_2)\big]$$

$$= \frac{1}{8}\{(X_1 + X_2) + (Y_1 + Y_2)\}$$

$$\times \mathbb{E}\left[\{(X_1 - X_2) - (Y_1 - Y_2)\}\,|\mathcal{F}(X_1 + X_2, Y_1 + Y_2)\right]$$

$$- \frac{1}{8}\{(X_1 + X_2) - (Y_1 + Y_2)\}$$

$$\times \mathbb{E}\left[\{(X_1 - X_2) + (Y_1 - Y_2)\}\,|\mathcal{F}(X_1 + X_2, Y_1 + Y_2)\right]$$

$$= \frac{1}{8} \{(X_1 + X_2) + (Y_1 + Y_2)\} \, \mathbb{E} \left[(X_1 - X_2) - (Y_1 - Y_2) \right]$$

$$- \frac{1}{8} \mathbb{E} \left[(X_1 - X_2) + (Y_1 - Y_2) \right] \{(X_1 + X_2) - (Y_1 + Y_2)\}$$

$$= 0$$

since X_1, Y_1, X_2 and Y_2 are all of zero mean. $\qquad\square$

We use this to prove

Theorem 2.2. $(M_n(a, b))_{n \in \mathbb{N}}$ *is a martingale adapted to the filtration* $(\mathcal{F}_n ([a, b[))_{n \in \mathbb{N}}$.

Proof. By subtracting the second row from the other two in the determinant (2) we find

$$M_2(a, b) = \frac{1}{2} \det \begin{bmatrix} 0 & -\left(X(\frac{a+b}{2}) - X(a)\right) & -\left(Y(\frac{a+b}{2}) - Y(a)\right) \\ 1 & X(\frac{a+b}{2}) & Y(\frac{a+b}{2}) \\ 0 & X(b) - X(\frac{a+b}{2}) & Y(b) - Y(\frac{a+b}{2}) \end{bmatrix}$$

$$= \frac{1}{2} \left(\left(X(\frac{a+b}{2}) - X(a) \right) \left(Y(b) - Y(\frac{a+b}{2}) \right) \right.$$

$$\left. - \left(Y(\frac{a+b}{2}) - Y(a) \right) \left(X(b) - X(\frac{a+b}{2}) \right) \right)$$

which shows that $M_2(a, b)$ is $\mathcal{F}_2 ([a, b[)$-measurable. Moreover

$$M_2(a, b) = \frac{1}{2} (X_1 Y_2 - X_2 Y_1)$$

where $X_1 = X(\frac{a+b}{2}) - X(a)$, $Y_1 = Y(\frac{a+b}{2}) - Y(a)$, $X_2 = X(b) - X(\frac{a+b}{2})$ and $Y_2 = Y(b) - Y(\frac{a+b}{2})$ satisfy the hypothesis of Theorem 1 (with $\sigma_1 = \sigma_2 = \frac{b-a}{2} \sigma^2$). Hence the conclusion of Theorem 1 is also satisfied:

$$\mathbb{E} \left[M_2(a, b) | \mathcal{F}_1 ([a, b[) \right] = \mathbb{E} \left[\frac{1}{2} (X_1 Y_2 - Y_1 X_2) | \mathcal{F} (X_1 + X_2, Y_1 + Y_2) \right]$$

$$= 0.$$

Similarly, replacing the interval $[a, b[$ by $[a + 2^{-n}(j-1)(b-a), a + 2^{-n} j(b-a)[$ in the preceding argument it follows that $M_2(a + 2^{-n}(j-1)(b-a), a + 2^{-n} j(b-a))$ is $\mathcal{F}_2([a + 2^{-n}(j-1)(b-a), a + 2^{-n} j(b-a)[)$ and that

$$\mathbb{E} \left[M_2(a + 2^{-n}(j-1)(b-a), a + 2^{-n} j(b-a)) \right.$$

$$\left. | \mathcal{F}_1([a + 2^{-n}(j-1)(b-a), a + 2^{-n} j(b-a)[) \right] = 0. \quad (4)$$

Since $\mathcal{F}_2([a + 2^{-n}(j-1)(b-a), a + 2^{-n}j(b-a)[) \subset \mathcal{F}_{n+1}([a, b[)$ it follows inductively from (3) that $M_{n+1}(a, b)$ is $\mathcal{F}_{n+1}([a, b[)$-measurable. Also, since

$$\mathcal{F}_n([a, b[) = \bigvee_{j=1}^{2^n} \mathcal{F}_1([a + 2^{-n}(j-1)(b-a), a + 2^{-n}j(b-a)[),$$

that is, the sigma-field generated by the $\mathcal{F}_1([a+2^{-n}(j-1)(b-a), a+2^{-n}j(b-a)[)$, $j = 1, 2, ..., 2^n$, and since, by the independent increments property of Brownian motion, each $M_2(a + 2^{-n}(j-1)(b-a), a + 2^{-n}j(b-a))$ is independent of all the sigma-fields $\mathcal{F}_1([a+2^{-n}(k-1)(b-a), a+2^{-n}k(b-a)[)$ for $k \neq j$, it follows from (3) and using (4) that

$$\mathbb{E}\left[M_{n+1}(a, b)\mid \mathcal{F}_n([a, b[)\right] - M_n(a, b)$$

$$= \sum_{j=1}^{2^n} \mathbb{E}\left[M_2(a + 2^{-n}(j-1)(b-a), a + 2^{-n}j(b-a))\mid \mathcal{F}_n([a, b[)\right]$$

$$= \sum_{j=1}^{2^n} \mathbb{E}\left[\mathbb{E}\left[M_2(a + 2^{-n}(j-1)(b-a), a + 2^{-n}j(b-a))\mid \right.\right.$$

$$\left.\left. \mathcal{F}_1([a + 2^{-n}(j-1)(b-a), a + 2^{-n}j(b-a)[)\right]\mid \mathcal{F}_n([a, b[)\right]$$

$$= 0. \qquad \qquad \Box$$

The martingale $(M_n)_{n \in \mathbb{N}}$ is uniformly bounded in L^2 norm and hence convergent in that norm.[16] We define the limit to be the Lévy area A_a^b. Lévy showed later[10] that it could be expressed alternatively as the Ito stochastic integral

$$A_a^b = \int_{a \leq x < b} (X(x) - X(a))dY(x) - (Y(x) - Y(a))dX(x)) \qquad (5)$$

$$= \int_{a \leq x < y < b} (dX(x)dY(y) - dY(x)dX(y)). \qquad (6)$$

To see this intuitively, we express the random polygonal area M_n using a triangulation by triangles with a common vertex $(X(a), Y(a))$ as

$$M_n(a, b) = \sum_{k=2}^{2^n} \left(\left\{X(a + 2^{-n}(k-1)(b-a)) - X(a)\right\}\right.$$

$$\left\{Y\left(a + 2^{-n}k(b-a)\right) - Y\left(a + 2^{-n}(k-1)(b-a)\right)\right\}$$

$$- \left\{Y(a + 2^{-n}(k-1)(b-a)) - Y(a)\right\}$$

$$\left.\left\{X\left(a + 2^{-n}k(b-a)\right) - X\left(a + 2^{-n}(k-1)(b-a)\right)\right\}\right).$$

By making the substitutions

$$X(a + 2^{-n}(k-1)(b-a)) - X(a)$$

$$= \sum_{j=1}^{k-1} \left(X(a + 2^{-n}j(b-a)) - X(a + 2^{-n}(j-1)(b-a)) \right)$$

$$Y(a + 2^{-n}(k-1)(b-a)) - Y(a)$$

$$= \sum_{j=1}^{k-1} \left(Y(a + 2^{-n}j(b-a)) - Y(a + 2^{-n}(j-1)(b-a)) \right)$$

we see that $M_n(a,b)$ can also be expressed in the form of the discrete approximation to the iterated Itô integral (6)

$$M_n(a,b) = 2^{-n}(b-a) \sum_{1 \leq j < k \leq 2^n} (\xi_j \eta_k - \eta_j \xi_k) \tag{7}$$

where $\xi_1, \xi_2, ...\xi_{2^n}, \eta_1, \eta_2, ..., \eta_{2^n}$ are the independent identically distributed Gaussian random variables of mean zero and variance σ^2 given by

$$\xi_j = \sqrt{\frac{2^n}{b-a}} \left(X(a + 2^{-n}j(b-a)) - X(a + 2^{-n}(j-1)) \right),$$

$$\eta_k = \sqrt{\frac{2^n}{b-a}} \left(Y(a + 2^{-n}k(b-a)) - Y(a + 2^{-n}(k-1)) \right).$$

The L^2-boundedness of the martingale $(M_n(a,b))_{n \in \mathbb{N}}$ follows from this.

3. An analog of Theorem 1 for canonical pairs

To see that an analog of Theorem 1 can be formulated, we first recall from[3] some facts about *canonical pairs*, that is, pairs of self-adjoint operators (p, q) satisfying the Heisenberg commutation relation

$$pq - qp = -iI$$

in the mathematically rigorous Weyl form

$$e^{ixp} e^{iyq} = e^{ixy} e^{iyq} e^{ixp}, \quad x, y \in \mathbb{R}.$$

For fixed $x, y \in \mathbb{R}$, if

$$U_z = e^{-\frac{1}{2}iz^2 xy} e^{izxp} e^{izyq} = e^{\frac{1}{2}iz^2 xy} e^{izyq} e^{izxp}, \quad z \in \mathbb{R},$$

then $(U_z)_{z \in \mathbb{R}}$ is a continuous one parameter unitary group whose self adjoint infinitesimal generator coincides with $xp + yq$ on the intersection of the domains of p and q. Thus we can *define* the self-adjoint operator $xp + yq$ for all real x and y by

$$e^{iz(xp+yq)} = e^{-\frac{1}{2}iz^2 xy} e^{izxp} e^{izyq}, \quad z \in \mathbb{R}.$$

By putting $z = 1$ we find that $e^{i(xp+yq)}$ is a well defined unitary operator and

$$e^{i(xp+yq)} = e^{-\frac{1}{2}ixy} e^{ixp} e^{iyq} = e^{\frac{1}{2}ixy} e^{iyq} e^{ixp} \qquad (8)$$

Given a quantum state to determine expectations, we can then define the *quasicharacteristic function*[3] $f_{p,q}$ of the pair (p, q) by

$$f_{(p,q)}(x, y) = \mathbb{E}\left[e^{i(xp+yq)}\right].$$

Note that in general $f_{(p,q)}$ is not nonnegative-definite and so is not the characteristic function of a probability distributionj on $\mathbb{R}^2$. The canonical pair (p, q) will be said to be *normally distributed with zero mean and variance* σ^2, abbreviated as $N(0, \sigma^2)$, if $f_{p,q}$ is given by

$$f_{(p,q)}(x, y) = e^{-\frac{\sigma^2}{4}(x^2+y^2)}. \qquad (9)$$

Thus p and q are individually normally distributed with zero mean and variance $\frac{1}{2}\sigma^2$; they share the variance equally. By the Heisenberg uncertainty principle, such a pair exists only if $\sigma^2 \geq 1$. Note that they are then stochastically independent in the sense that the quasicharacteristic function factorises as

$$f_{(p,q)}(x, y) = f_{(p,q)}(x, 0) f_{(p,q)}(0, y) = f_p(x) f_q(y)$$

where f_p and f_q are the individual characteristic functions of p and q.

Two canonical pairs (p_1, q_1) and (p_2, q_2) commute with eachother if, by definition, the corresponding pairs of one-parameter unitary groups commute. Then, for $(x_1, y_1, x_2, y_2) \in \mathbb{R}$

$$e^{i(x_1 p_1 + y_1 q_1)} e^{i(x_2 p_2 + y_2 q_2)} = e^{i(x_2 p_2 + y_2 q_2)} e^{i(x_1 p_1 + y_1 q_1)}$$

and we may define the unitary operator $e^{i(x_1 p_1 + y_1 q_1 + x_2 p_2 + y_2 q_2)}$ to be either of these, and with it the *joint quasicharacteristic function*

$$f_{(p_1,q_1),(p_2,q_2)}(x_1, y_1, x_2, y_2) = \mathbb{E}\left[e^{i(x_1 p_1 + y_1 q_1 + x_2 p_2 + y_2 q_2)}\right].$$

The commuting pairs (p_1, q_1) and (p_2, q_2) are *stochastically independent* if this factorizes as

$$f_{(p_1,q_1),(p_2,q_2)}(x_1, y_1, x_2, y_2) = f_{(p_1,q_1)}(x_1, y_1) f_{(p_2,q_2)}(x_2, y_2).$$

Thus the joint quasi-characteristic function of two stochastically independent $N(0, \sigma^2)$ canonical pairs is

$$f_{(p_1,q_1),(p_2,q_2)}(x_1, y_1, x_2, y_2) = e^{-\frac{\sigma^2}{4}(x_1^2 + y_1^2 + x_2^2 + y_2^2)}. \tag{10}$$

For example if $\sigma^2 > 1$ we can realise an $N(0, \sigma^2)$ canonical pair (p, q) in the thermal equilibrium state of the harmonic oscillator Hamiltonian $\frac{1}{2}(p^2 + q^2)$ with density operator

$$\rho_{(p,q)} = \left(\text{tr}\left(e^{-\frac{\beta}{2}(p^2+q^2)} \right) \right)^{-1} e^{-\frac{\beta}{2}(p^2+q^2)}$$

at inverse temperature β where

$$\coth \beta = \frac{\sigma^2}{4}.$$

Similarly, we can realise stochasticly independent $N(0, \sigma^2)$ pairs (p_1, q_1) and (p_2, q_2) in the thermal equilibrium state

$$\rho_{(p_1,q_1),(p_2,q_2)} = \left(\text{tr}\left(e^{-\beta\frac{1}{2}(p_1^2+q_1^2+p_2^2+q_2^2)} \right) \right)^{-1} e^{-\beta\frac{1}{2}(p_1^2+q_1^2+p_2^2+q_2^2)}$$

of the two-dimensional oscillator provided that $\sigma^2 > 1$. A canonical pair (p, q) of minimal variance $\sigma^2 = 1$ is realised as the ground state of the Hamiltonian $\frac{1}{2}(p^2 + q^2)$. For two independent such pairs (p_1, q_1) and (p_2, q_2) we can take the ground state of the two-dimensional oscillator Hamiltonian $\frac{1}{2}(p_1^2 + q_1^2 + p_2^2 + q_2^2)$.

Alternatively we may realise arbitrarily many independent $N(0, \sigma^2)$ canonical pairs $(p_n, q_n), n = 1, 2, \dots$ as follows. First consider the case $\sigma^2 = 1$ of minimum variance. Following[15] or[4] we define the Fock space $\Gamma(\mathfrak{h})$ over a Hilbert space $\mathfrak{h}$ as the closed linear span of the exponential vectors $e(f), f \in \mathfrak{h}$, which satisfy

$$\langle e(f), e(g) \rangle = \exp \langle f, g \rangle$$

and notice that, for each direct sum decomposition $\mathfrak{h} = \mathfrak{h}_1 \oplus \mathfrak{h}_2 \oplus \cdots \oplus \mathfrak{h}_m$, there is a unique tensor product decomposition

$$\Gamma(\mathfrak{h}_1 \oplus \mathfrak{h}_2 \oplus \cdots \oplus \mathfrak{h}_m) = \Gamma(\mathfrak{h}_1) \otimes \Gamma(\mathfrak{h}_2) \otimes \cdots \otimes \Gamma(\mathfrak{h}_m) \tag{11}$$

in which each exponential vector splits:

$$e(f_1 \oplus f_2 \oplus \cdots \oplus f_m) = e(f_1) \otimes e(f_2) \otimes \cdots \otimes e(f_m).$$

In particular the vacuum vector $e(0_\mathfrak{h})$ splits as the product vector of the component vacuum vectors

$$e(0_\mathfrak{h}) = e(0_{\mathfrak{h}_1}) \otimes e(0_{\mathfrak{h}_2}) \otimes \cdots \otimes e(0_{\mathfrak{h}_m}). \tag{12}$$

The Weyl operators $W_1(f), f \in \mathfrak{h}$, are the unitary operators on $\Gamma(\mathfrak{h})$ which act on exponential vectors as

$$W_1(f)e(g) = e^{-\frac{1}{4}\|f\|^2 - \frac{1}{\sqrt{2}}\langle f,g \rangle} e(\frac{1}{\sqrt{2}}f + g) \tag{13}$$

so that they have vacuum expectation value

$$\mathbb{E}\left[W_1(f)\right] = \langle e(0), W_1(f)e(0) \rangle = e^{-\frac{1}{4}\|f\|^2}. \tag{14}$$

They satisfy the Weyl relation

$$W_1(f)W_1(g) = e^{-\frac{1}{2}i\mathrm{Im}\langle f,g \rangle} W_1(f + g). \tag{15}$$

In particular $W_1(f)$ commutes with $W_1(g)$ whenever f and g are mutually orthogonal or more generally whenever $\langle f, g \rangle$ is real. Corresponding to the splitting (11) we have

$$W_1(f_1 \oplus f_2 \oplus \cdots \oplus f_m) = W_1(f_1) \otimes W_1(f_2) \otimes \cdots \otimes W_1(f_m) \tag{16}$$

Thus it follows from (15) that, for each fixed unit vector $f \in \mathfrak{h}$, $(W(ixf))_{x\in\mathbb{R}}$ and $(W(xf))_{x\in\mathbb{R}}$ are continuous one parameter unitary groups which satisfy

$$W(ixf)W(yf) = e^{-\frac{1}{2}ixy}W((ix + y)f),$$
$$W(yf)W(ixf) = e^{\frac{1}{2}ixy}W((ix + y)f) = e^{ixy}W(ixf)W(yf)$$

and we can define a canonical pair (p,q) by

$$e^{ixp} = W_1(ixf), \ e^{ixq} = W_1(xf). \tag{17}$$

By (14) we have

$$\mathbb{E}\left[e^{i(xp+yq)}\right] = \mathbb{E}\left[W((ix + y)f)\right] = e^{-\frac{1}{4}\|(ix+y)f\|^2} = e^{-\frac{1}{4}|ix+y|^2}$$
$$= e^{-\frac{1}{4}(x^2+y^2)}$$

so (p,q) is normally distributed with zero mean and variance 1. More generally for orthonormal vectors $f_1, f_2, ..., f_m$ the corresponding canonical pairs $(p_1,q_1), (p_2,q_2), ..., (p_m,q_m)$ constructed in this way commute with eachother and satisfy

$$\mathbb{E}\left[e^{i\Sigma_{j=1}^m(x_j p_j + y_j q_j)}\right] = \mathbb{E}\left[W\left(\Sigma_{j=1}^m (ix_j + y_j) f_j\right)\right]$$
$$= e^{-\frac{1}{4}\left\|\Sigma_{j=1}^m(ix_j+y_j)f_j\right\|^2} = e^{-\frac{1}{4}\Sigma_{j=1}^m\|(ix_j+y_j)f_j\|^2}$$
$$= e^{-\frac{1}{4}\Sigma_{j=1}^m(x_j+y_j)^2} = \prod_{j=1}^m e^{-\frac{1}{4}(x_j+y_j)^2}$$
$$= \prod_{j=1}^m \mathbb{E}\left[e^{i(x_j p_j + y_j q_j)}\right]$$

so they are stochastically independent.

In the case when $\sigma^2 > 1$ we set $\sigma^2 = \alpha^2 + \beta^2$ where α and β are positive real numbers satisfying $\alpha^2 - \beta^2 = 1$ and define Weyl operators $W(f), f \in \mathfrak{h}$ in the tensor product $\Gamma(\mathfrak{h}) \otimes \overline{\Gamma(\mathfrak{h})}$ of the Fock space $\Gamma(\mathfrak{h})$ with its dual by

$$W(f) = W_1(\alpha f) \otimes \overline{W_1(\beta f)}$$

where $W_1(f)$ is the Fock Weyl operator (13). These continue to satisfy the Weyl relation (15), whereas instead of (14) we have the double vacuum expectation

$$\mathbb{E}\left[W(f)\right] = \left\langle e(0) \otimes \overline{e(0)}, W(f)e(0) \otimes \overline{e(0)} \right\rangle = e^{-\frac{\sigma^2}{4}\|f\|^2}.$$

The splitting property (16) holds if we make the tensor product identification

$$\Gamma(\mathfrak{h}_1 \oplus \mathfrak{h}_2 \oplus \cdots \oplus \mathfrak{h}_m) \otimes \overline{\Gamma(\mathfrak{h}_1 \oplus \mathfrak{h}_2 \oplus \cdots \oplus \mathfrak{h}_m)}$$
$$= (\Gamma(\mathfrak{h}_1) \otimes \Gamma(\mathfrak{h}_2) \otimes \cdots \otimes \Gamma(\mathfrak{h}_m)) \otimes \overline{\Gamma(\mathfrak{h}_1) \otimes \Gamma(\mathfrak{h}_2) \otimes \cdots \otimes \Gamma(\mathfrak{h}_m)}$$
$$= (\Gamma(\mathfrak{h}_1) \otimes \Gamma(\mathfrak{h}_2) \otimes \cdots \otimes \Gamma(\mathfrak{h}_m)) \otimes \left(\overline{\Gamma(\mathfrak{h}_1)} \otimes \overline{\Gamma(\mathfrak{h}_2)} \otimes \cdots \otimes \overline{\Gamma(\mathfrak{h}_m)}\right)$$
$$= \left(\Gamma(\mathfrak{h}_1) \otimes \overline{\Gamma(\mathfrak{h}_1)}\right) \otimes \left(\Gamma(\mathfrak{h}_2) \otimes \overline{\Gamma(\mathfrak{h}_2)}\right) \otimes \cdots \otimes \left(\Gamma(\mathfrak{h}_m) \otimes \overline{\Gamma(\mathfrak{h}_m)}\right)$$
$$\tag{18}$$

where at the last stage the identification is made by appropriately permuting product vectors. Modulo this identification the double vacuum vector splits as

$$e(0_{\mathfrak{h}_1 \oplus \mathfrak{h}_2 \oplus \cdots \oplus \mathfrak{h}_m}) \otimes \overline{e(0_{\mathfrak{h}_1 \oplus \mathfrak{h}_2 \oplus \cdots \oplus \mathfrak{h}_m})}$$
$$= \left(e(0_{\mathfrak{h}_1}) \otimes \overline{e(0_{\mathfrak{h}_1})}\right) \otimes \left(e(0_{\mathfrak{h}_2}) \otimes \overline{e(0_{\mathfrak{h}_2})}\right) \otimes \cdots \otimes \left(e(0_{\mathfrak{h}_m}) \otimes \overline{e(0_{\mathfrak{h}_m})}\right).$$
$$\tag{19}$$

Thus we can define a canonical pair by (17) but with the Weyl operator $W_1(f)$ replaced by $W(f)$., Then (p, q) is $N(0, \sigma^2)$. Similarly, for orthonormal $f_1, f_2, ..., f_m$ the canonical pairs $(p_1, q_1), (p_2, q_2), ..., (p_m, q_m)$ constructed in this way satisfy

$$\mathbb{E}\left[e^{i\Sigma_{j=1}^m (x_j p_j + y_j q_j)}\right] = e^{-\frac{\sigma^2}{4}\Sigma_{j=1}^m (x_j^2 + y_j^2)} = \Pi_{j=1}^m e^{-\frac{\sigma^2}{4}(x_j^2 + y_j^2)}$$
$$= \Pi_{j=1}^m f_{(p_j, q_j)}(x_j, y_j)$$

and are thus independent and $N(0, \sigma^2)$.

Now let $\mathcal{N}$ denote the von Neumann algebra

$$\mathcal{N} = \{W(f), f \in \mathfrak{h}\}''$$

generated by the Weyl operators $W(f)$ with $f \in \mathfrak{h}$, and for each sub-Hilbert space $\mathfrak{h}_1$ of $\mathfrak{h}$, denote by $\mathcal{N}_{\mathfrak{h}_1}$ the von Neumann algebra

$$\mathcal{N}_{\mathfrak{h}_1} = \{W(f), f \in \mathfrak{h}_1\}''.$$

In general conditional expectations with reasonable properties do not exist in quantum probability. But, essentially because of the splitting property (19), in the (double) vacuum state there is a good notion of conditional expectation from $\mathcal{N}$ onto $\mathcal{N}_{\mathfrak{h}_1}$. Namely, there exists[4] a unique linear map $\mathbb{E}[\,.\,|\mathcal{N}_{\mathfrak{h}_1}]$ from $\mathcal{N}$ onto $\mathcal{N}_{\mathfrak{h}_1}$ having the two properties

-

$$\mathbb{E} = \mathbb{E} \circ \mathbb{E}[\,.\,|\mathcal{N}_{\mathfrak{h}_1}],$$

- For arbitrary $U \in \mathcal{N}$ and $V \in \mathcal{N}_{\mathfrak{h}_1}$

$$\mathbb{E}[UV|\mathcal{N}_{\mathfrak{h}_1}] = \mathbb{E}[U|\mathcal{N}_{\mathfrak{h}_1}]V, \ \mathbb{E}[VU|\mathcal{N}_{\mathfrak{h}_1}] = V\mathbb{E}[U|\mathcal{N}_{\mathfrak{h}_1}]. \quad (20)$$

In fact $\mathbb{E}[\,.\,|\mathcal{N}_{\mathfrak{h}_1}]$ acts on each Weyl operator $W(f)$ by

$$\mathbb{E}[W(f)|\mathcal{N}_{\mathfrak{h}_1}] = e^{-\frac{\sigma^2}{4}\|f-f_1\|^2}W(f_1)$$

where f_1 is the projection of the vector $f \in \mathfrak{h}$ on the subspace $\mathfrak{h}_1$.[4]

The following further properties are consequences of these.

- For $U \in \mathcal{N}_{\mathfrak{h}_1}$

$$\mathbb{E}[U|\mathcal{N}_{\mathfrak{h}_1}] = U$$

- For $U \in \mathcal{N}'_{\mathfrak{h}_1} \cap \mathcal{N} = \mathcal{N}_{\mathfrak{h}_1^{\perp}}$, the relative commutant of $\mathcal{N}'_{\mathfrak{h}_1}$ in $\mathcal{N}$,

$$\mathbb{E}[U|\mathcal{N}_{\mathfrak{h}_1}] = \mathbb{E}[U]I$$

- For $\mathfrak{h} = \mathfrak{h}_1 \oplus \mathfrak{h}_2 \oplus \mathfrak{h}_3$ if U is independent of $\mathfrak{h}_3$ in the sense that $U \in \mathcal{N} \cap \mathcal{N}'_{\mathfrak{h}_3} = \mathcal{N}_{\mathfrak{h}_1 \oplus \mathfrak{h}_2}$ then

$$\mathbb{E}[U|\mathcal{N}_{\mathfrak{h}_2 \oplus \mathfrak{h}_3}] = \mathbb{E}[U|\mathcal{N}_{\mathfrak{h}_2}] \quad (21)$$

- For $\mathfrak{h}_1 \subset \mathfrak{h}_2$,

$$\mathbb{E}[\,.\,|\mathcal{N}_{\mathfrak{h}_1}] = \mathbb{E}[\mathbb{E}[\,.\,|\mathcal{N}_{\mathfrak{h}_2}]|\mathcal{N}_{\mathfrak{h}_1}]$$

These properties extend in the natural way to unbounded self-adjoint elements whose domains include the (double) vacuum vector, which are affiliated rather than belonging to the relevant von Neumann algebra, by parametric differentiation at 0 of the corresponding one parameter unitary group. Products of two such unbounded self-adjoint elements such as $p_1 q_2 -$

$q_1 p_2$ below can be encompassed by the usual techniques for dealing with such products in quantum stochastic calculus such as separating the matrix elements between exponential vectors of such products by moving the first term of the product to the other side of the inner product ([15]).

Alternatively, in the case $\sigma^2 > 1$, when the double vacuum $\Omega = e(0) \otimes \overline{e(0)}$ is a cyclic and separating vector for $\mathcal{N}$, the extension to unbounded operators is accomplished more simply by replacing such operators by the corresponding vectors given by their actions on Ω, as explained in detail in.[8] Then in particular if ψ is the vector corresponding in this way to the affiliated operator U, the vector corresponding to $\mathbb{E}\left[U | \mathcal{N}_{\mathfrak{h}_1^c}\right]$ is given by the projection of ψ onto the closure of the subspace $\mathcal{N}_{\mathfrak{h}_1}\Omega$. It can then be checked that the properties of conditional expectations high-lighted above continue to hold for unbounded affiliated operators.

Now let (p_1, q_1) and (p_2, q_2) be stochastically independent $N(0, \sigma^2)$ canonical pairs. Then

$$p_+ = \frac{p_1 + p_2}{\sqrt{2}}, q_+ = \frac{q_1 + q_2}{\sqrt{2}},$$

$$p_- = \frac{p_1 - p_2}{\sqrt{2}}, q_- = \frac{q_1 - q_2}{\sqrt{2}}$$

are likewise independent $N(0, \sigma^2)$ canonical pairs since

$$\mathbb{E}\left[e^{i(x_1 p_+ + x_2 p_- + y_1 q_+ + y_2 q_-)}\right]$$

$$= \mathbb{E}\left[e^{i(x_1 \frac{p_1+p_2}{\sqrt{2}} + x_2 \frac{p_1-p_2}{\sqrt{2}} + y_1 \frac{q_1+q_2}{\sqrt{2}} + y_2 \frac{q_1-q_2}{\sqrt{2}})}\right]$$

$$= \mathbb{E}\left[e^{i(\frac{x_1+x_2}{\sqrt{2}} p_1 + \frac{x_1-x_2}{\sqrt{2}} p_2 + \frac{y_1+y_2}{\sqrt{2}} q_1 + \frac{y_1-y_2}{\sqrt{2}} q_2)}\right]$$

$$= e^{-\frac{\sigma^2}{4}\left(\left(\frac{x_1+x_2}{\sqrt{2}}\right)^2 + \left(\frac{x_1-x_2}{\sqrt{2}}\right)^2 + \left(\frac{y_1+y_2}{\sqrt{2}}\right)^2 + \left(\frac{y_1-y_2}{\sqrt{2}}\right)^2\right)}$$

$$= e^{-\frac{\sigma^2}{4}(x_1^2 + y_1^2 + x_2^2 + y_2^2)}.$$

Moreover the canonical pairs (p_+, q_+) and (p_-, q_-) generate the same von Neumann algebra

$$\mathcal{N}\left\{(p_1, q_1), (p_2, q_2)\right\} = \left\{e^{ixp_1}, e^{ixq_1}, e^{ixp_2}, e^{ixq_2} : x \in \mathbb{R}\right\}''$$

as do (p_1, q_1) and (p_2, q_2), since for example, because of commutativity

$$e^{ixp_+} = e^{ix\frac{p_1}{\sqrt{2}}} e^{ix\frac{p_2}{\sqrt{2}}}, e^{ixp_1} = e^{ix\frac{p_+}{\sqrt{2}}} e^{ix\frac{p_-}{\sqrt{2}}}.$$

The pair (p_+, q_+) generates a von Neumann subalgebra $\mathcal{N}\{p_+, q_+\}$ whose relative commutant within $\mathcal{N}\{(p_1, q_1), (p_2, q_2)\}$ is the von Neumann subalgebra $\mathcal{N}\{p_-, q_-\}$ generated by (p_-, q_-). In particular the pair (p_-, q_-) is independent of $\mathcal{N}\{p_+, q_+\}$.

Theorem 3.1. *The conditional expectation* $\mathbb{E}\left[p_1 q_2 - q_1 p_2 \,|\, \mathcal{N}\{p_+, q_+\}\right]$ *vanishes.*

Proof. The proof is essentially parallel to that of Theorem 1, but it exploits commutativity properties of the canonical pairs (p_1, q_1) and (p_2, q_2). Thus, since p_1 commutes with q_2 and q_1 commutes with p_2,

$$p_1 q_2 = \frac{1}{4}\left\{(p_1 + q_2)^2 - (p_1 - q_2)^2\right\},$$

$$q_1 p_2 = \frac{1}{4}\left\{(q_1 + p_2)^2 - (q_1 - p_2)^2\right\}$$

and so

$$p_1 q_2 - q_1 p_2 = \frac{1}{4}\left\{(p_1 + q_2)^2 - (p_1 - q_2)^2 - (q_1 + p_2)^2 + (q_1 - p_2)^2\right\}$$

$$= \frac{1}{4}\left\{(p_1 + q_2)^2 - (q_1 + p_2)^2 - \left((p_1 - q_2)^2 - (q_1 - p_2)^2\right)\right\}$$

$$= \frac{1}{4}\left\{(p_1 + q_2 + q_1 + p_2)(p_1 + q_2 - q_1 - p_2)\right.$$

$$\left. - (p_1 - q_2 + q_1 - p_2)(p_1 - q_2 - q_1 + p_2)\right\}$$

since $p_1 + q_2$ commutes with $q_1 + p_2$ and $p_1 - q_2$ commutes with $q_1 - p_2$. Since

$$p_1 + q_2 + q_1 + p_2 = \sqrt{2}\,(p_+ + q_+),$$

$$p_1 - q_2 - q_1 + p_2 = \sqrt{2}\,(p_+ - q_+)$$

both are affiliated to $\mathcal{N}\{p_+, q_+\}$, and using (20), we obtain

$$\mathbb{E}\left[p_1 q_2 - q_1 p_2 \,|\, \mathcal{N}\{p_+, q_+\}\right]$$

$$= \mathbb{E}\left[(p_1 + q_2 + q_1 + p_2)(p_1 + q_2 - q_1 - p_2) \,|\, \mathcal{N}\{p_+, q_+\}\right]$$

$$- \mathbb{E}\left[(p_1 - q_2 + q_1 - p_2)(p_1 - q_2 - q_1 + p_2) \,|\, \mathcal{N}\{p_+, q_+\}\right]$$

$$= (p_1 + q_2 + q_1 + p_2)\,\mathbb{E}\left[p_1 + q_2 - q_1 - p_2 \,|\, \mathcal{N}\{p_+, q_+\}\right]$$

$$- \mathbb{E}\left[p_1 - q_2 + q_1 - p_2 \,|\, \mathcal{N}\{p_+, q_+\}\right](p_1 - q_2 - q_1 + p_2)$$

$$= (p_1 + q_2 + q_1 + p_2)\,\mathbb{E}\left[p_1 + q_2 - q_1 - p_2\right]$$

$$- \mathbb{E}\left[p_1 - q_2 + q_1 - p_2\right](p_1 - q_2 - q_1 + p_2)$$

since

$$p_1 + q_2 - q_1 - p_2 = \sqrt{2}\,(p_- - q_-),$$

$$p_1 - q_2 + q_1 - p_2 = \sqrt{2}\,(p_- + q_-)$$

are both affiliated to $\mathcal{N}\{p_-, q_-\}$. But

$$\mathbb{E}[p_1] = \mathbb{E}[p_2] = \mathbb{E}[q_1] = \mathbb{E}[q_2] = 0.$$

Hence $\mathbb{E}[p_1 q_2 - q_1 p_2 \,|\, \mathcal{N}\{p_+, q_+\}] = 0.$ □

4. Quantum Lévy area

Now let P and Q be a canonical pair of quantum Brownian motions, satisfying the commutation relation (1). Then[2] for each nonempty real interval $[a, b[$, the operators

$$p_{[a,b[} = \frac{1}{\sqrt{b-a}}\left(P(b) - P(a)\right), q_{[a,b[} = \frac{1}{\sqrt{b-a}}\left(Q(b) - Q(a)\right)$$

form an $N(0, \sigma^2)$ canonical pair, and for disjoint intervals $[a, b[$ and $[c, d[$ the corresponding pairs $\left(p_{[a,b[}, q_{[a,b[}\right)$ and $\left(p_{[c,d[}, q_{[c,d[}\right)$ are independent. P and Q can be realised in terms of corresponding Weyl operators by

$$e^{ixP(s)} = W(ix\chi_{]0,s]}), \; e^{iyQ(t)} = W(y\chi_{]0,t]}),$$

where $\chi_{]0,s]}$ is the indicator function of the interval $]0, s]$ regarded as an element of the Hilbert space $L^2(\mathbb{R}_+)$. Similarly the canonical pairs $\left(p_{[a,b[}, q_{[a,b[}\right)$ are given in terms of the L^2-normalised indicator functions

$$\hat{\chi}_{]a,b]} = \frac{1}{\sqrt{b-a}}\chi_{]a,b]}$$

by

$$e^{ixp_{[a,b[}} = W(ix\hat{\chi}_{]a,b]}), e^{ixq_{[a,b[}} = W(x\hat{\chi}_{]a,b]}).$$

Note that in this realisation

$$P(t) = \alpha P_1(t) \otimes \bar{I} + \beta I \otimes \overline{P_1(t)},$$
$$Q(t) = \alpha Q_1(t) \otimes \bar{I} + \beta I \otimes \overline{Q_1(t)},$$

where P_1 and Q_1 are the usual Fock space momentum and position Brownian motions corresponding to variance $\sigma^2 = 1$.

For each interval $[a, b[$ we construct a filtration of von Neumann algebras $\mathcal{N}_1([a, b[) \subset \mathcal{N}_2([a, b[) \subset \mathcal{N}_3([a, b[) \subset \cdots$ generated by the increments of P and Q over the subintervals comprising the successive dyadic dissections of the time interval $[a, b[$, so that

$$\mathcal{N}_1\left([a, b[\right) = \left\{p_{[a,b[}, q_{[a,b[}\right\}'',$$

$$\mathcal{N}_2\left([a, b[\right) = \left\{\left(p_{[a,a+\frac{b-a}{2}[}, q_{[a,a+\frac{b-a}{2}[}\right), \left(p_{[a+\frac{b-a}{2},b[}, q_{[a+\frac{b-a}{2},b[}\right)\right\}''$$

$$\ldots\ldots$$

$$\mathcal{N}_{n+1}\left([a, b[\right) = \left\{\left(p_{[a+2^{-n}(j-1)(b-a),a+2^{-n}j(b-a)[},\right.\right.$$
$$\left.\left.q_{[a+2^{-n}(j-1)(b-a),a+2^{-n}j(b-a)[}\right); \; j = 1, 2, ..., 2^n\right\}''$$

$$\ldots\ldots$$

Quantum conditional expectations $\mathbb{E}\left[|\mathcal{N}_n([a, b[)]\right]$ exist from

$$\{P, Q\}'' = \left\{W(f) : f \in L^2(\mathbb{R}_+)\right\}$$

onto each $\mathcal{N}_n([a, b[)$. We imitate the construction of the martingale $(M_n(a, b))_{n=1,2,...}$ by defining quantum analogs $(L_n(a, b))_{n=1,2,...}$ by

$$L_1(a, b) = 0$$

$$L_2(a, b) = \frac{1}{2} \det \begin{bmatrix} 1 & P(a) & Q(a) \\ 1 & P(\frac{a+b}{2}) & Q(\frac{a+b}{2}) \\ 1 & P(b) & Q(b) \end{bmatrix}$$

$$= \frac{1}{2}\left\{\left(P(\frac{a+b}{2}) - P(a)\right)\left(Q(b) - Q(\frac{a+b}{2})\right)\right.$$
$$\left. - \left(Q(\frac{a+b}{2}) - Q(a)\right)\left(P(b) - P(\frac{a+b}{2})\right)\right\}$$

$$= \frac{b-a}{2}\left\{p_{[a,\frac{a+b}{2}[}q_{[\frac{a+b}{2},b[} - q_{[a,\frac{a+b}{2}[}p_{[\frac{a+b}{2},b[}\right\} \tag{22}$$

and for $n > 2$

$$L_{n+1}(a, b) = L_n(a, b) + \sum_{j=1}^{2^n} L_2(a+2^{-n}(j-1)(b-a), a+2^{-n}j(b-a)). \tag{23}$$

Theorem 4.1. $(L_n(a, b))_{n \in \mathbb{N}}$ *is a martingale with respect to the filtration of von Neumann algebras* $(\mathcal{N}_n([a, b[))_{n \in \mathbb{N}}$ *in the sense that each* $L_n(a, b)$ *is affiliated to* $\mathcal{N}_n([a, b[)$, *and for each* $n = 1, 2, ...$

$$\mathbb{E}\left[L_{n+1}(a, b)|\mathcal{N}_n[a, b[\right] = L_n(a, b). \tag{24}$$

Proof. (22) shows that $L_2(a, b)$ is affiliated to $\mathcal{N}_2([a, b[)$. Moreover Theorem 3 implies that

$$\mathbb{E}\left[L_2(a, b)|\mathcal{N}_1[a, b[\right] = 0 = L_1(a, b).$$

Replacing the interval $[a, b[$ by $[a + 2^{-n}(j - 1)(b - a), a + 2^{-n}j(b - a)[$, it follows that $L_2(a + 2^{-n}(j - 1)(b - a), a + 2^{-n}j(b - a))$ is affiliated to $\mathcal{N}_2([a + 2^{-n}(j - 1)(b - a), a + 2^{-n}j(b - a)[)$ and that

$$\mathbb{E}\left[L_2(a + 2^{-n}(j - 1)(b - a), a + 2^{-n}j(b - a))\right.$$
$$\left|\mathcal{N}_1([a + 2^{-n}(j - 1)(b - a), a + 2^{-n}j(b - a)[)\right] = 0.$$

Since $\mathcal{N}_2([a + 2^{-n}(j - 1)(b - a), a + 2^{-n}j(b - a)[) \subset \mathcal{N}_{n+1}([a, b[)$ it follows inductively from (23) that $M_{n+1}(a, b)$ is affiliated to $\mathcal{N}_{n+1}([a, b[)$.

Also, since for each $j = 1, 2, ..., 2^n$, $L_2(a + 2^{-n}(j - 1)(b - a), a + 2^{-n}j(b - a))$ is independent of the von Neumann algebra generated by all $p_{[a+2^{-n}(k-1)(b-a), a+2^{-n}k(b-a)[}$ and $q_{[a+2^{-n}(k-1)(b-a), a+2^{-n}kj(b-a)[}$ for $k \neq j$, by (21),

$$\mathbb{E}\left[L_2(a + 2^{-n}(j - 1)(b - a), a + 2^{-n}j(b - a))\middle|\mathcal{N}_n([a, b[)\right]$$
$$= \mathbb{E}\left[L_2(a + 2^{-n}(j - 1)(b - a), a + 2^{-n}j(b - a))\right|$$
$$\mathcal{N}_1([a + 2^{-n}(j - 1)(b - a), a + 2^{-n}j(b - a)[)\right]$$
$$= 0.$$

Hence (24) follows from (23). □

5. L^2-boundedness of the quantum martingale $(L_n(a, b))_{n \in \mathbb{N}}$

In this section we assume that $\sigma^2 > 1$. Because the Fock vacuum is not separating, it is difficult to formulate a satisfactory L^2-martingale convergence theorem in the case $\sigma^2 = 1$. In fact in this case the quantum Lévy area is identically zero in the same probabilistic sense that the Fock number process Λ is zero.

Theorem 5.1. *The quantum martingale $(L_n(a, b))_{n \in \mathbb{N}}$ is bounded in the sense that each $\|L_n(a, b)\Omega\| \leq (b - a)\alpha\beta$ where Ω is the cyclic and separating double vacuum vector $e(0) \otimes \overline{e(0)}$.*

Proof. We introduce the independent $N(0, \sigma^2)$ canonical pairs

$$p_j = p_{[a+2^{-n}(j-1)(b-a), a+2^{-n}j(b-a)[}, \quad q_j = q_{[a+2^{-n}(j-1)(b-a), a+2^{-n}j(b-a)[},$$

where $j = 1, 2, ..., 2^n$. Corresponding to (7) we can then write

$$L_n(a, b) = \frac{b - a}{2^n} \sum_{1 \leq j\,k \leq 2^n} (p_j q_k - q_j p_k).$$

We realise the Fock space $\Gamma(\mathfrak{h})$ with $\mathfrak{h} = L^2(\mathbb{R}_+)$ as the infinite direct sum

$$\Gamma(\mathfrak{h}) = \oplus_{N=0}^{\infty} \mathfrak{h}^{(N)}$$

of N-particle subspaces $\mathfrak{h}^{(N)}$ each consisting of the symmetric sector of the N-fold tensor product of $\mathfrak{h}$ with itself,

$$\mathfrak{h}^{(N)} = \left(\otimes_{j=1}^{N} \mathfrak{h}\right)_{\text{sym}}.$$

Each exponential vector $e(f)$ is realised as

$$e(f) = \left(1, f, \frac{1}{\sqrt{2!}} f \otimes f, \frac{1}{\sqrt{3!}} f \otimes f \otimes f, ...\right).$$

We express the canonical pairs (p_j, q_j) in terms of corresponding creation and annihilation operators

$$a_j^{\dagger} = \frac{1}{\sqrt{2}}(p_j + iq_j), \ a_j = \frac{1}{\sqrt{2}}(p_j - iq_j)$$

so that

$$p_j q_k - q_j p_k = i \left(a_j^{\dagger} a_k - a_k^{\dagger} a_j\right).$$

The creation and annihilation operators $a_k^{\dagger}$ and a_k are given in terms of corresponding Fock creation and annihilation operators $\hat{a}_k^{\dagger}$ and $\hat{a}_k$ by

$$a_k^{\dagger} = \alpha^2 \hat{a}_k^{\dagger} \otimes I + \beta^2 I \otimes \overline{\hat{a}_k}, \ a_k = \alpha^2 \hat{a}_k \otimes I + \beta^2 I \otimes \overline{\hat{a}_k^{\dagger}}.$$

$\hat{a}_k$ annihilates the Fock vacuum $e(0)$ while $\hat{a}_k^{\dagger}$ "creates" the normalised indicator function $\hat{\chi}_k = \hat{\chi}_{[a+2^{-n}(k-1)(b-a), a+2^{-n}k(b-a)[}$ of the interval $[a + 2^{-n}(k-1)(b-a), a + 2^{-n}k(b-a)[$. Hence a_k act on the double vacuum $\Omega = e(0) \otimes \overline{e(0)}$ as

$$a_k \Omega = e(0) \otimes \beta^2 \overline{f_k}$$

where the unit vector f_k is the one-particle vector

$$f_k = (0, \hat{\chi}_k, 0, ...) \in \Gamma(\mathfrak{h})$$

which is orthogonal to the Fock vacuum $e(0) = (1, 0, 0, ...)$. For $j \neq k$, since $\hat{\chi}_j$ is orthogonal to $\hat{\chi}_k$, the Fock annihilation operator $\hat{a}_j$ also annihilates f_k. consequently Consequently $a_j^{\dagger}$ acts on the vector $e(0) \otimes \overline{f_k}$ as

$$a_j^{\dagger} \left(e(0) \otimes \overline{f_k}\right) = \alpha^2 f_j \otimes \overline{f_k}$$

so that

$$a_j^{\dagger} a_k \Omega = a_j^{\dagger} \left(e(0) \otimes \beta^2 \overline{f_k}\right)$$
$$= \alpha^2 \beta^2 f_j \otimes \overline{f_k}.$$

Similarly $a_k^\dagger a_j \Omega$ is the orthogonal vector

$$a_k^\dagger a_j \Omega = \alpha^2 \beta^2 f_k \otimes \overline{f_j}.$$

Also, for $(j, k) \neq (j', k')$, $a_j^\dagger a_k \Omega$ and $a_k^\dagger a_j \Omega$ are orthogonal to $a_{j'}^\dagger a_{k'} \Omega$ and $a_{k'}^\dagger a_{j'} \Omega$. Thus we have

$$\left\| \sum_{1 \leq j < k \leq 2^n} (p_j q_k - q_j p_k) \, \Omega \right\|^2 = \left\| \sum_{1 \leq j < k \leq 2^n} \left(a_j^\dagger a_k - a_k^\dagger a_j \right) \Omega \right\|^2$$

$$= \sum_{1 \leq j < k \leq 2^n} \left\| \left(a_j^\dagger a_k - a_k^\dagger a_j \right) \Omega \right\|^2$$

$$= \sum_{1 \leq j < k \leq 2^n} \left(\left\| a_j^\dagger a_k \Omega \right\|^2 + \left\| a_k^\dagger a_j \Omega \right\|^2 \right)$$

$$= 2^n (2^n - 1) \alpha^2 \beta^2.$$

It follows that

$$\| L_n \Omega \|^2 = \left\| \frac{b - a}{2^n} \sum_{1 \leq j < k \leq N} \left(a_j^\dagger a_k - a_k^\dagger a_j \right) \Omega \right\|^2 \leq (b - a)^2 \alpha^2 \beta^2 . \qquad \square$$

The L^2-bounded martingale $(L_n)_{n \in \mathbb{N}}$ is seen to be convergent in the L^2 sense as follows. The bounded sequence of vectors $(L_n \Omega)_{n \in \mathbb{N}}$ must contain a weakly convergent subsequence $(L_{n_k} \Omega)_{k \in \mathbb{N}}$. Denote its limit by Ψ and define an operator L affiliated to $\mathcal{N}$ on the dense domain $\{W\Omega : W \in \mathcal{N}'\}$ by

$$LW\Omega = W\Psi.$$

Then for arbitrary $n \in \mathbb{N}$,

$$\langle L_n \Omega, L\Omega \rangle = \lim_k \langle L_n \Omega, L_{n_k} \Omega \rangle .$$

But $\langle L_n \Omega, L_{n_k} \Omega \rangle = \langle L_n \Omega, L_n \Omega \rangle$ for $n_k \geq n$. It follows that, for all $n \in \mathbb{N}$, $\mathbb{E}\left[L \, | \mathbb{N}_n \right] = L_n$.

Acknowledgement

The author thanks the referee for drawing his attention to a number of mistypes.

References

1. M Capitaine and C Donati-Martin, The Lévy area process for the free Brownian motion, Journal of Functional Analysis **179**, 153-169 (2000).
2. A M Cockroft and R L Hudson, Quantum mechanical Wiener processes, J. Multivariate Anal. **7**, 107-124 (1977).
3. C D Cushen and R L Hudson, A quantum mechanical central limit theorem, J. Applied Prob. **8**, 454-469 (1971).
4. D E Evans and J T Lewis, *Dilations of irreversible evolutions in algebraic quantum theory*, Comm. Dublin Inst. Adv. Studies A **24** (1977).
5. P K Friz and N B Victoir, *Multidimensional stochastic processes as rough paths*, Cambridge University Press (2010)
6. H Goldstein, *Classical mechanics*, Addison-Wesley, New York (1959).
7. The Itô-Nisio theorem , quadratic Wiener functionals and 1-solitons, Stochastic Processes and their Applications **120**, 605-621 (2010).
8. R L Hudson and J M Lindsay, A noncommutative martingale representation theorem for non-Fock quantum Brownian motion, J. Funct. Anal **61**, 202-221 (1985).
9. K Itô, Stochastic integral, Proc. Imp. Acad. Tokyo **20**, 519-524 (1944).
10. P Lévy, Wiener's random function and other Laplacian functions, pp 171-187, in Proc. 2nd Berkeley Symposium Math Statistics and Probability 1950, University of California Press (1951).
11. P Lévy, Le mouvement Brownien plan, Amer. Jour. Math. **62**, 487-550 (1940).
12. T Lyons, Differential equations driven by rough signals, Rev. Mat. Iberoamericana **14**, 215-310 (1998).
13. T Lyons and Z Qian, Flow of diffeomorphisms induced by a multiplicative functional, Probab. Theory and Related Fields **112**, 91-119 (1998).
14. J Ortmann, Functionals of free Brownian bridge, arXiv:1107.0218v1 [math.PR]
15. K R Parthasarathy, *An introduction to quantum stochastic calculus*, Birkhäuser (1993).
16. D Williams, *Diffusions, Markov processes and martingales* Volume I, Wiley (1979).

ON COMPUTATIONAL COMPLEXITY OF QUANTUM ALGORITHM FOR FACTORING

SATOSHI IRIYAMA* and MASANORI OHYA

*Department of Information Sciences, Tokyo University of Science,
2641 Yamazaki, Noda City, Chiba, 2788510, Japan
* E-mail: iriyama@is.noda.tus.ac.jp*

IGOR V. VOLOVICH

*Steklov Mathematical Institute, Russian Academy of Science
Gubkin St. 8, Moskow, 119991, Russia
E-mail: volovich@mi.ras.ru*

In 1994, Shor discovered the quantum algorithm which solves a prime factoring in a polynomial time of input data. However, he did not show how to create the unitary operator for the quantum algorithm, so that the proof of computational complexity is not complete. In this paper, we discuss an upper bound of the computational complexity of factoring algorthm, and show a quantum algorithm for factoring is not always effective even if the Oracle is given.

Keywords: Quantum algorithm; Prime factoring; Computational complexity.

1. Introduction

In 1994, Shor proposed a quantum algorithm for factoring[3,4] and discussed its computational complexity and probability of correct result. It is concluded that Shor's quantum algorithm can solve factoring in polynomial time of input size. However, the following problems still remain:

(1) How to define the computational complexity of quantum algorithm
(2) How to construct the unitary operator calculating modular
(3) How much is the computational complexity for modular exponentiation in the case of superposition

Moreover, Accardi pointed out that there exists a classical probabilistic algorithm to solve factoring with same probability as Shor's quantum algorithm.[1]

In this paper, we say that a quantum algorithm is *effective* if it requires a polynomial steps of input size. And we define the computational complexity of quantum algorithm as the number of fundamental gates such as NOT gate, controlled gate as so on. A product of unitary operators is also a unitary operator. Then we have to decompose the unitary operator in the algorithm into the fundamental gates.

For the second problem, we discuss how to make the unitary operator computing modular. In this talk, we propose two strategies to construct it. One is the way to construct directly, and the other is QTM approach. In both ways, we have to solve factoring before construct the unitary operator. Therefore constructing unitary operator is equivalent to solving factoring.

For the third problem, we compute the computational complexity of modular exponentiation. In Shor's quantum algorithm, one has to prepare $\left[\log N^2\right]$ different controlled gates such as $U_C\left(U_{y,N}^{2^j}\right)$, $(j = 0, 1, \cdots, \left[\log N^2\right] - 1)$ where $U_{y,N}$ is the unitary operator computing modular defined below. Using this unitary operators with n control qubits, we can compute modular exponentiation. Then we consider how much is the computational complexity of $U_C\left(U_{y,N}^{2^j}\right)$. If we define the computational complexity of $U_C\left(U_{y,N}^{2^j}\right)$ as 2^j, the computational complexity of Shor's quantum algorithm becomes N^2. This implies that Shor's quantum algorithm for factoring is not finished in polynomial time of $\log N$.

2. Quantum algorithm

First, we explain foundations of quantum algorithm.

A quantum algorithm is constructed by the following steps:

(1) Prepare a Hilbert space $\mathcal{H} = \mathbb{C}^{\otimes n}$
(2) Construct an initial state $|\psi_{in}\rangle \in \mathcal{H}$.
(3) Construct unitary operators U to solve the problem.
(4) Apply them for the initial state and obtain a result state $|\psi_{out}\rangle = U|\psi_{in}\rangle$.
(5) If necessary, amplify the probability of correct result.
(6) Measure an observable with the result state.

In the first step, we define the Hilbert space depending on the problem. Let $\mathbb{C}^2$ be a Hilbert space spanned by $|0\rangle = \begin{pmatrix} 1 \\ 0 \end{pmatrix}$ and $|1\rangle = \begin{pmatrix} 0 \\ 1 \end{pmatrix}$, a normalized vector $|\psi\rangle = \alpha|0\rangle + \beta|1\rangle$ on this space is called a qubit. Since we can use a superposition of $|0\rangle$ and $|1\rangle$ as an initial state vector, the quantum algorithm is more effective than classical one.

One can apply Hadamard transformation

$$U_H = \frac{1}{\sqrt{2}} \begin{pmatrix} 1 & 1 \\ 1 & -1 \end{pmatrix}$$

to create a superposition. Hadamard transformation has a very important role in a quantum algorithm.

Applying $(U_H)^{\otimes n}$ to the vector $|\psi\rangle = |0\rangle \otimes \cdots \otimes |0\rangle \in (\mathbb{C}^2)^{\otimes n}$, we have

$$(U_H)^{\otimes n} |\psi\rangle = \frac{1}{\sqrt{2^n}} \sum_{i=0}^{2^n-1} |e_i\rangle,$$

where $\{|e_i\rangle\}$ is a complete orthonormal system of $(\mathbb{C}^2)^{\otimes n}$ defined as

$$|e_0\rangle = |0\rangle \otimes \cdots \otimes |0\rangle$$
$$|e_1\rangle = |1\rangle \otimes |0\rangle \otimes \cdots \otimes |0\rangle$$
$$\cdots$$
$$|e_{2^n-1}\rangle = |1\rangle \otimes \cdots \otimes |1\rangle$$

One can represent any positive integers less than 2^n by $|e_i\rangle$.

Here we introduce unitary gates, which are NOT gate, Controlled gate We call these gates fundamental gates. We can also construct AND and OR gate by considering the product of them and some implementations.[11] The NOT gate U_{NOT} is defined on a Hilbert space $\mathbb{C}^2$ as

$$U_{NOT} = |1\rangle \langle 0| + |0\rangle \langle 1|.$$

Controlled gate $U_C(U)$ for a unitary operator U is given by

$$U_C(U) = |0\rangle \langle 0| \otimes I + |1\rangle \langle 1| \otimes U$$

We can easily check that these gates are unitary.

2.1. *Computational complexity of quantum algorithm*

In order to discuss the computational complexity of quantum algorithm, we introduced fundamental gates above. Here we define the computational complexity of quantum algorithm as the number of fundamental gates in it.

For example, we say the computational complexity is n if the unitary operator U is constructed by a product of n fundamental gates.

Definition 2.1. We say the algorithm is effective if the computational complexity is a polynomial of $\log N$ where N is a size of input.

Moreover, we defined a generalized quantum Turing machine(GQTM) using quantum channel and density operator. Based on GQTM we defined some language classes. BGQPP is the language class which is recognized a GQTM in polynomial time with a halting probability 1/2. We proved that the class NP in included by BGQPP.[11,14]

3. Quantum algorithm for factorization

Shor's factoring algorithm[3,4] is based on Simon's period finding.[5,6]

Let p and q be two prime numbers, and $N = pq$. Shor's quantum factoring is the following:

(1) Prepare an initial state.
(2) Modular exponentiation.
(3) Quantum Fourier transform.
(4) Measurement.
(5) Computation of the order at the classical computer. If it is odd we can obtain the factoring. Otherwise, go back to Step1 again.

Step 1: Put the first register in the uniform superposition of states representing numbers k. Choose $y \neq 1$ such that $g.c.d\,(N, y) = 1$ and put it the second register. The quantum computer will be in the state

$$|\psi_{in}\rangle = \frac{1}{\sqrt{N^2}} \sum_{k=0}^{N^2-1} |k\rangle \otimes |1\rangle .$$

Step 2: Compute y^k (mod N) in the second register by $U_{y,N}$. For all $x \in \{0, 1, \cdots, N-1\}$ it works as

$$U_{y,N} |x\rangle = |xy \bmod N\rangle .$$

This leaves the quantum computer in the state

$$|\psi_2\rangle = \frac{1}{\sqrt{N^2}} \sum_{k=0}^{N^2-1} |k\rangle \otimes |y^k \bmod N\rangle .$$

Step 3: Perform the quantum Fourier transform on the first register, mapping $|k\rangle$ to

$$\frac{1}{\sqrt{N^2}} \sum_{l=0}^{N^2-1} e^{2\pi i k l/N^2} |l\rangle .$$

The quantum computer will be in the state

$$|\psi_{out}\rangle = \frac{1}{N^2} \sum_{k=0}^{N^2-1} \sum_{l=0}^{N^2-1} e^{2\pi i k l/N^2} |l\rangle \otimes |y^k \bmod N\rangle .$$

Step 4: Make the measurement on both registers $|l\rangle$ and $|y^k \pmod{N}\rangle$. To find the period r we will need only the value of $|l\rangle$ in the first register but for clarity of computations we make the measurement on the both registers. The probability $P(l, y^k \pmod{N})$ that the quantum computer halts in a particular state $|l\rangle \otimes |y^k \pmod{N}\rangle$ is

$$P(l, y^k \pmod{N}) \geq \frac{1}{3r^2}$$

We will use the Theorem[3,4] which shows that the probability $P(l, y^k \pmod{N})$ is large if residue of $rl \pmod{N^2}$ is small.

Step 5: From the measurement above, we obtain the pair of l and $y^k \pmod{N}$ with the probability more than $\frac{1}{3r^2}$ where r is the order. Here we consider the following joint events

$$\{y \text{ is coprime to } N\} \cap \{\text{obtain } r\}$$

Let P be a uniform measure on the set $\{0, 1, \ldots, N-1\}$, i.e.,

$$P(x) \equiv \frac{1}{N}, \ \forall x \in \{0, 1, \ldots, N-1\}$$

From number theory, we have

$$P(\{y \text{ is coprime to } N\}) \geq \frac{1}{\log N}$$

Then the probability of the event $\{\text{we obtain } r\}$ is

$$P(\{\text{obtain } r\}) \geq \frac{\varphi(r)}{3r} = C \left(\geq \frac{1}{3} \text{ if } r \text{ is enough large} \right)$$

where $0 < C < 1$ is a constant, because there are $r\varphi(r)$ states $|l\rangle \otimes |y^k \pmod{N}\rangle$ which enable us to compute r, where $\varphi(r)$ is called Euler's function, i.e.,

$$\varphi(r) = \#\{0 \leq a < r; g.c.d.(a, r) = 1\}$$

Therefore we have

$$P(\{y \text{ is coprime to } N\} \cap \{\text{obtain } r\}) \geq \frac{C}{\log N} = p_q$$

Accardi pointed out[1] that a classical probabilistic algorithm can solve factoring with same probability as quantum one

$$\frac{1}{2\log N} = p_c$$

4. How to construct the unitary operator

Let $U_{y,N}$ be the unitary operator of quantum factoring algorithm such that for a state vector $|x\rangle$ ($x \in \{0, 1, \ldots, N-1\}$) on $\left(\mathbb{C}^2\right)^{\otimes n}$

$$U_{y,N}|x\rangle = |xy \bmod N\rangle$$

where $n = [\log N]$. Here, we consider how to construct this unitary operator. In this study we discuss the two strategy: the unitary operator is given by (1) a direct transformation from input to the output or (2) a product of fundamental gates(unitary quantum Turing machine).

In the case of (1) the computational complexity of the unitary operator is just 1 because this unitary operator is given by the following simple form

$$U_{y,N} = \sum_x |xy \bmod N\rangle \langle x|$$

however, we have to consider how to construct it.

In the case of (2) we have to discuss the stationarity for a superposition input.

Definition 4.1. We say a QTM is *stationary* if there exists a positive integer t such that for all $s < t$ a halting probability of QTM is 0 and it halts at time t with probability 1.

4.1. A direct transformation

To construct the unitary operator, we have to know the answer of $xy \bmod N$ for all x. Before the all calculations are finished, we inevitably solve the following classical algorithm(order finding):

Step 1: Obtain the number z_1 such that $|z_1\rangle \langle 1|$

Step 2: Repeat the Step3 from $i = 1$ until we obtain the number z_r such that $|1\rangle \langle z_r|$

Step 3: Obtain the number z_i such that $|z_{i+1}\rangle \langle z_i|$

Step 4: $r + 1$ is an order

Therefore, if we take this approach to construct the unitary operator, we have to solve the order finding before the quantum algorithm with probability 1.

4.2. QTM approach

Let M be a classical Turing machine such that for an input x it outputs $M(x)$. From Deutch's theorem, there exists a (unitary)QTM which works as same as the classical one.[2] However, it is not clear that even if there exist QTM for a classical input, we do not know the existence for a superposition input.

Let $f : \Sigma \to \Sigma$ be a partial recursive function

$$f(x) = \begin{cases} f(g(x)) & h(x) = 1 \\ x & h(x) = 0 \end{cases}$$

where $g : \Sigma \to \Sigma$ and $h : \Sigma \to \{0,1\}$ are computable functions, and M_f be a CTM that computes $f(x)$ for the input x. We write as

$$M_f(x) = f(x).$$

We don't know the steps exactly for computation $M_f(x)$ before the computation has done.

Here, we first consider the QTM $M_{Q,f}$ compute $f(x)$ for the input x, and its unitary operator is $U_{Q,f}$. We can construct $M_{Q,f}$ according to M following the Deutch's Theorem.

For certain input x, let $T(M_f(x))$ be the computational complexity of M_f with input x. Here,one can consider the another input $y \neq x$, and the computational complexity is

$$T(M_f(y)) \neq T(M_f(x))$$

generally.

Then, we consider a superposition

$$|\psi\rangle = \frac{1}{\sqrt{2}}(|x\rangle + |y\rangle).$$

After $T(M_f(x))$ steps, the state of QTM becomes

$$|\psi'\rangle = U_{Q,f}^{T(M_f(x))}|\psi\rangle = \frac{1}{\sqrt{2}}(|f(x)\rangle + |y'\rangle)$$

where $|y'\rangle$ is an intermediate state of computation. While for the same input, after $T(M_f(y))$ steps we obtain

$$|\psi''\rangle = U_{Q,f}^{T(M_f(y))}|\psi\rangle = \frac{1}{\sqrt{2}}(|x'\rangle + |f(y)\rangle)$$

where $|x'\rangle$ is an intermediate state of computation not same as $|f(x)\rangle$. By the unitarity of the transition of QTM, after the computation is finished, the state goes back to the initial state again. Therefore, we have to know the exact number of steps to obtain the correct result (see more discussions in the book and papers[11,14,16]).

For all input including a superposition, we call a QTM is stationary if there exists a step t such that for all x

$$T\left(M_f\left(x\right)\right) = t$$

Modulo calculation $f\left(x\right) = x \bmod N$ is a partial recursive function written as

$$f\left(x\right) = \begin{cases} f\left(x - N\right) & \text{if } x \geq N \\ x & \text{otherwise} \end{cases}$$

First we compare x and N, then we calculate $f\left(x - N\right)$ if $x \geq N$. We cannot know how many repetition there before we calculate. To construct a stationary QTM, we have to know how many repetition happens for all x, namely we have to know a_x such that

$$x = a_x N + b_x, \ 0 \leq b_x < N - 1$$

By looping Lemma,[9] we can construct a stationary QTM if we know a looping time for each inputs.

However, the computational complexity to obtain all a_x is the same as the above discussion. We again inevitably solve the order finding.

As a conclusion of this section we say that

The computational complexity of constructing the unitary operator $U_{y,N}$ is equal to factoring.

5. Computational complexity of Shor's quantum factorization

Here, we assume the modulo calculation is given. It means that we have the direct transition from input to the output discussed the above.

We can use a Controlled-U gate $U_C\left(U\right)$ which works as the following:

$$U_C\left(U\right)|1\rangle \otimes |x\rangle = |1\rangle \otimes U\,|x\rangle$$
$$U_C\left(U\right)|0\rangle \otimes |x\rangle = |0\rangle \otimes I\,|x\rangle$$

(e.g. $U_C\left(U_{NOT}\right)$ is a known as a Controlled-NOT gate)

The input superposition is given by

$$|\psi_{in}\rangle = \frac{1}{\sqrt{N^2}} \sum_{k=0}^{N^2-1} |k\rangle \otimes |1\rangle$$

where k is represented by a binary form. Using k as a controlled bit, we prepare $\left[\log\left(N^2\right)\right]$ controlled gates

$$U_C\left(U_{y,N}^{2^j}\right),\ j = 0, 1, 2, \cdots, \left[\log\left(N^2\right)\right] - 1$$

where $U_{y,N}$ is a unitary operator computing $xy \bmod N$.

Applying a product of these operators, we obtain

$$U_C\left(U_{y,N}\right) \cdot U_C\left(U_{y,N}^2\right) \cdots U_C\left(U_{y,N}^{2^{\left[\log\left(N^2\right)\right]-1}}\right) |\psi_{in}\rangle$$

$$= \frac{1}{\sqrt{N^2}} \sum_{k=0}^{N^2-1} |k\rangle \otimes \left|y^k \bmod N\right\rangle$$

Since

$$\sum_{j=0}^{N^2-1} 2^j = N^2 - 1$$

the computational complexity of this process is

$$T\left(U_C\left(U_{y,N}\right) \cdot U_C\left(U_{y,N}^2\right) \cdots U_C\left(U_{y,N}^{2^{\left[\log\left(N^2\right)\right]-1}}\right)\right)$$

$$= T\left(U_{y,N}\right) + T\left(U_{y,N}^2\right) + \cdots + T\left(U_{y,N}^{2^{\left[\log\left(N^2\right)\right]-1}}\right)$$

$$= N^2 - 1$$

where we count the computational complexity of $U_{y,N}$ as 1.

Finally, we assume the computational complexities of $U_C\left(U_{y,N}^{2^j}\right)$ is 1 for all $j = 0, \cdots, \left[\log\left(N^2\right)\right] - 1$.

Therefore the computational complexity becomes

$$T\left(U_C\left(U_{y,N}\right) \cdot U_C\left(U_{y,N}^2\right) \cdots U_C\left(U_{y,N}^{2^{\left[\log\left(N^2\right)\right]-1}}\right)\right)$$

$$= T\left(U_{y,N}\right) + T\left(U_{y,N}^2\right) + \cdots + T\left(U_{y,N}^{2^{\left[\log\left(N^2\right)\right]-1}}\right)$$

$$= \left[\log\left(N^2\right)\right]$$

because $j = 0, \cdots, \left[\log\left(N^2\right)\right] - 1$.

6. Conclusion

In this study, we discussed the computational complexity of quantum algorithm for factorization. To construct the unitary operator $U_{y,N}$, we have to mention the computational complexity to construct it In this case, when we construct it, we have to solve the factorization. Moreover, we discussed the computational complexity of the Shor's quantum algorithm assuming the unitary operator is given. From the above discussion, even if the computational complexity of the unitary operator is 1. If we define the computational complexity of $U_C \left(U_{y,N}^{2^j} \right)$ as 1 for all $j = 0, j = 0, \cdots, \left[\log \left(N^2 \right) \right] - 1$. the computational complexity of quantum factorization becomes $\log \left(N^2 \right)$.

We give more precise discussions in the paper,[17] and propose the new quantum factoring algorithm which can solve it in polynomial time.[18,19]

References

1. L.Accardi, Complexity considerations quantum computation, Volterra preprint, 2010
2. D.Deutsch, Quantum theory, the Church-Turing Principle and the universal quantum computer, Proc. R. Soc. Lond. A, 400:97, 1985
3. P.W.Shor, Algorithms for quantum computation: discrete logarithms and factoring, Proc. of 35th Annu. Symp. on Fou. of Comp. Sci., IEEE Press, Los Alamitos, CA, 1994
4. P.W.Shor, Polynomial-time algorithms for prime factorization and discrete logarithms on a quantum computer, SIAM J. Comp., 26(5):1484-1509, 1997
5. D.Simon, On the power of quantum computation, Proc. of 35th Annu. Symp. on Fou. of Comp. Sci., IEEE Press, Los Alamitos, CA, 1994
6. D.Simon, On the power of quantum computation, SIAM J. Comp., 26(5):1474-1483, 1997
7. M.Ohya and I.V.Volovich, Quantum computing and chaotic amplification, J. opt. B, 5,No.6 639-642, 2003.
8. M.Ohya and I.V.Volovich, New quantum algorithm for studying NP-complete problems, Rep.Math.Phys., 52, No.1,25-33 2003.
9. E. Bernstein and U. Vazirani.: Quantum complexity theory, in: Proc. of the 25th Annual ACM Symposium on Theory of Computing, ACM, New York, pp.11-22, 1993, SIAM Journal on Computing, 26, 1411, 1997
10. L.Accardi and M.Ohya, A Stochastic Limit Approach to the SAT Problem,Open Systems and Information dynamics, 11,1-16, 2004.
11. M.Ohya and I.V.Volovich, Mathematical Foundation of Quantum Information and Quantum Computation, Springer Velrag, 2011.
12. S.Iriyama and M.Ohya, Rigorous Estimation of the Computational Complexity for OV SAT Algorithm, Open System and Information Dynamics, 15, 2, 173-187, 2008.
13. S.Iriyama, T.Miyadera and M.Ohya, Note on a Universal Quantum Turing Machine, Physics Letter A, 372, 5120-5122, 2008.
14. S.Iriyama and M.Ohya, Language Classes Defined by Generalized Quantum Turing Machine, Open System and Information Dynamics 15:4, 383-396,

2008.

15. S.Iriyama, M.Ohya and I.V.Volovich, Generalized Quantum Turing Machine and its Application to the SAT Chaos Algorithm, QP-PQ:Quantum Prob. White Noise Anal., Quantum Information and Computing, 19, World Sci.Publishing, 204-225, 2006.

16. S.Iriyama and M.Ohya, The problem to construct Unitary Quantum Turing Machine computing partial recursive functions, TUS preprint, 2009.

17. S.Iriyama, M.Ohya and I.V.Volovich, On Computational Complexity of Shor's Quantum Factoring Algorithm, TUS preprint, 2012

18. S.Iriyama, M.Ohya and I.V.Volovich, On Quantum Algorithm for Binary Search and Its Computational Complexity, TUS preprint, 2012

19. K.Goto, S.Iriyama, M.Ohya, On Quantum Algorithm for Prime Factoring and Its Computational Complexity, in preparation, 2012

PREQUANTUM CLASSICAL STATISTICAL FIELD THEORY: DERIVATION OF GAUSSIANITY OF PROBABILITY DISTRIBUTIONS

ANDREI KHRENNIKOV

International Center for Mathematical Modelling
in Physics and Cognitive Sciences
Linnaeus University, S-35195, Växjö, Sweden

We developed a purely field model of microphenomena – prequantum classical statistical field theory (PCSFT). This model reproduces important probabilistic predictions of QM including correlations for entangled systems. The presence of the sufficiently strong background field is a fundamental element of PCSFT. In the absence of such a field ("fluctuations of vacuum") PCSFT can reproduce correlations only for quantum systems in factorizable states. By taking into account a random background (of the white noise type) we are able to construct the classical field representation in the general case. This has been done under the additional assumption that prequantum random fields are Gaussian. In previous works on PCSFT, Gaussianity of prequantum fields was invented as a purely mathematical assumption – to simplify calculation of correlations. In this paper we present the derivation of Gaussianity by representing the probability distributions of prequantum fields as the limiting distributions of wave-pulse Poissonian processes.

1. Introduction

One of the crucial differences between mathematical models of classical and quantum mechanics (QM)* is the use of the tensor product of the state spaces of subsystems as the state space of the corresponding composite quantum system. (To describe an ensemble of classical composite systems one uses random variables taking values in the Cartesian product of the state spaces of subsystems.) We show that, nevertheless, it is possible to establish a natural correspondence between the classical and quantum probabilistic descriptions of composite systems. Quantum averages for composite systems (including entangled) can be represented as averages with

---

*See books[1,2] for discussions on general foundational aspects of QM.

respect to classical random fields. (It is essentially what Albert Einstein was dreamed of.) QM is represented as classical statistical mechanics, but with infinite-dimensional phase space. We call our classical random field model for quantum phenomena *prequantum classical statistical field theory* (PCSFT).[3-14]

We stress that the presence of the sufficiently strong background field is a fundamental element of PCSFT. In the absence of such a field ("fluctuations of vacuum") PCSFT can reproduce correlations only for quantum systems in factorizable states. By taking into account a random background (of the white noise type) we are able to construct the classical field representation in the general case. This has been done under the additional assumption that prequantum random fields are Gaussian. In previous works on PCSFT,[3-14] Gaussianity of prequantum fields was invented as a purely mathematical assumption – to simplify calculation of correlations. In this paper we present the derivation of Gaussianity by representing the probability distributions of prequantum fields as limiting distributions of *wave-pulse Poissonian processes*.

We recall that by PCSFT[3-14] quantum systems are represented by classical Gaussian random fields. Correlations of quantum observables A_1 and A_2 on a composite system $S = (S_1, S_2)$ are represented as correlations of quadratic forms, $f_{A_1}(\phi_1), f_{A_2}(\phi_2)$, of components of the prequantum random field $\omega \to \phi(\omega) = (\phi_1(\omega), \phi_2(\omega))$ representing S at the subquantum level. (Here ω is a random parameter.)

2. Classical representation of quantum correlations

Take a Hilbert space H as the space of states of classical random fields. To simplify considerations, we shall work with real Hilbert spaces.

Consider a probability distribution P on H having zero average (it means that $\int_H (y, \phi) dP(\phi) = 0$ for any $y \in M$) and the covariance operator D:

$$(Dy_1, y_2) = \int_H (y_1, \phi)(y_2, \phi) dP(\phi), y_1, y_2 \in M. \tag{1}$$

The P can be considered as the probability distribution of an H-valued random variable – random field (signal) (the terminology which matches better the case $H = L_2(\mathbf{R}^3)$). We remark that a covariance operator does not determine the random signal uniquely. However, in the Gaussian case each D determines uniquely the Gaussian measure with zero mean value.

We recall that the dispersion of the random variable ϕ (with zero average) is given by

$$\sigma_\phi^2 = E\|\phi(\omega)\|^2 = \sum_{k=1}^n E|\phi_k(\omega)|^2.$$

3. Operator representation of wave function of composite system

In this section we show that the wave function of a composite system has an operator representation which is useful in coupling quantum and classical correlations at the subquantum level, see works[3-14].

Let H be a real Hilbert space. We denote the space of bounded self-adjoint operators acting in H by the symbol $L_s(H)$.

Let H_1 and H_2 be two Hilbert spaces. The tensor product $H = H_1 \otimes H_2$ is isomorphic to the space of Hilbert-Schmidt operators acting from H_2 to H_1. We shall use this isomorphism and represent any $\Psi \in H$ by the corresponding operator $\widehat{\Psi}$.

We remark that $\widehat{\Psi}\widehat{\Psi}^* : H_1 \to H_1$ and $\widehat{\Psi}^*\widehat{\Psi} : H_2 \to H_2$ and these operators are self-adjoint and positively defined. Consider operator $\rho = \Psi \otimes \Psi : H_1 \otimes H_2 \to H_1 \otimes H_2$ and the operators $\rho^{(1)} \equiv \mathrm{Tr}_{H_2}\rho$ and $\rho^{(2)} \equiv \mathrm{Tr}_{H_1}\rho$. If the vector Ψ is normalized by 1, then ρ is the density operator corresponding to the pure state Ψ and the operators $\rho^{(i)} \equiv \rho^{S_i}, i = 1, 2$, are the reduced density operators. These density operators describe quantum states of subsystems $S_i, i = 1, 2$, of a composite quantum system $S = (S_1, S_2)$. For any $\Psi \in H_1 \otimes H_2$, the following equalities hold:

$$\rho^{(1)} = \widehat{\Psi}\widehat{\Psi}^*, \ \rho^{(2)} = \widehat{\Psi}^*\widehat{\Psi}.$$

For any pair of operators $\widehat{A}_j \in L_s(H_j), j = 1, 2$, the following equality holds, see works[3-14],

$$\mathrm{Tr}\widehat{\Psi}\widehat{A}_2\widehat{\Psi}^*\widehat{A}_1 = \langle\widehat{A}_1 \otimes \widehat{A}_2\rangle_\Psi \equiv (\widehat{A}_1 \otimes \widehat{A}_2\Psi, \Psi). \tag{2}$$

It will play a fundamental role in representation of quantum correlations as classical correlations of quadratic forms of the prequantum random field.

Let the state vectors of systems S_1 and S_2 belong to Hilbert spaces H_1 and H_2, respectively. Then by QM the state vector Ψ of the composite system $S = (S_1, S_2)$ belongs to $H = H_1 \otimes H_2$. We remark that the interpretation of the state vector $\Psi \in H$ of a composite system is not as straightforward as for a single system. It is known that, in general, a pure state Ψ of a composite system does not determine pure states for its components. This viewpoint matches well our approach. We shall interpret a normalized vector $\Psi \in H$ not as the state vector of a concrete composite system $S = (S_1, S_2)$, but as one of blocks of the covariance operator for the prequantum random field $\omega \to \phi(\omega) = (\phi_1(\omega), \phi_2(\omega))$ describing $S = (S_1, S_2)$.

3.1. *From classical to quantum correlations*

Let ϕ_1 and ϕ_2 be two random vectors, in Hilbert spaces H_1 and H_2, respectively. Consider Cartesian product of these Hilbert spaces: $\mathbf{H} = H_1 \times H_2$(don't mix with $H = H_1 \otimes H_2$) and the random vector $\omega \to \phi(\omega) = (\phi_1(\omega), \phi_2(\omega)) \in \mathbf{H}$ such that: a) its expectation $E\phi = 0$; b) its dispersion $\sigma^2(\phi) = E||\phi||^2 < \infty$. Take its covariance operator D which is determined by the symmetric (positive) bilinear form: $(Du, v) = E(u, \phi)(v, \phi)$, where vectors $u, v \in \mathbf{H}$. This operator has the block structure $D = \begin{pmatrix} D_{11} & D_{12} \\ D_{21} & D_{22} \end{pmatrix}$, where $D_{ii} : H_i \to H_i, D_{ij} : H_j \to H_i$.

Let $\widehat{A}_i \in L_s(H_i), i = 1, 2$. It determines the quadratic function on the Hilbert space $H_i : f_{A_i}(\phi_i) = (\widehat{A}_i \phi_i, \phi_i)$. Such quadratic functionals are prequantum physical variables corresponding to quantum observables.

For any Gaussian random vector $\phi = (\phi_1, \phi_2)$ having zero average and any pair of operators $\widehat{A}_i \in L_s(H_i), i = 1, 2$, the following equality takes place[3-14]:

$$\langle f_{A_1}, f_{A_2} \rangle_\phi \equiv E f_{A_1}(\phi_1) f_{A_2}(\phi_2)$$

$$= (\mathrm{Tr} D_{11} \widehat{A}_1)(\mathrm{Tr} D_{22} \widehat{A}_2) + 2\mathrm{Tr} D_{12} \widehat{A}_2 D_{21} \widehat{A}_1. \tag{3}$$

We also remark[3-14] that $\mathrm{Tr} D_{ii} \widehat{A}_i = E f_{A_i}(\phi_i), i = 1, 2$. Thus $E f_{A_1} f_{A_2} = E f_{A_1} E f_{A_2} + 2\mathrm{Tr} D_{12} \widehat{A}_2 D_{21} \widehat{A}_1$. Now take an arbitrary pure state of a composite system $S = (S_1, S_2)$, a normalized vector $\Psi \in H$. Consider a Gaussian vector random field such that $D_{12} = \widehat{\Psi}$. By operator equality (2) the last summand in the right-hand side of (3) is equal to the QM-average. Hence, we obtain $\frac{1}{2}E(f_{A_1} - E f_{A_1})(f_{A_2} - E f_{A_2}) = (\widehat{A}_1 \otimes \widehat{A}_2 \Psi, \Psi) \equiv \langle \widehat{A}_1 \otimes \widehat{A}_2 \rangle_\Psi$, or, for the covariance of two classical random vectors f_{A_1}, f_{A_2}, we have:

$$\frac{1}{2}\mathrm{cov}\,(f_{A_1}, f_{A_2}) = \langle \widehat{A}_1 \otimes \widehat{A}_2 \rangle_\Psi, \tag{4}$$

see works[3-14].

Operators D_{ii} are responsible for averages of random variables $\omega \to f(\phi_i(\omega))$, i.e., depending only on one of components of the vector random field ϕ. In particular, $E f_{A_i}(\phi_i) = \mathrm{Tr} D_{ii} \widehat{A}_i$.

We shall construct such a random field that these averages will match those given by QM. It is natural to take the covariance operator $\tilde{D}_\Psi = \begin{pmatrix} \widehat{\Psi}\widehat{\Psi}^* & \widehat{\Psi} \\ \widehat{\Psi}^* & \widehat{\Psi}^*\Psi \end{pmatrix}$. However, in general this operator is not positively defined and, hence, it cannot serve as a covariance operator. In the series of works[3−14] it was proposed to modify aforementioned operator and consider $D_\Psi = \begin{pmatrix} \widehat{\Psi}\widehat{\Psi}^* + \epsilon I & \widehat{\Psi} \\ \widehat{\Psi}^* & \widehat{\Psi}^*\Psi + \epsilon I \end{pmatrix}$, where $\epsilon > 0$ is sufficiently large.[3−14] We remark that white noise is a Gaussian random variable with zero average and the unit covariance operator I. Thus additional terms in diagonal blocks are related to the white noise background. The situation is tricky: in general it is impossible (in the classical mathematical model) to separate this noisy background from a random prequantum field. We cannot consider a random field with the covariance operator D_Ψ as the sum of two signals, e.g., an electron signal and the background signal. For some states (entangled states), the matrix with $\epsilon = 0$ is not positively defined. We discuss this point in more detail:

Suppose now that $\phi(\omega)$ is a random vector with the covariance operator D_Ψ. Then

$$\langle \widehat{A}_1 \rangle_\Psi = E f_{A_1}(\phi_1(\omega)) - \epsilon \mathrm{Tr} \widehat{A}_1, \tag{5}$$

$$\langle \widehat{A}_2 \rangle_\Psi = E f_{A_2}(\phi_2(\omega)) - \epsilon \mathrm{Tr} \widehat{A}_2. \tag{6}$$

These relations for averages and relation (4) for the correlation provide coupling between QM and PCSFT.

4. Gaussian random field as approximation of wave-pulse process

A priori there are no physical reasons to assume that prequatum random fields are Gaussian. In a series of previous papers on PCSFT, we used Gaussian fields by purely mathematical reasons: to simplify mathematical computations. Gaussian integrals are easier for computation.

What can be physical reasons for usage of Gaussian probability distributions for prequantum random fields?

To find an answer to this question, we shall explore the analogy between PCSFT and the classical signal theory (radiophysics).

4.1. *Pulse processes*

An important class of random signals is given by pulse processes

$$\xi(s) = \sum_m a_m F(s - s_m), \tag{7}$$

where s_m is the instant of generation of the mth pulse, a_m is its amplitude and $F(s)$ describes the form of a pulse. As always, it is assumed that $|F(s)| \to 0, |s| \to \infty$, sufficiently quickly. For example, $\xi(s) = \sum_m a_m \delta(s - s_m)$, the sum of δ-pulses.

The random parameters a_m and s_m of the random process $\xi(s)$ satisfy the following natural assumptions:

a). All amplitudes $\{a_m\}$ and instances $\{s_m\}$ of generation of pulses are independent; amplitudes are equally distributed and instances of time are equally distributed as well

$$w(a_1 \ldots, a_m, \ldots; s_1, \ldots, s_m, \ldots)\Pi_m da_m ds_m$$

$$= \Pi_m w_a(a_m) da_m w_s(s_m) ds_m.$$

b). The probability of appearance of a pulse in the interval $[s, s+ds]$ does not depend on s and it is proportional ds, i.e., $w_s(s)ds = n_1 ds, n_1 = \text{const}.$

The latter implies that the probability of appearance of n pulses in the interval T is given by the Poisson distribution: $P(n) = \frac{\bar{n}^n e^{-\bar{n}}}{n!}, \bar{n} = n_1 T$. Set $\bar{a} = \bar{a}_m = \int_{-\infty}^{+\infty} a_m w_a(a_m) da_m$, the amplude mean value (we assumed that amplitudes of all pulses are equally distributed, so the average does not depend on m) and set $\overline{a^2} = \overline{a_m^2} = \int_{-\infty}^{+\infty} a_m^2 w_a(a_m) da_m$. Then the dispersion of $\xi(s)$ is given by $\sigma_\xi^2 \equiv \sigma_{\xi(s)}^2 = n_1 \overline{a^2} \int_{-\infty}^{+\infty} F^2(\theta) d\theta$. It does not depend on s (a feature of stationary of $\xi(s)$). We consider the case of symmetric fluctuations of amplitudes, i.e., $\bar{a} = 0$. In this case $\bar{\xi}(s) = 0$ and $\sigma_{\xi(s)}^2 = E\xi^2(s)$.

It can be proved that the density of the probability distribution $w(\xi)$ of the process $\xi(s)$ can be approximated (under some conditions) by the density of the Gaussian distribution with zero mean value and the dispersion which is equal to the dispersion of $\xi(s)$:

$$w(\xi) \approx \frac{1}{\sqrt{2\pi\sigma_\xi^2}} e^{-\frac{\xi^2}{2\sigma_\xi^2}} + \ldots$$

(As a consequence of stationarity of $\xi(s)$, its probability distribution does not depend on s.) If $\int_{-\infty}^{+\infty} F(\theta) d\theta = 1$ and the pulse duration is of the magnitude $\delta > 0$, then the approximation is good under the condition

$$n_1\delta \gg 1 \tag{8}$$

This means that the number of pulses during the interval δ must be very large. In other words, we can say: at the moment s,the process $\xi(s)$ is composed of pulses which appeared in the interval $[s - \delta, s]$, because earlier pulses died to the moment s. Thus the $n_1\delta$ is the mean value of summands $x_m = a_m F(s - s_m)$ composing $\xi(s)$ at each fixed s. The condition (8 can be interpreted in the following: the distribution $w(\xi)$ of the pulse process $\xi(s)$ is close to the Gaussian distribution if a huge number of pulses is mixed in $\xi(s)$ at each moment s. If the frequency of generation of pulses is low then the distribution of $\xi(s)$ would depend essentially on the form of an individual pulse and the Gaussian approximation would not be useful.

Thus the Gaussian law can be used to describe (of course, approximately) random pulse processes for sources of pulses which produce pulses with very high frequency (comparing with the average duration of pulses).

4.2. Wave-pulses

In this section we proceed on the physical level of rigorousness and consider random fields given by probability distributions on the space $H = L_2(\mathbf{R}^3)$.

Consider now a source of wave-pulses, in (7) the coefficients belong not to the field of complex numbers $\mathbf{C}$, but to the space of square integrable fields $L_2(\mathbf{R}^3)$

$$\phi(s) = \sum_m \phi_m F(s - s_m), \tag{9}$$

where $\phi_m \equiv \phi_m(x, \omega)$ are random fields and $s_m \equiv s_m(\omega)$ are random variables. At each instant of time s_m, a spatial wave $\phi_m(x)$ is produced. In the abstract formalism the coefficients ϕ_m of randomly generated pulses belong to complex Hilbert space H. We can repeat previous considerations for such random fields. We again assume that all random fields ϕ_m have the same probability distribution on H, say μ. The only restriction is that its mean value equals to zero. The covariation operator of ν is denoted by D_ν. Consider now the probability distribution, say w_ϕ of the wave-pulses process $\phi(s)$. It is possible to prove that $\phi(s)$ is a stationary process and its probability distribution can be approximated by the Gaussian probability distribution ν_{D_ϕ}, where D_ϕ is the covariance operator of $\phi(s)$.

$$D_\phi = \left(n_1 \int_{-\infty}^{+\infty} F^2(\theta)d\theta\right) D_\mu \tag{10}$$

For the pulse-function $F(s) = 1/\delta$, if $s \in [0, \delta]$, and 0, if $s \notin [0, \delta]$, we obtain $\int_{-\infty}^{+\infty} F(\theta)d\theta = 1$ and $\int_{-\infty}^{+\infty} F^2(\theta)d\theta = 1/\delta$. Hence,

$$D_\phi = \frac{n-1}{\delta}D_\mu \tag{11}$$

If the condition (8) holds, i.e., if the wave-pulses are produced densely on the duration interval δ, then

$$\sigma^2_{\omega_\phi} = \sigma^2_{\mu_{D_\phi}} = \frac{n_1^2 \mathrm{Tr}D_\mu}{n_1 \delta} << 1 \tag{12}$$

for

$$\mathrm{Tr}D_\mu \sim \frac{1}{n_1^2} \tag{13}$$

Thus if

$$\sigma_\mu \sim \frac{1}{n_1}, \tag{14}$$

then, instead of the wave-pulses process $\phi(s)$, we can consider a Gaussian random field with the covariance operator D_ϕ), where D_ϕ is given by (11). If the prequantum random field is really produced (at the subquantum time scale) as a stationary process of the wave-pulses type, then it is *mathematically* justified to operate everywhere with Gaussian random field. Of course, the same Gaussian random field approximates a huge variety wave-pulses processes.

At the moment PCSFT does not provide an estimate of the subquantum time scale. It is clear that it is sufficiently fine, see paper[8] for discussion. By approaching this time scale we will be able to monitor the wave-pulse structure of the prequantum random field.

Conclusion. *Quantum averages and correlations can be represented through quadratic forms of Gaussian random variables. Gaussianity at the subquantum level can be derived from composition of prequantum random fields through wave-pulse processes.*

This paper was written under support of the QBIC-grant of Tokyo University of Science and the grant Mathematical Modeling of complex hierarchic systems of Linnaeus University. Tha author would like to thank M. Ohya, N. Watanabe, M. Aasano, Y. Tanaka for hospitality and fruitful discussions. The author also would like to thank for critical discussions L. Accardi, I. Basieva, A. Zeilinger, G. Weihs, H. De Raedt, A. Plotnitsky, I. Volovich, S. Polyakov, V. Belavkin and R. Belavkin.

References

1. L. Accardi, *Urne e Camaleoni: Dialogo sulla Realta, le Leggi del Caso e la Teoria Quantistica*, Il Saggiatore, Rome 1997.
2. A. Khrennikov, *Contextual approach to quantum formalism*, Springer, Berlin-Heidelberg-New York, 2009.
3. A. Khrennikov, *J. Phys. A: Math. Gen.* **38** 9051 (2005).
4. A. Khrennikov, *Found. Phys. Letters* **18** 637 (2006).
5. A. Khrennikov, *Physics Letters A* **357** 171 (2006).
6. A. Khrennikov, *Found. Phys. Lett.* **19** 299 (2006).
7. A. Khrennikov, *Nuovo Cimento* B **121** 505 (2006).
8. A. Khrennikov, *Int. J. Theor. Phys.*, **47** 114-124 (2008).
9. A. Khrennikov, *EPL* **88** 40005.1 (2009).
10. A. Khrennikov, *EPL* **90** 40004 (2010).
11. A. Khrennikov, *J. of Russian Laser Research* **31** 191 (2010).
12. A. Khrennikov, *J. Math. Phys.* **48** 013512 (2007).
13. A. Khrennikov, M. Ohya, and N. Watanabe, *J. Russian Laser Research* **31** 462 (2010).
14. A. Khrennikov, M. Ohya, N. Watanabe, *Int. J. Quantum Information* **9** 281 (2011).

A SURVEY ON EXTENSIONS OF HILBERT C^*-MODULES

BISERKA KOLAREC

Department of Informatics and Mathematics,
Faculty of Agriculture, University of Zagreb
Svetošimunska cesta 25, 10000 Zagreb, Croatia
* E-mail: bkudelic@agr.hr

We survey the notion of extensions of Hilbert C^*-modules and discuss some of the basic results. They are obtained mostly as generalizations of results from the theory of extensions of C^*-algebras. Additionally, we establish an equivalence between categories of Hilbert C^*-modules and C^*-algebras and propose an alternative way of lifting results on extensions from C^*-algebra setting to Hilbert C^*-module setting.

1. Preliminaries and introduction

Hilbert C^*-modules are generalizations of Hilbert spaces.

A (right) Hilbert C^* -module V over a C^*-algebra A (a **Hilbert A-module**, for short) is a right A-module V with an "inner product" $(\cdot \mid \cdot) : V \times V \to A$ (which is A-linear in the second and conjugate linear in the first variable) such that V is a Banach space with the norm $\|v\| = \| (v \mid v) \|^{1/2}$ (see Ref. 8,11). V is a **full** Hilbert A-module if the (closed) ideal in A generated by the elements $(v_1 \mid v_2)$, $v_1, v_2 \in V$ is equal to A.

Let us have a Hilbert A-module V and a Hilbert B-module W. A map $\Phi : V \to W$ is a **morphism of modules** or φ-**morphism** if there is a homomorphism $\varphi : A \to B$ such that $(\Phi(v_1) \mid \Phi(v_2)) = \varphi((v_1 \mid v_2))$ for all $v_1, v_2 \in V$.

Example 1. Let I be an ideal of a C^*-algebra A. An ideal submodule V_I of V associated to an ideal I is defined by $V_I = VI = \{vb : v \in V, b \in I\}$. Let us denote by $\Pi : V \to V/V_I$ and $\pi : A \to A/_I$ canonical quotient maps. The quotient V/V_I has a natural Hilbert $A/_I$-module structure with the operation of right multiplication and the inner product given by: $\Pi(v)\pi(a) = \Pi(va)$, $(\Pi(v_1) \mid \Pi(v_2)) = \pi((v_1 \mid v_2))$. Obviously, Π is a π-morphism.

A definition of an extension (see Ref. 2,3) of a Hilbert C^*-module is given inside of the category whose objects are full Hilbert C^*-modules and whose morphisms are morphisms of modules.

We say that a Hilbert C^*-module W is an **extension** of a (full) Hilbert C^*-module V if we have a short exact sequence of Hilbert C^*-modules and morphisms of modules

$$0 \longrightarrow V \xrightarrow{\ \Phi\ } W \xrightarrow{\ \Pi\ } V/_{\Phi(V)} \longrightarrow 0.$$

Recall that a sequence of morphisms is exact if the kernel of the second morphism equals the image of the first morphism. There is also a short exact sequence of C^*-algebras and homomorphisms associated to the above short exact sequence

$$0 \longrightarrow A \xrightarrow{\ \varphi\ } B \xrightarrow{\ \pi\ } B/_{\varphi(A)} \longrightarrow 0.$$

So, a Hilbert B-module W is an extension of a Hilbert A-module V if $\Phi : V \to W$ is a φ-morphism, $\varphi(A)$ is an ideal in B and $\Phi(V)$ is an ideal submodule of W associated to $\varphi(A)$ i.e. $\Phi(V) = W\varphi(A)$. W is an **essential extension** of V if $\varphi(A)$ is an essential ideal in B.

We know from the theory of extensions of C^*-algebras that there is a maximal (essential) extension of a C^*-algebra A. Namely, this is the extension of A by the multiplier algebra $M(A)$ of A. There are: a homomorphism $\gamma : A \to M(A)$ defined by $\gamma(a)b = ab$, $b \in A$ and a unique homomorphism $\lambda : B \to M(A)$ such that $\lambda|_A = id_A$ and $\lambda\varphi = \gamma$.

Likewise, there is the maximal essential extension of a Hilbert A-module V by a **multiplier module** $M(V)$. Precisely, $M(V) = \mathbf{B}(A, V)$; it consists of all adjointable maps from A to V with the inner product $(r \mid s) = r^*s$. $M(V)$ is a Hilbert C^*-module analogue of a multiplier C^*-algebra. Namely, we can define a strict topology on V such that V is strictly dense in $M(V)$ and that $M(V)$ is a strict completion of V, see Ref. 2. Further, a map $\Gamma : V \to M(V)$ defined as $\Gamma(v) = l_v$ (where $l_v : A \to V$ is given by $l_v(a) = va$) is a γ-morphism and $M(V)$ is a Hilbert $M(A)$-module. Because A is an essential ideal in $M(A)$, $M(V)$ is an essential extension of V. If a Hilbert B-module W is another extension of V and $\lambda : B \to M(A)$ is a unique homomorphism of C^*-algebras such that $\lambda|_A = id_A$, then there is a unique λ-morphism $\Lambda : W \to M(V)$ such that $\Lambda\Phi = \Gamma$ (see Theorem 1.1

from Ref. 2). Let us denote by $Q(V)$ the quotient $M(V)/_{\Gamma(V)}$. After the identification of $\Gamma(V)$ with V (via $l_v \leftrightarrow v$), we have $Q(V) = M(V)/_V$. We can form a short exact sequence

$$0 \longrightarrow V \xrightarrow{\ \Gamma\ } M(V) \xrightarrow{\ \Pi_m\ } Q(V) \longrightarrow 0.$$

We know from Ref. 4 that there is a unique homomorphism $\delta : B/_{\varphi(A)} \to M(A)/_A$ called the Busby invariant of an extension B of A that makes the following diagram commutative:

$$
\begin{array}{ccccccccc}
0 & \longrightarrow & A & \xrightarrow{\ \varphi\ } & B & \xrightarrow{\ \pi\ } & B/_{\varphi(A)} & \longrightarrow & 0 \\
& & \| & & \downarrow{\scriptstyle \lambda} & & \downarrow{\scriptstyle \delta} & & \\
0 & \longrightarrow & A & \xrightarrow{\ \gamma\ } & M(A) & \xrightarrow{\ \pi_m\ } & M(A)/_A & \longrightarrow & 0.
\end{array}
$$

On the level of Hilbert modules we define a morphism $\Delta : W/_{\Phi(V)} \to Q(V)$ by $\Delta(\Pi(w)) = \Pi_m(\Lambda(w))$. This definition is good because $\Pi(w) = 0$ implies $w = \Phi(v)$ for some $v \in V$, hence $\Lambda(w) = \Lambda(\Phi(v)) = \Gamma(v)$ and, finally, $\Pi_m(\Lambda(w)) = \Pi_m(\Gamma(v)) = 0$ as required. The definition of the map Δ ensures that the next diagram is commutative:

$$
\begin{array}{ccccccccc}
0 & \longrightarrow & V & \xrightarrow{\ \Phi\ } & W & \xrightarrow{\ \Pi\ } & W/_{\Phi(V)} & \longrightarrow & 0 \\
& & \| & & \downarrow{\scriptstyle \Lambda} & & \downarrow{\scriptstyle \Delta} & & \\
0 & \longrightarrow & V & \xrightarrow{\ \Gamma\ } & M(V) & \xrightarrow{\ \Pi_m\ } & Q(V) & \longrightarrow & 0.
\end{array}
$$

By Theorem 2.6 from Ref. 3 there is a bijective correspondence between the set of equivalence classes of full extensions W of V and the set of all morphisms $\Delta : W/_{\Phi(V)} \to Q(V)$. The morphism Δ is called the **Busby invariant** corresponding to the extension W of V.

Further, for an arbitrary Hilbert C^*-module Z (that we put on the place of the quotient $W/_{\Phi(V)}$ in the first row of the above diagram) and a morphism $\Delta : Z \to Q(V)$ there is an extension W such that Δ is the Busby invariant for W, cf. Ref. 3. To be precise, this extension W equals the pullback $M(V) \oplus_{Q(V)} Z$. The construction of $M(V) \oplus_{Q(V)} Z$ goes the same way in which the pullback of C^*-algebras is constructed, see Ref. 3. We have the following commutative diagram:

$$
\begin{array}{ccc}
W & \xrightarrow{\;\Pi\;} & Z \\
\Big\downarrow{\scriptstyle\Lambda} & & \Big\downarrow{\scriptstyle\Delta} \\
M(V) & \xrightarrow{\;\Pi_m\;} & Q(V).
\end{array}
$$

Like in the case of C^*-algebras, the pullback $W = M(V) \oplus_{Q(V)} Z$ has the following universal property: for an arbitrary Hilbert C^*-module U and a coherent pair of morphisms $\Theta : U \to Z$ and $\Psi : U \to M(V)$ (i. e. such that $\Delta\Theta = \Pi_m\Psi$) there is a unique morphism $\Omega : U \to W$ such that $\Pi\Omega = \Theta$ and $\Lambda\Omega = \Psi$.

2. Main results

Definition 1. Let us have Hilbert C^*-modules V and V_1 and let W and W_1 be extensions of V and V_1, respectively. A morphism $\mathcal{E} : W \to W_1$ is called a **morphism between extensions** if we have a commutative diagram of Hilbert C^*-modules and morphisms of modules

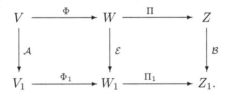

Obviously, a morphism $\Lambda : W \to M(V)$ from above is one example of a morphism between extensions.

In order to establish necessary and sufficient conditions for the existence of a morphism between extensions, we extend the diagram from the definition by adding maximal extensions of V and V_1. We know by Ref. 1, Proposition 1 that a surjective morphism $\Phi : V_1 \to V_2$ can be extended to a unique morphism $\overline{\Phi} : M(V_1) \to M(V_2)$ that extends Φ in the sense that $\overline{\Phi}(l_v) = l_{\Phi(v)}$ for all $v \in V_1$. We are able to extend Φ even further, to a morphism $\tilde{\Phi} : Q(V_1) \to Q(V_2)$; it is defined by $\tilde{\Phi}(l + V_1) = \overline{\Phi}(l) + V_2$, see Ref. 6.

Next theorem gives the criterion for the existence of a morphism between extensions.

Theorem 1 (Theorem 2.4 from Ref. 6). *Let V and V_1 be full Hilbert C^*-modules. Let Δ and Δ_1 be the Busby invariants of extensions W and W_1 of V and V_1, respectively. Let us have a surjective morphism $\mathcal{A} : V \to V_1$ and a morphism $\mathcal{B} : Z \to Z_1$. A unique morphism $\mathcal{E} : W \to W_1$ between extensions exists if and only if $\tilde{\mathcal{A}}\Delta = \Delta_1 \mathcal{B}$.*

Proof. We add maximal extensions to given ones and draw the following 3D diagram with exact rows:

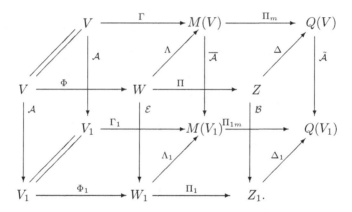

Suppose that $\tilde{\mathcal{A}}\Delta = \Delta_1 \mathcal{B}$. We can extract the following commutative diagram from the above diagram:

$$
\begin{array}{ccc}
W & \xrightarrow{\mathcal{B}\Pi} & Z_1 \\
\downarrow{\overline{\mathcal{A}}\Lambda} & & \downarrow{\Delta_1} \\
M(V_1) & \xrightarrow{\Pi_{1m}} & Q(V_1).
\end{array}
$$

Recall that W_1 is the pullback

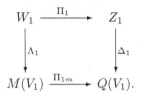

Notice that we have a coherent pair of morphisms $\mathcal{B}\Pi$: $W \to Z_1$ and $\overline{\mathcal{A}}\Lambda$: $W \to M(V_1)$. So, by the universal property for the pullback W_1, there is a unique morphism $\mathcal{E}$: $W \to W_1$ such that $\Pi_1\mathcal{E} = \mathcal{B}\Pi$ and $\Lambda_1\mathcal{E} = \overline{\mathcal{A}}\Lambda$. It remains to see that $\mathcal{E}\Phi = \Phi_1\mathcal{A}$. This follows from $\Lambda_1\mathcal{E}\Phi = \overline{\mathcal{A}}\Lambda\Phi = \overline{\mathcal{A}}\Gamma = \Gamma_1\mathcal{A} = \Lambda_1\Phi_1\mathcal{A}$ by canceling out Λ_1 (we can do that because Λ_1 is injective, see Ref. 3).

Conversely, let us have a morphism $\mathcal{E}$: $W \to W_1$ such that $\mathcal{E}\Phi = \Phi_1\mathcal{A}$ and $\Pi_1\mathcal{E} = \mathcal{B}\Pi$. We form the 3D diagram as before. The front face of the 3D diagram is commutative: indeed, $\mathcal{B}\Pi\Phi = \Pi_1\mathcal{E}\Phi = \Pi_1\Phi_1\mathcal{A}$. We also have $\Lambda_1\mathcal{E} = \overline{\mathcal{A}}\Lambda$, namely $\Lambda_1\mathcal{E}\Phi = \Lambda_1\Phi_1\mathcal{A} = \Gamma_1\mathcal{A} = \overline{\mathcal{A}}\Gamma = \overline{\mathcal{A}}\Lambda\Phi$. Further,

$$\Delta_1\mathcal{B}\Pi = \Delta_1\Pi_1\mathcal{E} = \Pi_{1m}\Lambda_1\mathcal{E} = \Pi_{1m}\overline{\mathcal{A}}\Lambda = \tilde{\mathcal{A}}\Pi_m\Lambda = \tilde{\mathcal{A}}\Delta\Pi$$

and the equality $\tilde{\mathcal{A}}\Delta = \Delta_1\mathcal{B}$ follows.   □

We consider now the question of describing morphisms out of an extension of a Hilbert C^*-module.

Definition 2. Let a Hilbert B-module W be an extension of a Hilbert A-module V and U be an arbitrary Hilbert C^*-module. A morphism Ω : $W \to U$ is called a **morphism out of extension**.

In order to study morphisms out of extensions in a general case, one has to define an idealizer of a Hilbert C^*-module, cf. Ref. 6. For C^*-algebras A and B and a homomorphism $\varphi : A \to B$, the idealizer of $\varphi(A)$ in B is defined as $I(\varphi(A); B) = \{b \in B : b\varphi(A) \subseteq \varphi(A), \varphi(A)b \subseteq \varphi(A)\}$.

Definition 3. Let V and W be Hilbert C^*-modules over C^*-algebras A and B, respectively. Let Φ : $V \to W$ be a φ-morphism for a morphism $\varphi : A \to B$ of C^*-algebras. An **idealizer of $\Phi(V)$ in W** is defined as

$I(\Phi(V); W)$
$= \{w \in W : (w \mid \Phi(v)) \in I(\varphi(A)), w\varphi(a) \in \Phi(V), a \in A, v \in V\}.$

It is easy to show that $\Phi(V)$ is an ideal submodule of $I(\Phi(V); W)$ associated to an ideal $\varphi(A) \subseteq I(\varphi(A); B)$ i. e. $\Phi(V) = I(\Phi(V); W)\varphi(A)$. If we have a submodule $W' \subseteq W$ such that $\Phi(V)$ is an ideal submodule of W', then $W' \subseteq I(\Phi(V))$. We define an **eigenmodule** $E(\Phi(V); W)$ as a quotient $I(\Phi(V); W)/_{\Phi(V)}$; it is a Hilbert C^*-module over an eigenalgebra $E(\varphi(A); B) = I(\varphi(A); B)/_{\varphi(A)}$. We shall write $I(\Phi(V))$, $E(\Phi(V))$ to shorten the notation $I(\Phi(V); W)$ and $E(\Phi(V); W)$, respectively. Notice that we have an exact sequence of Hilbert C^*-modules

$$0 \longrightarrow \Phi(V) \longrightarrow I(\Phi(V)) \longrightarrow E(\Phi(V)) \longrightarrow 0.$$

We denote the Busby invariant for extension of $\Phi(V)$ by its idealizer by Δ_Φ. We have:

Theorem 2 (Ref. 6, Theorem 2.7). *Let Δ be the Busby invariant of an extension W of a full Hilbert A-module V. Given a Hilbert C^*-module U over a C^*-algebra D, there is a bijective correspondence between a morphism $\Omega : W \to U$ and a pair of morphisms (Θ, Ψ), where $\Theta : V \to U$ and $\Psi : Z \to E(\Theta(V); U)$ are such that $\Delta_\Theta \Psi = \tilde{\Theta} \Delta$.*

Proof. Suppose that a morphism $\Omega : W \to U$ is given. We define $\Theta : V \to U$ by $\Theta = \Omega\Phi$. We know from the definition of an extension that $\Phi(V)$ is an ideal submodule in W associated to an ideal $\varphi(A) \subseteq B$, i. e. $\Phi(V) = W\varphi(A)$. Then $\Theta(V) = \Omega(\Phi(V)) = \Omega(W\varphi(A)) = \Omega(W)\omega(\varphi(A)) = \Omega(W)\theta(A)$. This equality shows that $\Theta(V)$ is an ideal submodule of $\Omega(W)$ associated to an ideal $\theta(A) \subseteq \omega(B)$. Hence, we conclude that Ω actually maps to $I(\Theta(V); U)$, i. e. $\Omega(W) \subseteq I(\Theta(V); U)$. We have the following 3D diagram with exact rows:

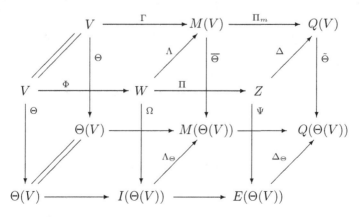

As a morphism $\Omega : W \to I(\Phi(V))$ exists, Theorem 1 ensures $\Delta_\Theta \Psi = \tilde{\Theta}\Delta$.
Conversely, if there is a pair (Θ, Ψ) of morphisms such that $\Delta_\Theta \Psi = \tilde{\Theta}\Delta$,
then by Theorem 1 there is a unique morphism $\Omega : W \to I(\Theta(V); U)$
associated to the given pair. $\square$

In case of a split extension of Hilbert C^*-modules we are able to obtain
even nicer result on the existence of morphisms out of extensions.

Definition 4. A Hilbert B-module W is a **split extension** of a Hilbert
A-module V if there is a morphism $\Sigma : Z \to W$ such that $\Pi\Sigma = id$, i. e. we
have the exact sequence of Hilbert C^*-modules and morphisms of modules:

$$0 \longrightarrow V \xrightarrow{\ \Phi\ } W \underset{\Sigma}{\overset{\Pi}{\rightleftarrows}} Z \longrightarrow 0.$$

As we work with full Hilbert C^*-modules, it is easy to see that the underlying exact sequence of C^*-algebras is split, so B equals the direct sum of
$\varphi(A)$ and $\sigma(C)$. This sum is orthogonal if $\sigma(C)$ is an ideal of B.

In case W is a split extension of V, it really splits into the direct sum of
$\Phi(V)$ and $\Sigma(Z)$. In general, this sum is not orthogonal, but it is if $\Sigma(Z)$ is an
ideal submodule of W associated to an ideal $\sigma(C) \subseteq B$. An easy observation
shows this (recall that $\Phi(V)$ is an ideal submodule of W associated to an
ideal $\varphi(A) \subseteq B$):

$$(\Phi(V) \mid \Sigma(Z)) = (W\varphi(A) \mid W\sigma(C)) = \varphi(A)(W \mid W)\sigma(C)$$
$$= \varphi(A)B\sigma(C) \subseteq \varphi(A)\sigma(C) = 0.$$

Famous T. A. Loring's theorem on morphisms out of a split extension
of a C^*-algebra claims the following.

Theorem 3. *(Ref. 9, 7.3.8) Let us have a split extension of a C^*-algebra
A and an arbitrary C^*-algebra D:*

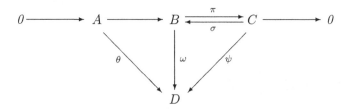

A homomorphism $\omega : B \to D$ is in a bijective correspondence with a pair

of homomorphisms (θ, ψ) *($\theta : A \to D$, $\psi : C \to D$) such that for all $a \in A$, $c \in C$ we have*

$$(C^*) \quad \psi(c)\theta(a) = \theta(\sigma(c)a).$$

In Ref. 7 we generalize Loring's theorem to full Hilbert C^*-modules and morphisms of modules. There we have a diagram:

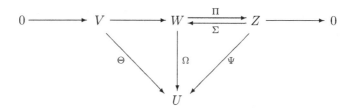

We want to establish a bijection between a morphism $\Omega : W \to U$ (ω-morphism for a homomorphism of C^*-algebras $\omega : B \to D$) and a pair of morphisms (Θ, Ψ) where $\Theta : V \to U$ is a θ-morphism for a homomorphism of C^*-algebras $\theta : A \to D$ and $\Psi : Z \to U$ is a ψ-morphism for a homomorphism $\psi : C \to D$. Loring's condition (C^*) naturally generalizes to:

$$(1) \quad (\Psi(z) \mid \Theta(v)) = \theta((\Sigma(z) \mid v)), \quad v \in V, z \in Z.$$

It turnes out that condition (1) alone is not sufficient to give a statement that generalizes Loring's theorem. Problem lies in the fact that condition (1) does not imply condition (C^*) which is necessary to establish a bijection between a homomorphism ω and a pair of homomorphisms (θ, ψ) on the C^*-algebra level. Namely, we can rewrite (1) as $(\Theta(v) \mid \Psi(z)) = \theta((v \mid \Sigma(z)))$ and get

$$\theta((v \mid \Sigma(z)))\psi(c) = (\Theta(v) \mid \Psi(z))\,\psi(c) = (\Theta(v) \mid \Psi(zc))$$
$$= \theta((v \mid \Sigma(zc))) = \theta((v \mid \Sigma(z))\,\sigma(c)).$$

However, elements of the form $(v \mid \Sigma(z))$, $v \in V$, $z \in Z$ do not generate a C^*-algebra A in general. This is easy to note in the extreme case when $\Sigma(Z)$ is an ideal submodule of W associated to an ideal $\sigma(C)$ in B. Then, the sum of V and $\Sigma(Z)$ is orthogonal, i. e. $\{(v \mid \Sigma(z)) : v \in V, z \in Z\} = \{0\}$.

We conclude that, besides condition (1), we need some condition that will ensure (C^*) on underlying C^*-algebras. Required condition is

$$(2) \quad \Psi(z)\theta(a) = \Theta(\Sigma(z)a), \quad z \in Z, a \in A.$$

Notice that this definition is good because $\Sigma(z)a \in V$ for all $z \in Z, a \in A$ thanks to the fact that V is an ideal submodule in W associated to an ideal $A \subseteq B$. We must only check that conditions (1) and (2) are not equivalent to (C^*). We illustrate that fact in the following example.

Example 2. Let us take a diagram

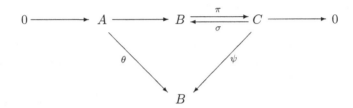

of C^*-algebras with the identity map $\theta : A \to B$. Suppose that (C^*) (it transforms to $\psi(c)a = \sigma(c)a$) holds. Now consider all C^*-algebras as Hilbert C^*-modules over themselves and take following morphisms of modules: $\Theta : A \to B$ given by $\Theta(a) = va$ for $v \in B$ such that $v^*v = 1$ and $\Psi = \psi = \sigma$, $\Sigma = \sigma$. It is easy to see that in this setting condition (2) does not hold. Similarly, we can see that the conditions (C^*) and (1) are not equivalent as well.

Now assume that a Hilbert C^*-module Z is full. Conditions (1) and (2) ensure (C^*): for all $z \in Z, a \in A$ we have

$$\psi((z \mid z))\theta(a) = (\Psi(z) \mid \Psi(z))\,\theta(a) = (\Psi(z) \mid \Psi(z)\theta(a)) =$$
$$\overset{(2)}{=} (\Psi(z) \mid \Theta(\Sigma(z)a)) \overset{(1)}{=} \theta((\Sigma(z) \mid \Sigma(z)a) =$$
$$= \theta((\Sigma(z) \mid \Sigma(z))\,a) = \theta(\sigma((z \mid z))a).$$

Theorem 4 (Ref. 7, Theorem 2.4). *Let V be a full Hilbert A-module, let a Hilbert B-module W be a full split extension of V, let morphism $\Phi :$ $V \to W$ be the inclusion map and let U be a Hilbert D-module. A morphism $\Omega : W \to U$ is in a bijective correspondence with a pair of morphisms (Θ, Ψ) where $\Theta : V \to U$ (a θ-morphism for a homomorphism of C^*-algebras $\theta : A \to D$) and $\Psi : Z \to U$ (a ψ-morphism for a homomorphism $\psi : C \to D$) are such that for all $v \in V, z \in Z, a \in A$ we have*

$$(1)\ (\Psi(z) \mid \Theta(v)) = \theta((\Sigma(z) \mid v))$$
$$(2)\ \Psi(z)\theta(a) = \Theta(\Sigma(z)a).$$

Proof. We have decomposition of W as the direct sum of V and $\Sigma(Z)$.

Suppose we have a pair of morphisms (Θ, Ψ) with given properties. As all modules are full, an extension of a C^*-algebra A is split. From the discussion preceeding this theorem we know that conditions (1) and (2) ensure condition (C^*) to hold for homomorphisms of C^*-algebras. By Loring's theorem we can associate a homomorphism $\omega : B \to D$ of C^*-algebras to a pair of homomorphisms (θ, ψ). It is given by $\omega(a + \sigma(c)) = \theta(a) + \psi(c)$. Quite similarly, we define a map $\Omega : W \to U$ by $\Omega(w) = \Omega(v + \Sigma(z)) = \Theta(v) + \Psi(z)$. It is easy to see that Ω is an ω-morphism.

Let us, on the other hand, have a morphism $\Omega : W \to U$, an ω-morphism for a homomorphism $\omega : B \to D$. By Loring's theorem, we know that ω is in a bijective correspondence with a pair (θ, ψ) of homomorphisms of C^*-algebras, where $\theta = \omega|_A$, $\psi = \omega\sigma$. We define morphisms of modules similarly: $\Theta = \Omega|_V$, $\Psi = \Omega\Sigma$. Obviously Θ is a θ-morphism, Ψ is a ψ-morphism. This pair (Θ, Ψ) satisfies conditions (1) and (2): namely, for all $v \in V, z \in Z, a \in A$

$$(\Psi(z) \mid \Theta(v)) = (\Omega(\Sigma(z)) \mid \Omega(v)) = \omega((\Sigma(z) \mid v)) = \theta((\Sigma(z) \mid v)),$$

$$\Psi(z)\theta(a) = \Omega(\Sigma(z))\omega(a) = \Omega(\Sigma(z)a) = \Theta(\Sigma(z)a). \qquad \square$$

3. Equivalence of categories of Hilbert C^*-modules and C^*-algebras

The results of the theory of extensions of Hilbert C^*-modules were obtained mostly as generalizations of facts from the theory of extensions of C^*-algebras, see Ref. 4,5. In this section we establish a natural equivalence between categories of Hilbert C^*-modules and C^*-algebras. It gives an alternative way to lift results on extensions from C^*-algebras to Hilbert C^*-modules.

By $\mathcal{M}$ we denote a category whose objects are pairs (V, A), where A is a C^*-algebra and V is a full Hilbert A-module. Morphisms from an object (V, A) to an object (W, B) are pairs (Φ, φ), where $\Phi : V \to W$ is a φ-morphism of modules for a homomorphism $\varphi : A \to B$.

By $\mathcal{A}$ we denote a category whose objects are pairs (C, p), where C is a C^*-algebra and p is a projection in the multiplier algebra $M(C)$ of C. The morphism between two objects (C, p) and (D, q) is a map $\psi : (C, p) \to (D, q)$ such that $\hat{\psi}(p) = q$, where $\hat{\psi} : M(C) \to M(D)$ is an extension of ψ to multiplier algebras.

How can we construct functors between the two categories?

A full (right) Hilbert A-module V has also a structure of a full left $\mathbf{K}(V)$-module. (Here $\mathbf{K}(V)$ denotes a C^*-algebra of all 'compact' operators on a Hilbert C^*-module V. Recall that $\mathbf{K}(V) = [\{F_{x,y} : x, y \in V\}]^-$, with $F_{x,y}(z) = x\,(y \mid z)_A$.) Namely, besides an inner product $(\cdot \mid \cdot)_A$ that takes values in A, one can naturally define an inner product $_{\mathbf{K}(V)}(x \mid y) = F_{x,y}$ with values in $\mathbf{K}(V)$. We additionally have

$$_{\mathbf{K}(V)}(x \mid y)\,z = F_{x,y}(z) = x\,(y \mid z)_A\,.$$

This property characterizes V as a $\mathbf{K}(V)$-A-imprimitivity bimodule (or a Morita equivalence bimodule). For details about imprimitivity bimodules or Morita equivalence we refer to Ref. 10. Notice that we are free to identify $\mathbf{K}(A)$ and A. Namely, every $a \in A$ induces the map $T_a \in \mathbf{K}(A)$ given by $T_a b = ab$. A map $a \mapsto T_a$ is an isomorphism of C^*-algebras $\mathbf{K}(A)$ and A. The fact that C^*-algebras $\mathbf{K}(V)$ and $\mathbf{K}(A)$ are Morita equivalent is equivalent to the existence of a C^*-algebra with complementary full corners isomorphic to $\mathbf{K}(V)$ and $\mathbf{K}(A)$, respectively (cf. Theorem 3.19 from Ref. 10. A C^*-algebra that implements the Morita equivalence of $\mathbf{K}(V)$ and $\mathbf{K}(A)$ is a linking algebra. Given a Hilbert A-module V, a **linking algebra** $\mathcal{L}(V)$ of V is defined as a matrix algebra of the form:

$$\mathcal{L}(V) = \begin{bmatrix} \mathbf{K}(V) & \mathbf{K}(A, V) \\ \mathbf{K}(V, A) & \mathbf{K}(A) \end{bmatrix}.$$

By Corollary 3.21 from Ref. 10, the linking algebra $\mathcal{L}(V)$ of a Hilbert A-module V is isomorphic to $\mathbf{K}(V \oplus A)$, the C^*-algebra of all 'compact' operators on a Hilbert C^*-module $V \oplus A$. Here, $V \oplus A = \{(v, a) : v \in V, a \in A\}$ is a Hilbert A-module with an inner product and a right A-action given by:

$$((v_1, a_1) \mid (v_2, a_2)) = (v_1 \mid v_2) + a_1^* a_2, \quad (v, a)a_1 = (va_1, aa_1).$$

Let

$$p = \begin{bmatrix} id_V & 0 \\ 0 & 0 \end{bmatrix}, \quad p^\perp = \begin{bmatrix} 0 & 0 \\ 0 & id_A \end{bmatrix},$$

p and $p^\perp$ are selfadjoint projections on $V \oplus A$. A C^*-algebra $V \oplus A$ is invariant for left and right multiplication by p and $p^\perp$ and therefore $p, p^\perp \in M(V \oplus A)$. Obviously, $p + p^\perp = id_{M(V \oplus A)}$. Furthermore, $p\mathcal{L}(V)p = \mathbf{K}(V)$ and $p^\perp \mathcal{L}(V)p^\perp = \mathbf{K}(A) = A$.

We construct functors between categories $\mathcal{M}$ and $\mathcal{A}$ as follows. To an object $(V, A) \in \mathcal{M}$ we associate a C^*-algebra $C = \mathbf{K}(V \oplus A)$ and take $p \in M(C)$ to be a projection $p : V \oplus A \to V$. A morphism $(\Phi, \varphi) : (V, A) \to$

(W, B) induces a homomorphism $\Theta : \mathbf{K}(V \oplus A) \to \mathbf{K}(W \oplus B)$ given by

$$\Theta(F_{(v_1, a_1), (v_2, a_2)}) = F_{(\Phi(v_1), \varphi(a_1)), (\Phi(v_2), \varphi(a_2))}.$$

In the other direction, for a pair $(C, p) \in \mathcal{A}$, we define $V = pCp$, $A = p^\perp C p^\perp$. A functor Θ' adjoins to a morphism $\psi : (C, p) \to (D, q)$ a pair $(\Psi, \varphi) : (V, A) \to (W, B)$, where $W = qDq$, $B = q^\perp D q^\perp$. Functors Θ and Θ' establish an equivalence between categories $\mathcal{M}$ and $\mathcal{A}$. Therefore, most results of the theory of extensions of Hilbert C^*-module can be pulled from the theory of extensions of C^*-algebras.

References

1. D. Bakić: Tietze extension theorem for Hilbert C^*-modules, *Proc. Amer. Math. Soc. 133(2) (2005), 441-448*
2. D. Bakić, B. Guljaš: Extensions of Hilbert C^*-modules, *Houston J. of Math. 30, no. 2 (2004), 537-558*
3. D. Bakić, B. Guljaš: Extensions of Hilbert C^*-modules II, *Glasnik Matematički, vol. 38(58)(2003), 341-357*
4. R. C. Busby: Double cenralizers and extensions of C^*-algebras, *Trans. Amer. Math. Soc. 132 (1968), 79-99*
5. S. Eilers, T. A. Loring, G. K. Pedersen: Morphisms of Extensions of C^*-algebras: Pushing Forward the Busby Invariant, *Adv. Math 147 (1999), 74-109*
6. B. Kolarec: Morphisms of extensions of Hilbert C^*-modules, *Glasnik Matematički, vol. 42(62)(2007), 401-409*
7. B. Kolarec: Morphisms out of a split extension of a Hilbert C^*-module, *Glasnik Matematički, vol. 41(61)(2006), 309-315*
8. E. C. Lance: *Hilbert C^*-modules - a toolkit for operator algebraist*, Lecture Note Series 210, Cambridge University Press, 1995.
9. T. A. Loring: *Lifting Solutions to Perturbing Problems in C^*-algebras*, Fields Institute Monographs, 1997.
10. I. Raeburn, D. P. Williams: *Morita Equivalence and Continuous-Trace C^*-algebras*, Math. Surveys and Monographs, vol. 60, 1998.
11. N. E. Wegge-Olsen: *K-theory and C^*-algebras: a friendly approach*, Oxford University Press, 1993.

AN ISOMETRY FORMULA FOR A NEW STOCHASTIC INTEGRAL

HUI-HSIUNG KUO

Department of Mathematics, Louisiana State University
Baton Rouge, LA 70803, USA
kuo@math.lsu.edu
http://www.math.lsu.edu/~kuo

ANUWAT SAE-TANG

Department of Mathematics, Faculty of Science
King Mongkut's University of Technology Thonburi
Bangkok, Thailand
anuwat.sae@kmutt.ac.th

BENEDYKT SZOZDA

Department of Mathematics, Louisiana State University
Baton Rouge, LA 70803, USA
benny@math.lsu.edu
http://www.math.lsu.edu/~benny

The isometry formula for the Itô integral is generalized to a new stochastic integral involving both adapted and instantly independent stochastic processes. The proof is rather elementary. An example is given to ilustrate the isometry formula.

Keywords: Brownian motion, Itô integral, isometry formula, adapted stochastic processes, instantly independent stochastic processes, conditional expectation, power series expansion, binomial expansion.

1. The Itô integral

Let B_t, $t \geq 0$, be a Brownian motion and $\{\mathcal{F}_t; t \geq 0\}$ a filtration such that (i) B_t is adapted to the filtration $\{\mathcal{F}_t; t \geq 0\}$, namely, B_t is measurable with respect to $\mathcal{F}_t$ for each t and (ii) $B_s - B_t$ and $\mathcal{F}_t$ are independent for any $t \leq s$.

Suppose $f(t), a \leq t \leq b$, is an $\int_a^b |f(t)|^2 \, dt < \infty$ almost surely. Then the well-known Itô integral

$$\int_a^b f(t) \, dB_t$$

is defined (see, e.g., chapters 4 and 5 of the book[4]). For such a stochastic process $f(t)$ with continuous sample paths, we have the equality

$$\int_a^b f(t) \, dB_t = \lim_{\|\Delta\| \to 0} \sum_{i=1}^n f(t_{i-1}) \left(B_{t_i} - B_{t_{i-1}} \right) \quad \text{in probability}, \qquad (1)$$

where $\Delta = \{a = t_0, t_1, t_2, \ldots, t_n = b\}$ is a partition of $[a, b]$ (see, e.g., Theorem 5.3.3 in the book[4]).

In particular, suppose $f(t)$ is an adapted stochastic process such that $E \int_a^b |f(t)|^2 \, dt < \infty$, then the convergence in Eq. (1) holds in the space $L^2(\Omega)$ and we have the isometry formula

$$\mathbb{E}\left[\left(\int_a^b f(t) \, dB_t \right)^2 \right] = \int_a^b \mathbb{E}\left[f(t)^2 \right] dt \qquad (2)$$

The purpose of this paper is to generalize this isometry formula to a new stochastic integral introduced by Ayed–Kuo.[1,2] We mention that the white noise generalization of this isometry formula is given by Theorem 13.16 in the book.[3] However, the white noise methods require too much background and strong conditions.

In our case for the new stochastic integral, not only the formulation of the isometry formula is very simple and natural, but also the proof is rather elementary by using the binomial expansion and applying conditional expectation with respect to four σ-fields at one time. More importantly, the new stochastic integral, contrary to white noise methods, has a direct probabilistic interpretation as a random variable.

2. A new stochastic integral

Let a Brownian motion B_t, $t \geq 0$, and a filtration $\{\mathcal{F}_t; t \geq 0\}$ be specified as in section 1. A stochastic process $\varphi(t)$ is called *instantly independent* with respect to the filtration $\{\mathcal{F}_t; t \geq 0\}$ if $\varphi(t)$ and $\mathcal{F}_t$ are independent for each t. For example, if φ is a measurable function on $\mathbb{R}$, the the stochastic process $\varphi(B_1 - B_t), 0 \leq t \leq 1$ is instantly independent. However, note that the stochastic process $\varphi(B_1 - B_t), t \geq 1$ is adapted.

The family of instantly independent stochastic processes can be regarded as a counterpart of the adapted stochastic processes in the Itô theory of stochastic integration. It is easy to check that if $\varphi(t)$ is both adapted and instantly independent, then $\varphi(t)$ must be a deterministic function.

Suppose $f(t), a \leq t \leq b$, is a continuous adapted stochastic process and $\varphi(t), a \leq t \leq b$, is a continuous instantly independent stochastic process. Ayed–Kuo[1,2] have defined a new stochastic integral of the stochastic process $f(t)\varphi(t)$ by

$$\int_a^b f(t)\varphi(t)\, dB_t = \lim_{\|\Delta\|\to 0} \sum_{i=1}^n f(t_{i-1})\varphi(t_i)\big(B_{t_i} - B_{t_{i-1}}\big), \qquad (3)$$

provided that the limit exists in probability. Observe that the evaluation points for the adapted factor are the left-endpoints of subintervals as in the Itô theory, while the evaluation points for the instantly independent factor are the right-endpoints of subintervals. Note that when $\varphi(t) = 1$, then this new integral reduces to the Itô integral of $f(t)$ in view of Eq. (1). Moreover, it can be checked that this new integral is well defined. For more information and properties on this new stochastic integral, see also the paper by Kuo–Sae-Tang–Szozda.[5] In particular, the analogue of martingale property for the Itô theory is the near-martingale property for the counterpart as shown by Kuo–Sae-Tang–Szozda.[5]

3. An isometry formula

We first prepare a lemma for the proof of the isometry formula.

Lemma 3.1. *Suppose f and φ are continuous functions on $\mathbb{R}$ such that $\int_a^b \mathbb{E}\left[f(B_t)^2 \varphi(B_b - B_t)^2 \right] dt < \infty$. Then*

$$\lim_{\|\Delta\|\to 0} \sum_{i=1}^n \mathbb{E}\left[f(B_{t_{i-1}})^2 \varphi(B_b - B_{t_i})^2 (B_{t_i} - B_{t_{i-1}})^2 \right]$$

$$= \int_a^b \mathbb{E}\left[f(B_t)^2 \varphi(B_b - B_t)^2 \right] dt \qquad (4)$$

where $\Delta = \{a = t_0, t_1, \ldots, t_n = b\}$ is a partition of $[a,b]$.

Proof. For simplicity, we use the notation:

$$f_i = f(B_{t_i}), \quad \varphi_i = \varphi(B_b - B_{t_i}), \quad \Delta B_i = B_{t_i} - B_{t_{i-1}}. \qquad (5)$$

Using the properties of conditional expectation and the fact that f_{i-1} and ΔB_i are $\mathcal{F}_{t_i}$-measurable, we have

$$\mathbb{E}\left[f_{i-1}^2 \varphi_i^2 (\Delta B_i)^2 \right] = \mathbb{E}\left[\mathbb{E}\left[f_{i-1}^2 \varphi_i^2 (\Delta B_i)^2 \big| \mathcal{F}_{t_i} \right] \right]$$
$$= \mathbb{E}\left[f_{i-1}^2 (\Delta B_i)^2 \, \mathbb{E}\left[\varphi_i^2 \big| \mathcal{F}_{t_i} \right] \right]. \qquad (6)$$

But φ_i is independent of $\mathcal{F}_{t_i}$, f_{i-1} and ΔB_i are independent, and φ_i and f_{i-1} are independent. Therefore,

$$
\begin{aligned}
\mathbb{E}\left[f_{i-1}^2(\Delta B_i)^2 \mathbb{E}\left[\varphi_i^2 \mid \mathcal{F}_{t_i}\right]\right] &= \mathbb{E}\left[f_{i-1}^2(\Delta B_i)^2 \mathbb{E}\left[\varphi_i^2\right]\right] \\
&= \mathbb{E}\left[\varphi_i^2\right] \mathbb{E}\left[f_{i-1}^2(\Delta B_i)^2\right] \\
&= \mathbb{E}\left[\varphi_i^2\right] \mathbb{E}\left[f_{i-1}^2\right] \Delta t_i \\
&= \mathbb{E}\left[\varphi_i^2 f_{i-1}^2\right] \Delta t_i.
\end{aligned} \tag{7}
$$

It follows from Eqs. (6) and (7) that

$$
\mathbb{E}\left[f_{i-1}^2 \varphi_i^2 (\Delta B_i)^2\right] = \mathbb{E}\left[\varphi_i^2 f_{i-1}^2\right] \Delta t_i.
$$

Upon summing up this equality over $i = 1, 2, \ldots, n$ and then taking the limit as $\|\Delta\| \to 0$, we immediately obtain Eq. (4). $\qquad \square$

Now, we can state the main result in this paper.

Theorem 3.1. *Let f and φ be C^1-functions on $\mathbb{R}$. Then*

$$
\begin{aligned}
\mathbb{E}\left[\left(\int_a^b f(B_t)\varphi(B_b - B_t)\, dB_t\right)^2\right] &= \int_a^b \mathbb{E}\left[f(B_t)^2 \varphi(B_b - B_t)^2\right] dt \\
&+ 2\int_a^b \int_a^t \mathbb{E}\left[f(B_s)\varphi'(B_b - B_s)f'(B_t)\varphi(B_b - B_t)\right] ds\, dt,
\end{aligned} \tag{8}
$$

provided that the integrals in the right-hand side exist.

Remark 3.1. Observe that if $\varphi(x) = 1$, then Eq. (8) reduces to Eq. (2) for the Itô integral. On the other hand, if $f(x) = 1$, then the double integral in Eq. (8) also disappears. Thus the isometry formula for purely instantly independent stochastic processes $\varphi(B_b - B_t)$ has the same form as the adapted stochastic processes as shown in Kuo–Sae-Tang–Szozda.[5]

4. Proof of the isometry formula

We will prove the isometry formula in Eq. (8) under stronger assumption that the functions f and φ can be represented by their Maclaurin series on the whole real line $\mathbb{R}$ with bounded derivatives f' and φ'. The general case can be proved by approximation.

For simplicity we will break the proof into several steps. The notation in Eq. (5) will be adopted throughout the proof.

Step 1. Use the definition of the new integral in Eq. (3) to write out the left-hand side of Eq. (8),

$$
\mathbb{E}\left[\left(\int_a^b f(B_t)\varphi(B_b - B_t)dB_t\right)^2\right]
$$

$$
= \lim_{\|\Delta\|\to 0}\left[\sum_{i=0}^n \mathbb{E}\left[f_{i-1}^2\varphi_i^2(\Delta B_i)^2\right] + 2\sum_{i=2}^n\sum_{j=1}^{i-1}\mathbb{E}\left[f_{i-1}\varphi_i f_{j-1}\varphi_j\Delta B_i\Delta B_j\right]\right].
$$
(9)

By Lemma 3.1, we have

$$
\lim_{\|\Delta\|\to 0}\sum_{i=0}^n \mathbb{E}\left[f_{i-1}^2\varphi_i^2(\Delta B_i)^2\right] = \int_a^b \mathbb{E}\left[f(B_t)^2\varphi(B_b - B_t)^2\right]dt.
$$

Thus we need to prove the following equality in order to complete the proof of the theorem,

$$
\lim_{\|\Delta\|\to 0}\sum_{i=2}^n\sum_{j=1}^{i-1}\mathbb{E}\left[f_{i-1}\varphi_i f_{j-1}\varphi_j\Delta B_i\Delta B_j\right]
$$

$$
= \int_a^b\int_a^t \mathbb{E}\left[f(B_s)\varphi'(B_b - B_s)f'(B_t)\varphi(B_b - B_t)\right]dsdt. \qquad (10)
$$

Step 2. For convenience, we denote

$$
D_{ij} \equiv f_{i-1}\varphi_i f_{j-1}\varphi_j\Delta B_i\Delta B_j, \quad 2 \le i \le n,\ 1 \le j \le i-1, \qquad (11)
$$

which appears in the left-hand side of Eq. (10). Evaluate expectation by conditioning to show that

$$
\mathbb{E}D_{ij} = \mathbb{E}\left[\mathbb{E}\left[f_{i-1}\varphi_i f_{j-1}\varphi_j\Delta B_i\Delta B_j\,\middle|\,\mathcal{F}_{t_i}\right]\right]
$$

$$
= \mathbb{E}\left[f_{i-1}f_{j-1}\Delta B_i\Delta B_j\,\mathbb{E}\left[\varphi_i\varphi_j\,\middle|\,\mathcal{F}_{t_i}\right]\right]. \qquad (12)
$$

Now, for each i and j as specified in Eq. (11), we can use the assumption on φ to get

$$
\mathbb{E}\left[\varphi_i\varphi_j\,\middle|\,\mathcal{F}_{t_i}\right] = \mathbb{E}\left[\varphi_i\varphi(B_b - B_{t_j})\,\middle|\,\mathcal{F}_{t_i}\right]
$$

$$
= \mathbb{E}\left[\varphi_i\sum_{n=0}^\infty\frac{\varphi^{(n)}(0)}{n!}(B_b - B_{t_j})^n\,\middle|\,\mathcal{F}_{t_i}\right]
$$

$$
= \sum_{n=0}^\infty\frac{\varphi^{(n)}(0)}{n!}\mathbb{E}\left[\varphi_i(B_b - B_{t_j})^n\,\middle|\,\mathcal{F}_{t_i}\right].
$$

Then we use the binomial expansion to rewrite the factor $(B_b - B_{t_j})^n$ as

$$(B_b - B_{t_j})^n = (B_b - B_{t_i} + B_{t_i} - B_{t_j})^n$$
$$= \sum_{k=0}^{n} \binom{n}{k} (B_b - B_{t_i})^k (B_{t_i} - B_{t_j})^{n-k}.$$

Therefore, we have

$$\mathbb{E}\left[\varphi_i\varphi_j \middle| \mathcal{F}_{t_i}\right]$$
$$= \sum_{n=0}^{\infty} \frac{\varphi^{(n)}(0)}{n!} \mathbb{E}\left[\varphi_i \sum_{k=0}^{n} \binom{n}{k} (B_b - B_{t_i})^k (B_{t_i} - B_{t_j})^{n-k} \middle| \mathcal{F}_{t_i}\right]$$
$$= \sum_{n=0}^{\infty} \sum_{k=0}^{n} \binom{n}{k} \frac{\varphi^{(n)}(0)}{n!} \mathbb{E}\left[\varphi_i (B_b - B_{t_i})^k (B_{t_i} - B_{t_j})^{n-k} \middle| \mathcal{F}_{t_i}\right]. \tag{13}$$

Step 3. Since $(B_{t_i} - B_{t_j})$ is measurable with respect to $\mathcal{F}_{t_i}$ and $(B_b - B_{t_i})$ is independent of $\mathcal{F}_{t_i}$, we see that

$$\mathbb{E}\left[\varphi_i (B_b - B_{t_i})^k (B_{t_i} - B_{t_j})^{n-k} \middle| \mathcal{F}_{t_i}\right]$$
$$= (B_{t_i} - B_{t_j})^{n-k} \mathbb{E}\left[\varphi_i (B_b - B_{t_i})^k \middle| \mathcal{F}_{t_i}\right]$$
$$= (B_{t_i} - B_{t_j})^{n-k} \mathbb{E}\left[\varphi_i (B_b - B_{t_i})^k\right]. \tag{14}$$

It follows from Eqs. (12), (13), and (14) that

$$\mathbb{E}D_{ij} = \sum_{n=0}^{\infty} \sum_{k=0}^{n} \binom{n}{k} \frac{\varphi^{(n)}(0)}{n!} \mathbb{E}\left[\varphi_i (B_b - B_{t_i})^k\right]$$
$$\times \mathbb{E}\left[f_{i-1}f_{j-1}\Delta B_i \Delta B_j (B_{t_i} - B_{t_j})^{n-k}\right]. \tag{15}$$

Step 4. Take expectation by conditioning to get

$$\mathbb{E}\left[f_{i-1}f_{j-1}\Delta B_i \Delta B_j (B_{t_i} - B_{t_j})^{n-k}\right]$$
$$= \mathbb{E}\left[f_{i-1}f_{j-1}\Delta B_j \mathbb{E}\left[\Delta B_i (B_{t_i} - B_{t_j})^{n-k} \middle| \mathcal{F}_{t_{i-1}}\right]\right]. \tag{16}$$

Put Eq. (16) into Eq. (15). Then apply the binomial expansion and use the fact that $(B_{t_{i-1}} - B_{t_j})^{n-k+l}$ is $\mathcal{F}_{t_{i-1}}$-measurable and ΔB_i is independent of $\mathcal{F}_{t_{i-1}}$ to show that

$$
\mathbb{E}\left[\Delta B_i (B_{t_i} - B_{t_j})^{n-k} \middle| \mathcal{F}_{t_{i-1}}\right]
$$
$$
= \mathbb{E}\left[\Delta B_i (B_{t_i} - B_{t_{i-1}} + B_{t_{i-1}} - B_{t_j})^{n-k} \middle| \mathcal{F}_{t_{i-1}}\right]
$$
$$
= \sum_{l=0}^{n-k} \binom{n-k}{l} \mathbb{E}\left[(\Delta B_i)^{l+1}(B_{t_{i-1}} - B_{t_j})^{n-k-l} \middle| \mathcal{F}_{t_{i-1}}\right]
$$
$$
= \sum_{l=0}^{n-k} \binom{n-k}{l} (B_{t_{i-1}} - B_{t_j})^{n-k-l} \mathbb{E}\left[(\Delta B_i)^{l+1}\right]. \tag{17}
$$

Step 5. Note that $\mathbb{E}\left[(\Delta B_i)^{l+1}\right] = 0$ for $l = 0$, $\mathbb{E}\left[(\Delta B_i)^{l+1}\right] = \Delta t_i$ for $l = 1$ and $\mathbb{E}\left[(\Delta B_i)^{l+1}\right] = o(\Delta t_i)$ for $l \geq 2$. Therefore, we have

$$
\sum_{l=0}^{n-k} \binom{n-k}{l} (B_{t_{i-1}} - B_{t_j})^{n-k-l} \mathbb{E}\left[(\Delta B_i)^{l+1}\right]
$$
$$
\approx (n-k)(B_{t_{i-1}} - B_{t_j})^{n-k-1} \Delta t_i. \tag{18}
$$

On the other hand, note that when $n = k$, the quantity in the left-hand side of Eq. (17) is zero because $\mathbb{E}\left[\Delta B_i\right] = 0$. By putting Eqs. (15), (16), (17), and (18) together and then changing the order of summations, we get

$$
\mathbb{E}D_{ij} \approx \sum_{n=1}^{\infty} \sum_{k=0}^{n-1} \binom{n}{k} \frac{\varphi^{(n)}(0)}{n!} (n-k) \mathbb{E}\left[\varphi_i(B_b - B_{t_i})^k\right]
$$
$$
\times \mathbb{E}\left[f_{i-1}f_{j-1}\Delta B_j (B_{t_{i-1}} - B_{t_j})^{n-k-1}\right] \Delta t_i.
$$

For simplicity, let K_{ij} denote

$$
K_{ij} = f_{i-1}f_{j-1}\Delta B_j (B_{t_{i-1}} - B_{t_j})^{n-k-1},
$$

Then we have

$$
\mathbb{E}D_{ij} \approx \sum_{n=1}^{\infty} \sum_{k=0}^{n-1} \binom{n}{k} \frac{\varphi^{(n)}(0)}{n!} (n-k) \mathbb{E}\left[\varphi_i(B_b - B_{t_i})^k\right] \mathbb{E}\left[K_{ij}\right] \Delta t_i. \tag{19}
$$

Step 6. Since f_{j-1} and ΔB_j are $\mathcal{F}_{t_j}$-measurable, we can apply expectation by conditioning to obtain

$$
\mathbb{E}K_{ij} = \mathbb{E}\left[\mathbb{E}\left[f_{i-1}f_{j-1}\Delta B_j (B_{t_{i-1}} - B_{t_j})^{n-k-1} \middle| \mathcal{F}_{t_j}\right]\right]
$$
$$
= \mathbb{E}\left[f_{j-1}\Delta B_j \mathbb{E}\left[f_{i-1}(B_{t_{i-1}} - B_{t_j})^{n-k-1} \middle| \mathcal{F}_{t_j}\right]\right]. \tag{20}
$$

Use the power series expansion of f to write the conditional expectation:

$$\mathbb{E}\left[f_{i-1}(B_{t_{i-1}} - B_{t_j})^{n-k-1}\big|\mathcal{F}_{t_j}\right]$$

$$= \mathbb{E}\left[\sum_{m=0}^{\infty}\frac{f^{(m)}(0)}{m!}(B_{t_{i-1}})^m(B_{t_{i-1}} - B_{t_j})^{n-k-1}\bigg|\mathcal{F}_{t_j}\right]$$

$$= \sum_{m=0}^{\infty}\frac{f^{(m)}(0)}{m!}\mathbb{E}\left[(B_{t_{i-1}})^m(B_{t_{i-1}} - B_{t_j})^{n-k-1}\big|\mathcal{F}_{t_j}\right]. \tag{21}$$

Apply the binomial expansion to express $(B_{t_{i-1}})^m$ as

$$(B_{t_{i-1}})^m = (B_{t_{i-1}} - B_{t_j} + B_{t_j})^m$$

$$= \sum_{l=0}^{m}\binom{m}{l}(B_{t_{i-1}} - B_{t_j})^l(B_{t_j})^{m-l}.$$

Hence (21) can be rewritten as

$$\mathbb{E}\left[f_{i-1}(B_{t_{i-1}} - B_{t_j})^{n-k-1}\big|\mathcal{F}_{t_j}\right]$$

$$= \sum_{m=0}^{\infty}\frac{f^{(m)}(0)}{m!}\mathbb{E}\left[\sum_{l=0}^{m}\binom{m}{l}(B_{t_j})^{m-l}(B_{t_{i-1}} - B_{t_j})^{n-k-1+l}\bigg|\mathcal{F}_{t_j}\right]$$

$$= \sum_{m=0}^{\infty}\sum_{l=0}^{m}\frac{f^{(m)}(0)}{m!}\binom{m}{l}(B_{t_j})^{m-l}\mathbb{E}\left[(B_{t_{i-1}} - B_{t_j})^{n-k-1+l}\right]. \tag{22}$$

It follows from Eqs. (19), (20), and (22) that

$$\mathbb{E}D_{ij} \approx \sum_{n=1}^{\infty}\sum_{k=0}^{n-1}\sum_{m=0}^{\infty}\sum_{l=0}^{m}\binom{n}{k}\frac{\varphi^{(n)}(0)}{n!}(n-k)\binom{m}{l}\frac{f^{(m)}(0)}{m!}$$

$$\times \mathbb{E}\left[\varphi_i(B_b - B_{t_i})^k\right]\mathbb{E}\left[(B_{t_{i-1}} - B_{t_j})^{n-k-1+l}\right]$$

$$\times \mathbb{E}\left[f_{j-1}(B_{t_j})^{m-l}\Delta B_j\right]\Delta t_i. \tag{23}$$

Step 7. Repeat the same arguments as those in Step 4 to show that

$$\mathbb{E}\left[f_{j-1}(B_{t_j})^{m-l}\Delta B_j\right]$$

$$= \mathbb{E}\left[f_{j-1}\mathbb{E}\left[(B_{t_j})^{m-l}\Delta B_j\big|\mathcal{F}_{t_{j-1}}\right]\right]$$

$$= \sum_{q=0}^{m-l}\binom{m-l}{q}\mathbb{E}\left[f_{j-1}(B_{t_{j-1}})^{m-l-q}\mathbb{E}\left[(\Delta B_j)^{q+1}\big|\mathcal{F}_{t_{j-1}}\right]\right]$$

$$= \sum_{q=0}^{m-l}\binom{m-l}{q}\mathbb{E}\left[f_{j-1}(B_{t_{j-1}})^{m-l-q}\right]\mathbb{E}\left[(\Delta B_j)^{q+1}\right]$$

$$\approx (m-l)\mathbb{E}\left[f_{j-1}(B_{t_{j-1}})^{m-l-q}\right]\Delta t_j, \tag{24}$$

where we have used the fact that $\mathbb{E}\left[(\Delta B_j)^{q+1}\right] \to 0$ as $\|\Delta\| \to 0$ for $q \neq 1$ and $\mathbb{E}\left[(\Delta B_j)^{q+1}\right] = \Delta t_j$ for $q = 1$.

It follows from Eqs. (23) and (24) that

$$
\begin{aligned}
\mathbb{E}D_{ij} & \\
\approx & \sum_{n=1}^{\infty}\sum_{k=0}^{n-1}\sum_{m=1}^{\infty}\sum_{l=0}^{m-1} \binom{n}{k}\frac{\varphi^{(n)}(0)}{n!}(n-k)\binom{m}{l}\frac{f^{(m)}(0)}{m!}(m-l) \\
& \times \mathbb{E}\left[\varphi_i(B_b - B_{t_i})^k\right] \mathbb{E}\left[(B_{t_{i-1}} - B_{t_j})^{n-k-1+l}\right] \\
& \times \mathbb{E}\left[f_{j-1}(B_{t_{j-1}})^{m-l-1}\right]\Delta t_i \Delta t_j \\
= & \Delta t_i \Delta t_j \sum_{n=1}^{\infty}\sum_{k=0}^{n-1}\sum_{m=1}^{\infty}\sum_{l=0}^{m-1}\binom{n}{k}\frac{\varphi^{(n)}(0)}{n!}(n-k)\binom{m}{l}\frac{f^{(m)}(0)}{m!}(m-l) \\
& \times \mathbb{E}\left[f_{j-1}\varphi_i(B_b - B_{t_i})^k(B_{t_{i-1}} - B_{t_j})^{n-k-1+l}(B_{t_{j-1}})^{m-l-1}\right]. \quad (25)
\end{aligned}
$$

Step 8. We can easily check the following equality

$$
\binom{n}{k}\frac{n-k}{n!} = \binom{n-1}{k}\frac{1}{(n-1)!}. \quad (26)
$$

By using Eq. (26), we can simplify Eq. (25) to

$$
\begin{aligned}
\mathbb{E}D_{ij} \approx \mathbb{E}\Bigg[& \Delta t_i \Delta t_j f_{j-1}\varphi_i \\
& \times \sum_{n=1}^{\infty}\sum_{k=0}^{n-1}\binom{n-1}{k}\frac{\varphi^{(n)}(0)}{(n-1)!}(B_b - B_{t_i})^k(B_{t_{i-1}} - B_{t_j})^{n-k-1} \\
& \times \sum_{m=1}^{\infty}\sum_{l=0}^{m-1}\binom{m-1}{l}\frac{f^{(m)}(0)}{(m-1)!}(B_{t_{i-1}} - B_{t_j})^l(B_{t_{j-1}})^{m-l-1}\Bigg],
\end{aligned}
$$

which, by rearranging the terms, can be further simplified as

$$
\begin{aligned}
\mathbb{E}D_{ij} \approx \mathbb{E}\Bigg[& \Delta t_i \Delta t_j f_{j-1}\varphi_i \\
& \times \sum_{n=1}^{\infty}\frac{\varphi^{(n)}(0)}{(n-1)!}\sum_{k=0}^{n-1}\binom{n-1}{k}(B_b - B_{t_i})^k(B_{t_{i-1}} - B_{t_j})^{n-k-1} \\
& \times \sum_{m=1}^{\infty}\frac{f^{(m)}(0)}{(m-1)!}\sum_{l=0}^{m-1}\binom{m-1}{l}(B_{t_{i-1}} - B_{t_j})^l(B_{t_{j-1}})^{m-l-1}\Bigg].
\end{aligned}
$$

Then apply the binomial expansion to show that

$$
\mathbb{E}D_{ij} \approx \mathbb{E}\left[\Delta t_i \Delta t_j f_{j-1}\varphi_i \sum_{n=1}^{\infty} \frac{\varphi^{(n)}(0)}{(n-1)!}(B_b - B_{t_i} + B_{t_{i-1}} - B_{t_j})^{n-1}\right.
$$

$$
\left. \times \sum_{m=1}^{\infty} \frac{f^{(m)}(0)}{(m-1)!}(B_{t_{i-1}} - B_{t_j} + B_{t_{j-1}})^{m-1}\right]
$$

$$
= \mathbb{E}\left[\Delta t_i \Delta t_j f_{j-1}\varphi_i \sum_{n=1}^{\infty} \frac{\varphi^{(n)}(0)}{(n-1)!}(B_b - B_{t_j} - \Delta B_i)^{n-1}\right.
$$

$$
\left. \times \sum_{m=1}^{\infty} \frac{f^{(m)}(0)}{(m-1)!}(B_{t_{i-1}} - \Delta B_j)^{m-1}\right]. \tag{27}
$$

Step 9. Now, note that

$$
f'(x) = \sum_{n=1}^{\infty} \frac{f^{(n)}(0)}{(n-1)!}x^{n-1}, \quad \varphi'(x) = \sum_{n=1}^{\infty} \frac{\varphi^{(n)}(0)}{(n-1)!}x^{n-1}.
$$

Therefore, Eq. (27) can be rewritten as

$$
\mathbb{E}D_{ij} \approx \mathbb{E}\left[\Delta t_i \Delta t_j f_{j-1}\varphi_i \varphi'(B_b - B_{t_j} - \Delta B_i)f'(B_{t_{i-1}} - \Delta B_j)\right],
$$

which yields the summation

$$
\sum_{i=1}^{n}\sum_{j=0}^{i-1} \mathbb{E}D_{ij}
$$

$$
\approx \sum_{i=1}^{n}\sum_{j=0}^{i-1} \mathbb{E}\left[f_{j-1}\varphi_i\varphi'(B_b - B_{t_j} - \Delta B_i)f'(B_{t_{i-1}} - \Delta B_j)\right]\Delta t_i \Delta t_j
$$

$$
\to \int_a^b \int_a^t \mathbb{E}\left[f(B_s)\varphi'(B_b - B_s)f'(B_t)\varphi(B_b - B_t)\right] ds\, dt, \tag{28}
$$

as $\|\Delta\| \to 0$. In view of Eqs. (11) and (28), we have proved Eq. (10). Thus the proof of the theorem is complete.

5. An example

Consider a new stochastic integral given by

$$
X = \int_0^1 e^{B_1}\, dB_t.
$$

We can compute $\mathbb{E}\left[X^2\right]$ in two ways. First we apply Theorem 3.1 for the functions $f(x) = \varphi(x) = e^x$ to get

$$\int_0^1 \mathbb{E}\left[f(B_t)^2\varphi(B_1 - B_t)\right]^2 dt = e^2,$$

$$\int_0^1 \int_0^t \mathbb{E}\left[f(B_s)\varphi'(B_b - B_s)f'(B_t)\varphi(B_b - B_t)\right] dsdt = \frac{1}{2}e^2.$$

Therefore, by Theorem 3.1 we have the value

$$\mathbb{E}\left[X^2\right] = 2e^2.$$

On the other hand, by Example 2.6 in Ayed–Kuo,[1] the random variable X is given by

$$X = e^{B_1}(B_1 - 1),$$

which leads to

$$\mathbb{E}\left[X^2\right] = \int_{-\infty}^{\infty} e^{2x}(x - 1)^2 \frac{1}{\sqrt{2\pi}} e^{-x^2/2}\, dx = 2e^2.$$

Thus we get the same value for $\mathbb{E}\left[X^2\right]$.

References

1. W. Ayed and H.-H. Kuo: An extension of the Itô integral, *Communications on Stochastic Analysis* **2**, no. 3 (2008) 323–333.
2. W. Ayed and H.-H. Kuo: An extension of the Itô integral: Toward a general theory of stochastic integration, *Theory of Stochastic Processes* **16(32)** no. 1 (2010) 17–28.
3. H.-H. Kuo: *White Noise Distribution Theory*, CRC Press, 1996.
4. H.-H. Kuo: *Introduction to Stochastic Integration*, Universitext, Springer, 2006.
5. H.-H. Kuo, A. Sae-Tang, and B. Szozda: A stochastic integral for adapted and instantly independent stochastic processes; A volume in honor of R. Elliott (accepted for publication, 2011)

INFINITE DIMENSIONAL LAPLACIANS ASSOCIATED WITH DERIVATIVES OF WHITE NOISE

KIMIAKI SAITÔ

Department of Mathematics, Meijo University
Shiogamaguchi 1-501, Tenpaku, Nagoya 468-8502, Japan
E-mail: ksaito@meijo-u.ac.jp

This paper is written based on the joint work with L. Accardi and U. C. Ji. In this paper we present a generalization of the result in Ref.7 on a relationship between the exotic Laplacians and the Lévy Laplacians in terms of higher order derivatives of white noise by a generalization of the Accardi-Smolyanov theorem.[5] Moreover, we give a relationship between exotic Laplacians, acting on higher order singular functionals, each other in terms of the second quantization of the adjoint operator of the higher order differentiation. These relationships imply an interesting result on an infinite dimensional stochastic process generated by the exotic Laplacian.

Keywords: Exotic Laplacians, deivatives of white noise, associated stochastic processes

1. Introduction

In Ref.22, P. Lévy introduced an infinite dimensional Laplacian, called the Lévy Laplacian, which is defined as the Cesàro mean of the second derivatives along a sequence of orthogonal axes. The Lévy Laplacian, the associated heat equation and stochastic processes have been studied by many authors from several different points of view, particularly in white noise theory (see Refs. 2, 4, 9, 10, 19-21, 23, 24, 27-29 and references cited therein). Recently, in Refs. 1 and 17, it was proved that the Lévy Laplacian can be considered as a particular Volterra–Gross Laplacian.

The exotic Laplacians were introduced in L. Accardi and O. Smolyanov in Ref.3 as natural generalizations of the Lévy Laplacian, defined by Cesàro means of higher order. In Ref.6, it was proved that the Exotic Laplacian also can be considered as a particular Volterra–Gross Laplacian and generates an infinite dimensional Brownian motion. A relationship between the exotic Laplacian and the Lévy Laplacian in terms of higher order derivatives of white noise was given in Ref.7. As a generalization, a relationship between exotic Laplacians each other also has been obtained in Ref.7.

The main purpose of this paper is to give a generalization of the result in Ref.7 on a relationship between the exotic Laplacians and the Lévy Laplacians in terms of higher order derivatives of white noise by generalizing the Accardi-Smolyanov theorem.[5] Moreover we give infinite dimensional stochastic processes associated with the exotic Laplacians by taking derivatives of an infinite dimensional Brownian motion obtained in Ref.17.

We introduce an operator which transfers white noise functionals into functionals of higher order derivatives of white noise. The operator is defined by the second quantization of the adjoint of the α-th differentiation. It gives an isomorphism from the space $(E)^*$ of white noise distributions onto itself. Generalizing the Accardi-Smolyanov result in Ref.5, we have a relationship between exotic Laplacians acting on distributions in $(E)^*$ each other by the above operator. This relationship is a generalization of the result obtained in Ref.7.

We also introduce an imbedding the space $(\mathcal{N}_{c,2a+1})$ of test functionals associated with the exotic trace of order $a \geq 0$ into the space of distributions in $(E)^*$. It is a quite important operator to give stochastic processes generated by exotic Laplacians. Based on the imbedding we construct a space of white noise distributions in $(E)^*$ which is isomorphic to $(\mathcal{N}_{c,2a+1})$. In the last section we give an infinite dimensional stochastic process associated with the exotic Laplacian of each order acting on $(\mathcal{N}_{c,2a+1})$ by taking the higher order derivative of the stochastic process generated by the Lévy Laplacian obtained in Ref.17.

2. Preliminaries

Let H be a complex Hilbert space with inner product $\langle \cdot, \cdot \rangle$ and $e \equiv \{e_k\}_{k=1}^{\infty}$ an orthonormal basis of H. Let $\{\ell_k\}_{k=1}^{\infty}$ be an arbitrary sequence such that

$$1 \leq \ell_1 \leq \ell_2 \leq \ell_3 \leq \ldots, \qquad \sum_{k=1}^{\infty} \ell_k^{-2} < \infty,$$

and A the densely defined selfadjoint operator on $(H, \{e_k\})$ by

$$A := \sum_{k=1}^{\infty} \ell_k \langle \xi, e_k \rangle e_k, \qquad \xi \in H.$$

We also assume that for any $\alpha > 0$ there exists $\beta > 0$ such that $(2[(k+1)/2]\pi)^{\alpha} \leq \ell_k^{\beta}$ for all $k \geq 1$, which is a necessary condition for the continuity of the higher order differential operator in Section 3.

Then A^{-1} is Hilbert-Schmidt, $\inf \operatorname{Spec}(A) \geq 1$ and, for each $p \in \mathbf{R}$, the p-norm is defined by

$$|\xi|_p^2 = |A^p \xi|_H^2 = \sum_{k=1}^{\infty} \ell_k^{2p} |\langle \xi, e_k \rangle|^2, \qquad \xi \in H.$$

For $p \geq 0$ one defines $E_p = \{\xi \in H \,;\, |\xi|_p < \infty\}$ and E_{-p} to be the completion of H with respect to $|\cdot|_{-p}$. Thus we obtain a chain of Hilbert spaces $\{E_p \,;\, p \in \mathbf{R}\}$ and the corresponding standard triple:

$$E := \operatorname*{proj\,lim}_{p \to \infty} E_p \subset H \subset E^* := \operatorname*{ind\,lim}_{p \to \infty} E_{-p},$$

which now depends on the triple $(H, \{e_k\}, \{\ell_k\})$.

Let J_e be the conjugation operator on H with respect to the e-basis defined by $J_e \xi = \sum_{n=1}^{\infty} \bar{\alpha}_n e_n$ for $\xi = \{\alpha_n\}_{n=1}^{\infty} \in H$. Then E, H and E^* are invariant under the action of J_e and their fixed point subspaces are denoted by $E_{\mathbf{R}}$, $H_{\mathbf{R}}$ and $E_{\mathbf{R}}^*$, respectively. Then we obtain a real standard triple $E_{\mathbf{R}} \subset H_{\mathbf{R}} \subset E_{\mathbf{R}}^*$. When no confusion is possible also the inner product on $H_{\mathbf{R}}$ and the canonical $\mathbf{C}$-bilinear forms on $E^* \times E$ and on $E_{\mathbf{R}}^* \times E_{\mathbf{R}}$ will be denoted by $\langle \cdot, \cdot \rangle$ again.

For each $p \in \mathbf{R}$ let $\Gamma(E_p)$ be the Fock space over the Hilbert space E_p, i.e.,

$$\Gamma(E_p) = \left\{ \phi = (f_n)_{n=0}^{\infty} \,;\, f_n \in E_p^{\hat{\otimes} n}, \; \|\phi\|_p^2 = \sum_{n=0}^{\infty} n! \, |f_n|_p^2 < \infty \right\}.$$

Then by identifying $\Gamma(H)$ with its dual space, we have a chain of Fock spaces:

$$\cdots \subset \Gamma(E_p) \subset \Gamma(E_0) = \Gamma(H) \subset \Gamma(E_{-p}) \subset \cdots$$

and a new triple

$$(E) = \operatorname*{proj\,lim}_{p \to \infty} \Gamma(E_p) \subset \Gamma(H) \subset (E)^* = \operatorname*{ind\,lim}_{p \to \infty} \Gamma(E_{-p}). \tag{1}$$

The canonical $\mathbf{C}$-bilinear form on $(E)^* \times (E)$, denoted by $\langle\!\langle \cdot, \cdot \rangle\!\rangle$, has the form:

$$\langle\!\langle \Phi, \phi \rangle\!\rangle = \sum_{n=0}^{\infty} n! \, \langle F_n, f_n \rangle, \qquad \Phi = (F_n) \in (E)^*, \quad \phi = (f_n) \in (E).$$

The *exponential vector* associated with $\xi \in H$ is defined by

$$\phi_\xi = \left(1, \xi, \frac{\xi^{\otimes 2}}{2!}, \dots, \frac{\xi^{\otimes n}}{n!}, \dots \right). \tag{2}$$

If $\xi, \eta \in E_p$, then $\langle\!\langle \phi_\xi, \phi_\eta \rangle\!\rangle = \exp\langle \xi, \eta \rangle$, therefore $\phi_\xi \in (E)$ for any $\xi \in E$. The S-*transform* of an element $\Phi \in (E)^*$ is the function $S\Phi : E \to \mathbb{C}$ defined by

$$S\Phi(\xi) = \langle\!\langle \Phi, \phi_\xi \rangle\!\rangle, \qquad \xi \in E.$$

Every element $\Phi \in (E)^*$ is uniquely specified by its S-transform $S\Phi$ since $\{\phi_\xi \, ; \, \xi \in E\}$ spans a dense subspace of (E).

A $\mathbb{C}$-valued function F defined on E is called a U-*functional* if F is Gâteaux entire and there exist constants $C, K \geq 0$ and $p \geq 0$ such that

$$|F(\xi)| \leq C \exp\left(K \, |\xi|_p^2 \right), \qquad \xi \in E.$$

Theorem 2.1.[26] *A $\mathbb{C}$-valued function F on E is the S-transform of an element of $(E)^*$ if and only if F is a U-functional.*

Remark 2.1. The Bochner-Minlos Theorem implies the existence of a probability measure μ on $E_{\mathbf{R}}^*$ such that

$$\int_{E_{\mathbf{R}}^*} e^{i\langle x, \xi \rangle} d\mu(x) = e^{-\frac{1}{2}\langle \xi, \xi \rangle}, \qquad \xi \in E_{\mathbf{R}}.$$

The Wiener-Itô-Segal isomorphism between $\Gamma(H)$ and $L^2(E^*, \mu)$ is the unitary isomorphism uniquely determined by the correspondence:

$$\phi_\xi = \left(1, \xi, \frac{\xi^{\otimes 2}}{2!}, \dots, \frac{\xi^{\otimes n}}{n!}, \dots \right) \quad \longleftrightarrow \quad \phi_\xi(x) = e^{\langle x, \xi \rangle - \langle \xi, \xi \rangle/2}, \quad \xi \in E.$$

The standard triple obtained from (1) through the Wiener-Itô-Segal isomorphism is denoted also by

$$(E) \subset L^2(E^*, \mu) \subset (E)^*,$$

which is referred to as the *Hida–Kubo–Takenaka space*. An element of (E) (resp. $(E)^*$) is called a test (resp. generalized) white noise functional.

3. Higher Order Derivatives of White Noise

In the following H will be the closed subspace of the Hilbert space $L^2([0,1])$ generated by the functions

$$e_{2k-1}(t) = e^{2\pi i k t}, \quad e_{2k}(t) = e^{-2\pi i k t}, \quad k = 1, 2, \dots, \tag{3}$$

that is, H is the subspace of $L^2([0,1])$ orthogonal to the constant functions. Then $\{e_n\}_{n=1}^\infty$ is an orthonormal basis of H. We denote ∂ the partial derivative in the t–variable and, for $\alpha \in \mathbb{N}$, $\xi \in E$, we will use indifferently the following notations:

$$\partial^\alpha \xi =: \partial_\alpha \xi =: \xi^{(\alpha)}.$$

Lemma 3.1. *For any $\alpha \in \mathbb{N}$ ∂_α on the basis (3) is an isomorphism from E onto itself.*

Proof. The linearity of ∂_α is obvious. Let $\alpha \in \mathbb{N}$ and $p \geq 0$ be given. Then the equalities

$$e^{(\alpha)}_{2k-1}(t) = (2\pi ik)^\alpha e_{2k-1}(t); \quad e^{(\alpha)}_{2k}(t) = (-2\pi ik)^\alpha e_{2k}(t), \quad k = 1, 2, 3, \ldots,$$
$$(4)$$

follow by direct computation. Moreover, for $\alpha = 0$, $\partial_\alpha = id$ and the proof is obvious. Suppose $\alpha \geq 1$. Then, if $\xi \in E$ is such that $\langle \xi, e_j \rangle = 0$ for all $j \in \mathbb{N}^*$ except a finite number of j, we have

$$\xi^{(\alpha)} = \sum_{j=1}^{\infty} \langle \xi, e_j \rangle e^{(\alpha)}_j = \sum_{k=1}^{\infty} \langle \xi, e_{2k-1} \rangle e^{(\alpha)}_{2k-1} + \sum_{k=1}^{\infty} \langle \xi, e_{2k} \rangle e^{(\alpha)}_{2k},$$

therefore, if $\beta = \beta(\alpha)$ is given in Section 1, then for any $p \geq 0$

$$\left| \xi^{(\alpha)} \right|_p^2 = \sum_{k=1}^{\infty} (2k\pi)^{2\alpha} \ell_{2k-1}^{2p} |\langle \xi, e_{2k-1} \rangle|^2 + \sum_{k=1}^{\infty} (2k\pi)^{2\alpha} \ell_{2k}^{2p} |\langle \xi, e_{2k} \rangle|^2$$

$$\leq \sum_{k=1}^{\infty} \ell_{2k-1}^{2(p+\beta)} |\langle \xi, e_{2k-1} \rangle|^2 + \sum_{k=1}^{\infty} \ell_{2k}^{2(p+\beta)} |\langle \xi, e_{2k} \rangle|^2$$

$$= \sum_{j=1}^{\infty} \ell_j^{2(p+\beta)} |\langle \xi, e_j \rangle|^2 = |\xi|_{p+\beta}^2.$$

Therefore ∂_α can be extended by continuity to the whole of E with preservation of the above inequality. This means that ∂_α is a continuous linear operator from E onto itself. The bijectivity of ∂_α is clear from its explicit form (4). In fact the inverse ∂_α^{-1} of ∂_α is given by

$$\partial_\alpha^{-1} \xi = \sum_{k=1}^{\infty} (2\pi ki)^{-\alpha} \langle \xi, e_{2k-1} \rangle e_{2k-1} + \sum_{k=1}^{\infty} (-2\pi ki)^{-\alpha} \langle \xi, e_{2k} \rangle e_{2k}$$

for $\xi \in E$. The linearity of ∂_α^{-1} is also obvious. It is also checked that for any $p \geq 0$

$$|\partial_\alpha^{-1}\xi|_p^2 = \sum_{k=1}^\infty (2k\pi)^{-2\alpha} \ell_{2k-1}^{2p} |\langle \xi, e_{2k-1} \rangle|^2 + \sum_{k=1}^\infty (2k\pi)^{-2\alpha} \ell_{2k}^{2p} |\langle \xi, e_{2k} \rangle|^2$$

$$\leq \sum_{j=1}^\infty \ell_j^{2p} |\langle \xi, e_j \rangle|^2 = |\xi|_p^2.$$

Thus the inverse is also continuous linear operator from E onto itself. $\square$

By similar arguments in the proof of Lemma 3.1 we can prove that there exists $\beta = \beta(\alpha)$ such that $\left|\xi^{(\alpha)}\right|_{-p} \leq |\xi|_{p-\beta}$, $\xi \in E$ for any $p \geq \beta$. It is also checked that $\left|\partial_\alpha^{-1}\xi\right|_{-p} \leq |\xi|_{-p}^2$, $\xi \in E$ for any $p \geq 0$. Then we have the following.

Lemma 3.2. *For any $\alpha \in \mathbb{N}$ the operator ∂_α is an isomorphism from E^* onto itself.*

For any two topological vector spaces $\mathfrak{X}$, $\mathfrak{Y}$ we denote $\mathcal{L}(\mathfrak{X}, \mathfrak{Y})$ the space of all continuous linear operators from $\mathfrak{X}$ into $\mathfrak{Y}$. The *second quantization* of an operator $T \in \mathcal{L}(E^*, E^*)$, denoted $\Gamma(T)$, is defined by $\Gamma(T)\phi_\xi = \phi_{T\xi}$ for $\xi \in E$. It is known (see Ref.25) that $\Gamma(T) \in \mathcal{L}((E)^*, (E)^*)$ and that, if $T \in \mathcal{L}(E, E)$, then $\Gamma(T) \in \mathcal{L}((E), (E))$. Then, from Lemma 3.1, we have the following.

Theorem 3.1. *The operator $\Gamma(\partial_\alpha^*)$ is an isomorphism from $(E)^*$ onto itself.*

4. A Relationship between Exotic Laplacians on White Noise Distributions

A function $F : E \to \mathbb{C}$ is said to be of class C^2 if it is twice (continuously) Fréchet differentiable, i.e. there exist two continuous maps

$$\xi \longmapsto F'(\xi) \in E^*, \quad \xi \longmapsto F''(\xi) \in \mathcal{L}(E, E^*), \quad \xi \in E,$$

such that for any $\eta \in E$

$$F(\xi + \eta) = F(\xi) + \langle F'(\xi), \eta \rangle + \frac{1}{2} \langle F''(\xi)\eta, \eta \rangle + \epsilon(\eta)$$

where the error terms satisfy $\lim_{t \to 0} \frac{\epsilon(t\eta)}{t} = 0$, $\eta \in E$. We denote $\widetilde{D}_\eta$ the Gateaux differentiation in the direction η, i.e.,

$$\widetilde{D}_\eta F(\xi) = \lim_{t \to 0} \frac{1}{t} [F(\xi + t\eta) - F(\xi)] = \frac{d}{dt} F(\xi + t\eta)\Big|_{t=0}.$$

It is known that, under general regularity conditions on F, one has

$$\tilde{D}_\eta F(\xi) = \langle F'(\xi), \eta \rangle .$$

(See Ref.15.) In the sense that, the existence of either side of the identity implies the existence of the othe one and the equality. The kernel theorem identifies $\mathcal{L}(E, E^*)$ with $E^* \otimes E^*$ (see Ref.11 and also Ref.25). Using this identification we will use indifferently the notations

$$\langle F''(\xi)\eta, \eta \rangle = \langle F''(\xi), \eta \otimes \eta \rangle = F''(\xi)(\eta, \eta) = \tilde{D}_\eta^2 F(\xi).$$

For $a \in \mathbb{N}$ arbitrary natural number, let $\mathrm{Dom}(\Delta_{c,2a+1})$ denote the set of all $\Phi \in (E)^*$ such that the limit

$$\tilde{\Delta}_{c,2a+1} S\Phi(\xi) := \lim_{N \to \infty} \frac{1}{N^{2a+1}} \sum_{k=1}^{N} \langle (S\Phi)''(\xi), e_k \otimes e_k \rangle$$

exists for each $\xi \in E$ is the S-transform of an element of $(E)^*$. The *exotic Laplacian* $\Delta_{c,2a+1}$ is defined on $\mathrm{Dom}(\Delta_{c,2a+1})$ by $\Delta_{c,2a+1}\Phi := S^{-1}(\tilde{\Delta}_{c,2a+1}S\Phi)$ for $\Phi \in \mathrm{Dom}(\Delta_{c,2a+1})$. The operator $\Delta_{c,1}$ is called the *Lévy Laplacian* and is also denoted by Δ_L.

Using the bijectivity of $\Gamma(\partial_\alpha^*)$ we introduce the *Lévy Laplacian* $\Delta_{L,\alpha}$ *of order* α, defined on the domain $\Gamma(\partial_\alpha^*)(\mathrm{Dom}(\Delta_L))$ of the derivatives of order α of white noise functionals, by

$$\Delta_{L,\alpha}\Phi = \Gamma(\partial_\alpha^*)\Delta_L\Gamma(\partial_\alpha^*)^{-1}\Phi, \quad \Phi \in \Gamma(\partial_\alpha^*)(\mathrm{Dom}(\Delta_L)) \tag{5}$$

The following theorem and lemma are obtained by Accardi and Smolyanov.[5]

Theorem 4.1.[5] *Let $p > 0$. If the limit*

$$\lim_{N \to \infty} \frac{1}{N^p} \sum_{n=1}^{N} a_n =: A_p(a)$$

exists for $a = (a_n)_{n=1}^{\infty} \in \mathbb{C}^\infty$, then

$$\lim_{N \to \infty} \frac{1}{N^{p+1}} \sum_{n=1}^{N} n a_n = \frac{p}{p+1} A_p(a).$$

Lemma 4.1.[5] *Suppose $p \geq 0$ and $a \in \mathbb{C}^\infty$. If $\dfrac{a_n}{n^p} \to 0$ as $n \to \infty$, then*

$$\frac{1}{n^{p+1}} \sum_{k=1}^{n} a(k) \to 0 \text{ as } n \to \infty.$$

Theorem 4.1 is extended to the following theorem.

Theorem 4.2. *Suppose $p > 0$ and $\alpha \in \mathbb{N}$. If*

$$\lim_{N \to \infty} \frac{1}{N^p} \sum_{n=1}^{N} a_n =: A_p(a) \tag{6}$$

exists for $a = (a_n)_{n=1}^{\infty} \in \mathbb{C}^{\infty}$, then

$$\lim_{N \to \infty} \frac{1}{N^{p+\alpha}} \sum_{n=1}^{N} n^{\alpha} a_n = \frac{p}{p+\alpha} A_p(a). \tag{7}$$

Proof. Suppose that (6) exists for $a = (a_n)_{n=1}^{\infty} \in \mathbb{C}^{\infty}$. By the Abel transform we see that

$$\frac{1}{N^{p+\alpha}} \sum_{k=1}^{N} k^{\alpha} a_k = \frac{1}{N^{p+\alpha}} \left[\left(\sum_{k=1}^{N} a_k \right) N^{\alpha} - \sum_{k=1}^{N-1} \{(k+1)^{\alpha} - k^{\alpha}\} \sum_{r=1}^{k} a_r \right]$$

$$= \frac{1}{N^p} \sum_{k=1}^{N} a_k - \frac{1}{N^{p+\alpha}} \sum_{k=1}^{N-1} \{(k+1)^{\alpha} - k^{\alpha}\} \sum_{r=1}^{k} a_r. \tag{8}$$

By (6) we have $\frac{1}{N^p} \left[\sum_{k=1}^{N} a_k - N^p A_p(a) \right] \to 0$ as $N \to \infty$. Therefore, by Lemma 4.1, we obtain the assertion as follows:

$$\lim_{N \to \infty} \frac{1}{N^{p+\alpha}} \sum_{k=1}^{N} k^{\alpha} a_k$$

$$= \lim_{N \to \infty} \frac{1}{N^p} \sum_{k=1}^{N} a_k - \lim_{N \to \infty} \frac{1}{N^{p+\alpha}} \sum_{k=1}^{N-1} \{(k+1)^{\alpha} - k^{\alpha}\} \sum_{r=1}^{k} a_r$$

$$= \lim_{N \to \infty} \frac{1}{N^p} \sum_{k=1}^{N} a_k - \lim_{N \to \infty} \frac{1}{N^{p+\alpha}} \sum_{k=1}^{N-1} k^p \{(k+1)^{\alpha} - k^{\alpha}\} A_p(a)$$

$$= A_p(a) - A_p(a) \lim_{N \to \infty} \frac{1}{N} \sum_{k=1}^{N-1} \left(\frac{k}{N} \right)^p \sum_{j=0}^{\alpha-1} {}_{\alpha}C_j \frac{k^j}{N^{\alpha-1}}$$

$$= A_p(a) - A_p(a)\alpha \int_0^1 x^{p+\alpha-1} dx$$

$$= \left(1 - \frac{\alpha}{p+\alpha} \right) A_p(a) = \frac{p}{p+\alpha} A_p(a). \qquad \square$$

Let Φ be a functional in $\mathrm{Dom}(\Delta_{c,2\alpha+1})$. Then by Theorem 4.2 we note that

$$\lim_{N \to \infty} \frac{1}{N^{2(\alpha+K)+1}} \sum_{n=1}^{N} S\Phi''(\xi^{(\alpha)})(e_n^{(\alpha)}, e_n^{(\alpha)})$$

$$= \lim_{N \to \infty} \frac{1}{N^{2(\alpha+K)+1}} \sum_{n=1}^{N} (-1)^{\alpha} (2n\pi)^{2\alpha} S\Phi''(\xi^{(\alpha)})(e_n, e_n)$$

$$= \frac{(-1)^{\alpha} (2\pi)^{2\alpha} (2K+1)}{2(\alpha+K)+1} \cdot \lim_{N \to \infty} \frac{1}{N^{2K+1}} \sum_{n=1}^{N} S\Phi''(\xi^{(\alpha)})(e_n, e_n).$$

$$= \frac{(-1)^{\alpha} (2\pi)^{2\alpha} (2K+1)}{2(\alpha+K)+1} \cdot S[\Delta_{c,2K+1}\Phi](\xi^{(\alpha)}).$$

Thus we have the following.

Theorem 4.3. *Let* $\alpha \in \mathbb{N}$ *and* $\Phi \in \mathrm{Dom}(\Delta_{c,2K+1})$. *Then* $\Gamma(\partial_\alpha^*)\Phi$ *is in* $\mathrm{Dom}(\Delta_{c,2(\alpha+K)+1})$ *and the equation*

$$\Delta_{c,2(\alpha+K)+1} \Gamma(\partial_\alpha^*)\Phi = \frac{(-1)^{\alpha} (2\pi)^{2\alpha} (2K+1)}{2(\alpha+K)+1} \Gamma(\partial_\alpha^*)\Delta_{c,2K+1}\Phi$$

holds.

Corollary 4.1. *Let* $\alpha \in \mathbb{N}$ *and* $\Phi \in \mathrm{Dom}(\Delta_L)$. *Then the equation*

$$\Delta_{c,2\alpha+1} \Gamma(\partial_\alpha^*)\Phi = \frac{(-1)^{\alpha} (2\pi)^{2\alpha}}{2\alpha+1} \Gamma(\partial_\alpha^*)\Delta_L\Phi = \frac{(-1)^{\alpha} (2\pi)^{2\alpha}}{2\alpha+1} \Delta_{L,\alpha}\Gamma(\partial_\alpha^*)\Phi \tag{9}$$

holds.

5. Exotic Traces and Exotic Triples

The Cesàro semi–norm of order a of $x \in E^*$ is defined by

$$|x|_{c,a}^2 := \lim_{N \to \infty} \frac{1}{N^a} \sum_{n=1}^{N} \overline{\langle x, e_n \rangle} \langle x, e_n \rangle$$

in the sense that, when the limit exists, the semi–norm is defined by the above limit. The Cesàro pre–scalar product of order a between $x, y \in E^*$ is defined, in the same sense, by

$$\langle x, y \rangle_{c,a} := \lim_{N \to \infty} \frac{1}{N^a} \sum_{n=1}^{N} \overline{\langle x, e_n \rangle} \langle y, e_n \rangle.$$

For all $a \in \mathbb{N}$ and $\lambda \in \mathbb{R}$, define $x_{\lambda,a} := \sum_{n=1}^{\infty} e^{2\pi i \lambda n} e_n^{(a)}$. Then, for all $a \in \mathbb{N}$ and $\lambda \in \mathbb{R}$, $x_{\lambda,a} \in E^*$. For all $a \in \mathbb{N}$ and $\lambda, \mu \in \mathbb{R} \setminus \mathbb{Q}$, we have

$$\langle x_{\lambda,p}, x_{\mu,p} \rangle_{c,2a+1} = \frac{(2\pi)^{2a}}{2a+1} \delta_{\lambda,\mu} \quad \text{(Kronecker's delta)}. \qquad (10)$$

Let $\mathcal{C}$ be a countable set in $\mathbb{R} \setminus \mathbb{Q}$. Then we know that, for each $a \in \mathbb{N}$, the set

$$\left\{ e_{a,\lambda} := x_{\lambda,a} \sqrt{2a+1}/(2\pi)^a \ : \ \lambda \in \mathcal{C} \right\} \qquad (11)$$

is orthonormal for the scalar product $\langle \, \cdot \, , \, \cdot \, \rangle_{c,2a+1}$. Therefore, the space

$$H_{c,2a+1}^{\circ} : = \text{linear span} \left\{ e_{a,\lambda}; \ \lambda \in \mathcal{C} \right\}$$

is a pre–Hilbert space for the inner product

$$\langle \xi, \eta \rangle_{c,2a+1} := \lim_{N \to \infty} \frac{1}{N^{2a+1}} \sum_{n=1}^{N} \overline{\langle \xi, e_n \rangle} \langle \eta, e_n \rangle, \qquad \xi, \eta \in H_{c,2a+1}^{\circ}.$$

Let $H_{c,2a+1}$ be the completion of the pre–Hilbert space ($H_{c,2a+1}^{\circ}$, $\langle \cdot, \cdot \rangle_{c,2a+1}$) which in general $H_{c,2a+1}$ will not be contained in E^*. Then $H_{c,2a+1}$ becomes an infinite dimensional separable Hilbert space whose inner product will still be denoted $\langle \cdot, \cdot \rangle_{c,2a+1}$ when no confusion is possible and, by construction, the set $\{e_{a,\lambda}\}_{\lambda \in \mathcal{C}}$ is an orthonormal basis of $H_{c,2a+1}$. By relabeling of the orthonormal basis $\{e_{a,\lambda}\}_{\lambda \in \mathcal{C}}$ of $H_{c,2a+1}$ we use the notation $\{e_{a,k}\}_{k=1}^{\infty}$.

The construction in Section 2 can be applied to the space $H_{c,2a+1}$ replacing the natural integers by an arbitrary sequence $(\ell_{c,k})$ such that

$$1 \le \ell_{c,1} \le \ell_{c,2} \le \ell_{c,3} \le \dots, \qquad \sum_{k=1}^{\infty} \ell_{c,k}^{-2} < \infty,$$

and introducing the densely defined selfadjoint operator on $(H_{c,2a+1}, \{e_{a,k}\})$

$$A_a := \sum_{k=1}^{\infty} \ell_{c,k} \langle \xi, e_{a,k} \rangle_{c,2a+1} e_{a,k}, \qquad \xi \in H_{c,2a+1}.$$

We also assume that for any $\alpha > 0$ there exists $\beta > 0$ such that $(2k\pi)^{\alpha} \le \ell_{c,k}^{\beta}$ for all $k \ge 1$. Then A_a^{-1} is Hilbert-Schmidt, $\inf \text{Spec}(A_a) \ge 1$ and p-norms are defined by

$$|\xi|_{c,a,p}^2 = |A_a^p \xi|_{c,2a+1}^2 = \sum_{k=1}^{\infty} \ell_{c,k}^{2p} |\langle \xi, e_{a,k} \rangle_{c,2a+1}|^2, \qquad \xi \in H_{c,2a+1}.$$

For $p \geq 0$ one defines $\mathcal{N}_{c,2a+1,p} = \{\xi \in H_{c,2a+1} ; |\xi|_{c,a,p} < \infty\}$ and $\mathcal{N}_{c,2a+1,-p}$ to be the completion of H with respect to $|\cdot|_{c,a,-p}$. Thus we obtain a chain of Hilbert spaces $\{\mathcal{N}_{c,2a+1,p} ; p \in \mathbf{R}\}$ and the corresponding standard triple:

$$\mathcal{N}_{c,2a+1} := \operatorname*{proj\,lim}_{p\to\infty} \mathcal{N}_{c,2a+1,p} \subset H_{c,2a+1} \subset \mathcal{N}^*_{c,2a+1} := \operatorname*{ind\,lim}_{p\to\infty} \mathcal{N}_{c,2a+1,-p}$$

which now depends on the triple $(H_{c,2a+1}, \{e_{a,k}\}, \{\ell_{c,k}\})$.

Using the duality $\langle \mathcal{N}^*_{c,2a+1}, \mathcal{N}_{c,2a+1} \rangle$ the usual trace τ_a on $H_{c,2a+1}$ is characterized by strong continuity and $\langle \tau_a, J_e z \otimes w \rangle = \langle z, w \rangle$ can be considered as an element in the dual of $E \otimes E$ and in fact identified to $\tau_a = \sum_{k=1}^\infty e^*_{a,k} J_e \otimes e^*_{a,k}$. It is easily checked that τ_a belongs to $\mathcal{N}_{c,2a+1,-1/2} \otimes \mathcal{N}_{c,2a+1,-1/2}$.

Remark 5.1. Let $a \in \mathbb{N}$. For $N \in \mathbb{N}_*$, let p_N be an operator on $\mathcal{N}^*_{c,2a+1}$ given by

$$p_N f = \sum_{k=1}^N \langle f, e_{a,k} \rangle_{c,2a+1} e_{a,k} \quad \text{for} \quad f \in \mathcal{N}^*_{c,2a+1}.$$

Then, for any $f \in \mathcal{N}_{c,2a+1}$, we have $p_N f \in E^*$. For any $f \in \mathcal{N}_{c,2a+1} \cap E^*$, there exists some $p > 1$ such that the p-norm $|f|_{-p}$ is estimated as follows:

$$|f|^2_{-p} = \sum_{\nu=1}^\infty \ell_\nu^{-2p} |\langle f, e_\nu \rangle|^2$$

$$= \frac{2a+1}{(2\pi)^{2a}} \sum_{\nu=1}^\infty \left(2\pi \left[\frac{\nu+1}{2} \right] \right)^{2a} \ell_\nu^{-2p} \left| \sum_{k=1}^\infty \langle f, e_{a,k} \rangle_{c,2a+1} e^{2\pi i \lambda_k \nu} \right|^2$$

$$\leq \frac{2a+1}{(2\pi)^{2a}} \sum_{\nu=1}^\infty \ell_\nu^{-2(p-q)} \sum_{k=1}^\infty \ell_{c,k}^2 |\langle f, e_{a,k} \rangle_{c,2a+1}|^2 \sum_{k=1}^\infty \ell_{c,k}^{-2}$$

$$= \frac{2a+1}{(2\pi)^{2a}} \sum_{\nu=1}^\infty \ell_\nu^{-2(p-q)} |f|^2_{c,a,1} \sum_{k=1}^\infty \ell_{c,k}^{-2} < \infty$$

for some $0 < q \leq p - 1$. By this estimation we see that for any $f \in \mathcal{N}_{c,2a+1}$ the sequence $(p_N f)_{N=1}^\infty$ converges to $\tilde{f}$ in E^*.

Let $\tilde{\mathcal{N}}_{c,2a+1} = \{\tilde{f}; f \in \mathcal{N}_{c,2a+1}\}$ with induced topology from $\mathcal{N}_{c,2a+1}$. Define the norm $|\tilde{f}|_{c,a,p}$ on $\tilde{\mathcal{N}}_{c,2a+1}$ by $|\tilde{f}|_{c,a,p} = |f|_{c,a,p}$ for any $f \in \mathcal{N}_{c,2a+1}$. Then $\mathcal{N}_{c,2a+1}$ is in E^* and isomorphic to $\tilde{\mathcal{N}}_{c,2a+1}$ by norms $|\cdot|_{c,a,p}$, $p \in \mathbb{R}$. This kind of imbedding is discussed in detail by K. Harada.[12]

6. Exotic HKT Spaces

For each $a \in \mathbb{N}$ and $p \in \mathbb{R}$ let $\Gamma(\mathcal{N}_{c,2a+1,p})$ be the Fock space over the Hilbert space $\mathcal{N}_{c,2a+1,p}$, i.e.,

$$\Gamma(\mathcal{N}_{c,2a+1,p})$$
$$= \left\{ \phi = (f_n)_{n=0}^{\infty} \,;\, f_n \in \mathcal{N}_{c,2a+1,p}^{\widehat{\otimes}n}, \, \| \phi \|_{c,a,p}^2 = \sum_{n=0}^{\infty} n! \, | f_n |_{c,a,p}^2 < \infty \right\}.$$

Then by identifying $\Gamma(H_{c,2a+1})$ with its dual space, we have a rigging of Fock spaces:

$$(\mathcal{N}_{c,2a+1}) = \operatorname*{proj\,lim}_{p \to \infty} \Gamma(\mathcal{N}_{c,2a+1,p}) \subset \cdots \subset \Gamma(\mathcal{N}_{c,2a+1,p})$$
$$\subset \Gamma(\mathcal{N}_{c,2a+1,0}) = \Gamma(H_{c,2a+1}) \subset \Gamma(\mathcal{N}_{c,2a+1,-p}) \subset \cdots$$
$$\subset (\mathcal{N}_{c,2a+1})^* = \operatorname*{ind\,lim}_{p \to \infty} \Gamma(\mathcal{N}_{c,2a+1,-p}). \tag{12}$$

The canonical $\mathbf{C}$-bilinear form on $(\mathcal{N}_{c,2a+1})^* \times (\mathcal{N}_{c,2a+1})$ takes then the form:

$$\langle\!\langle \Phi, \phi \rangle\!\rangle_{c,2a+1} = \sum_{n=0}^{\infty} n! \, \langle F_n, f_n \rangle_{c,2a+1} \,,$$

$$\Phi = (F_n) \in (\mathcal{N}_{c,2a+1})^*, \quad \phi = (f_n) \in (\mathcal{N}_{c,2a+1}).$$

The exponential vectors ϕ_ξ are defined as in (2). The $S_{c,2a+1}$-*transform* of an element $\Phi \in (\mathcal{N}_{c,2a+1})^*$ is defined by

$$S_{c,2a+1}\Phi(\xi) = \langle\!\langle \Phi, \phi_\xi \rangle\!\rangle_{c,2a+1} \,, \quad \xi \in \mathcal{N}_{c,2a+1}.$$

Using similar methods in Lemmas 3.1 and 3.2, we have the following:

Lemma 6.1. For any $\alpha \in \mathbb{N}$, the operator ∂_α is an isomorphism from $\mathcal{N}_{c,2\alpha+1}$ onto itself.

Lemma 6.2. For any $\alpha \in \mathbb{N}$, the operator ∂_α is an isomorphism from $\mathcal{N}_{c,2\alpha+1}^*$ onto itself.

Remark 6.1. By same argument in Remark 3.2, the Bochner-Minlos Theorem admits the existence of a probability measure $\mu_{c,2a+1}$ on $\mathcal{N}_{c,2a+1,\mathbf{R}}^*$ such that

$$\int_{\mathcal{N}_{c,2a+1,\mathbf{R}}^*} e^{i\langle x,\xi \rangle_{c,2a+1}} d\mu_{c,2a+1}(x) = e^{-\frac{1}{2}\langle \xi,\xi \rangle_{c,2a+1}}, \quad \xi \in \mathcal{N}_{c,2a+1,\mathbf{R}}.$$

The standard triple consisting of white noise distributions obtained from (12) through the Wiener-Itô-Segal isomorphism is denoted also by

$$(\mathcal{N}_{c,2a+1}) \subset \Gamma(H_{c,2a+1}) \equiv L^2(\mathcal{N}^*_{c,2a+1,\mathbf{R}}, \mu_{c,2a+1}) \subset (\mathcal{N}_{c,2a+1})^*.$$

We call $(\mathcal{N}_{c,2a+1})^*$ the *exotic Hida-Kubo-Takenaka (HKT) space of order* $2a+1$.

7. Exotic Laplacians on Exotic HKT Spaces

We introduce an isomorphism i_a from $(\mathcal{N}_{c,2a+1})$ onto $(\widetilde{\mathcal{N}}_{c,2a+1})$ given by

$$i_a\left((f_n)\right) = (\widetilde{f_n})$$

for $(f_n) \in (\mathcal{N}_{c,2a+1})$, Then we have the following.

Theorem 7.1. *Any element* $\varphi \in (\widetilde{\mathcal{N}}_{c,2a+1})$ *is in* $\mathrm{Dom}(\Delta_{c,2a+1})$. *Moreover, if* $\varphi = (\widetilde{f_n})_{n=0}^{\infty}$, *then we have*

$$\Delta_{c,2a+1}\varphi = \left((n+2)(n+1)\tau_a\widehat{\otimes}^2 f_{n+2} \right).$$

Proof. For the proof, we refer to Ref.17. □

By Theorem 7.1, we define an operator $\overline{\Delta}_{c,2a+1}$ on $(\mathcal{N}_{c,2a+1})$ by $\overline{\Delta}_{c,2a+1} = i_a^{-1}\Delta_{c,2a+1}i_a$. Then this operator is a continuous linear operator from $(\mathcal{N}_{c,2a+1})$ into itself. For each $t \in \mathbf{R}$ there exists a unique operator $G_t \in \mathcal{L}((\mathcal{N}_{c,2a+1}), (\mathcal{N}_{c,2a+1}))$ such that

$$\langle\!\langle G_t\phi_\xi, \phi_\eta\rangle\!\rangle_{c,2a+1} = e^{\frac{t}{2}\langle\tau_a, \xi\otimes\xi\rangle + \langle\xi, \eta\rangle_{c,2a+1}}, \qquad \xi, \eta \in \mathcal{N}_{c,2a+1}.$$

In fact, for any $\phi = (f_n) \in (\mathcal{N}_{c,2a+1})$, $G_t\phi$ is given by

$$G_t\phi = \left(\sum_{m=0}^{\infty} \frac{(n+2m)!}{n!m!} \left(\frac{t}{2}\right)^m (\tau_a^{\otimes m}\widehat{\otimes}_{2m}f_{n+2m}) \right).$$

Theorem 7.2. $\{G_t; t \in \mathbf{R}\}$ *becomes a regular one–parameter group of operators acting on* $(\mathcal{N}_{c,2a+1})$ *with infinitesimal generator* $\frac{1}{2}\overline{\Delta}_{c,2a+1}$.

The proof is a simple modification of the proof of Theorem 4.3 in Ref.17.

Remark 7.1. By Theorem 7.2 and Corollary 4.1 we have
$$G_t = i_a^{-1}\Gamma(\partial_a^*)e^{\frac{t}{2}\Delta_L}\Gamma(\partial_a^*)^{-1}i_a \text{ on } (\mathcal{N}_{c,2a+1}).$$

8. Infinite Dimensional Stochastic Processes Generated by Exotic Laplacians

Let $\{\mathbf{X}_t; t \geq 0\}$ be a $(\mathcal{N}_{c,2a+1})$-valued stochastic process. Then we can write the process in the form $\mathbf{X}_t = (\mathbf{X}_{t,n})$. The expectation $E[\mathbf{X}_t]$ of $\mathbf{X}_t$, if it exists in $(\mathcal{N}_{c,2a+1})$, is given by $E[\mathbf{X}_t] = (E[\mathbf{X}_{t,n}])$. For $\eta \in \mathcal{N}_{c,2a+1}$ let $\mathbf{T}_\eta$ be a translation operator defined on $(\mathcal{N}_{c,2a+1})^*$ by

$$S_{c,2a+1}[\mathbf{T}_\eta \Phi](\xi) = S_{c,2a+1}\Phi(\xi + \eta), \qquad \Phi \in (\mathcal{N}_{c,2a+1})^*.$$

Then we have

$$\mathbf{T}_\eta \Phi = \left(\sum_{k=0}^{\infty} \frac{(n+k)!}{n!k!} \left\langle F_{n+k}, \eta^{\otimes k} \right\rangle_{c,2a+1} \right). \tag{13}$$

Therefore, for given $z \in \mathcal{N}_{c,2a+1}^*$ and $\Phi \in (\mathcal{N}_{c,2a+1})^*$, it is natural to define $\mathbf{T}_z \Phi$ by (13) whenever the right hand side of (13) is well defined as an element in $(\mathcal{N}_{c,2a+1})^*$. By a similar method of the proof of Theorem 4.2.3 in[25] we have $\mathbf{T}_z \in \mathcal{L}((\mathcal{N}_{c,2a+1}), (\mathcal{N}_{c,2a+1}))$ for all $z \in \mathcal{N}_{c,2a+1}^*$.

Let $\{B_k(t); t \geq 0\}$, $k = 1, 2, \ldots$, be an infinite sequence of independent one dimensional Brownian motions and $\{\mathbf{B}_a(t); t \geq 0\}$ an infinite dimensional stochastic process defined by

$$\mathbf{B}_a(t) = \sum_{k=1}^{\infty} B_k(t) e_{a,k}, \qquad t \geq 0. \tag{14}$$

Then since $E[\|\mathbf{B}_a(t)\|_{c,2a+1,-p}^2] = t \sum_{k=1}^{\infty} \ell_{c,k}^{-2p} < \infty$, for any $p \geq 1$, we have $\mathbf{B}_a(t) \in \mathcal{N}_{c,2a+1}^*$ (a.e.) for all $t \geq 0$. From (13) and direct computation of the $S_{c,2a+1}$-transform of $E[\mathbf{T}_{\mathbf{B}_a(t)}\phi]$ for $\phi \in (\mathcal{N}_{c,2a+1})$, we have the following.

Theorem 8.1. *Let $\phi \in (\mathcal{N}_{c,2a+1})$. Then the equality*

$$G_t \phi = E[\mathbf{T}_{\mathbf{B}_a(t)}\phi]$$

holds for $t \geq 0$.

Theorem 8.1 means that $\{\mathbf{B}_a(t)\}_{t \geq 0}$ is a stochastic process generated by $\frac{1}{2}\overline{\Delta}_{c,2a+1}$. By Corollary 4.1 and Theorem 8.1 we have the following.

Theorem 8.2. *For $a \in \mathbb{N}$ the $\mathcal{N}_{c,2a+1}^*$-valued stochastic process*

$$\mathbf{B}_0(t)^{(a)} = \sum_{n=1}^{\infty} B_n(t) e_{0,n}^{(a)}$$

is generated by $\frac{c}{2}\overline{\Delta}_{c,2a+1}$, $c = \dfrac{(-1)^a (2\pi)^{2a}}{2a+1}$, acting on $(\mathcal{N}_{c,2a+1})$, where $\mathbf{B}_0(t)^{(a)}$ is the a-th derivative of the $\mathcal{N}_{c,1}^*$-valued stochastic process $\mathbf{B}_0(t)$.

The proofs of theorems and detail discussions in this paper will be written in Ref.8.

Acknowledgments

The author would like to express his deep thanks to Professor Accardi for the invitation to talk in this stimulating conference. The author was supported by JSPS Grant-in-Aid Scientific Research 21540151.

References

1. L. Accardi, A. Barhoumi and H. Ouerdiane: *A quantum approach to Laplace operators*, Infin. Dimen. Anal. Quantum Probab. Rel. Top. **9** (2006), 215–248.
2. L. Accardi, P. Gibilisco and I. V. Volovich: *The Lévy Laplacian and the Yang-Mills equations*, Russ. J. Math. Phys. **2** (1994), 235-250.
3. L. Accardi and O. G. Smolyanov: *On Laplacians and Traces*, Conferenze del Seminario di Matematica dell'Universit'a di Bari, N. **250**, 1993.
4. L. Accardi and O. G. Smolyanov: *The trace formulae for Lévy-Gaussian measures and their applications*, in "Mathematical Approach to Fluctuations Vol. II (T. Hida, ed.)" pp. 31–47, World Scientific, 1995.
5. L. Accardi and O. G. Smolyanov: *Generalized Levy Laplacians and Cesàro Means*, Doklady Akademii Nauk **424** (2009), 583-587.
6. L. Accardi, U. C. Ji and K. Saitô: *Exotic Laplacians and assosiated stochastic processes*, Infin. Dimen. Anal. Quantum Probab. Rel. Top. **12** (2009), 1–19.
7. L. Accardi, U. C. Ji and K. Saitô: *Infinite dimensional Laplacians and derivatives of white noise*, Infin. Dimen. Anal. Quantum Probab. Rel. Top. **14** (2011), 1–14.
8. L. Accardi, U. C. Ji and K. Saitô: *Infinite dimensional Laplacians and derivatives of white noise II*, (2012) Preprint.
9. D. M. Chung, U. C. Ji and K. Saitô: *Cauchy problems associated with the Lévy Laplacian in white noise analysis*, Infin. Dimen. Anal. Quantum Probab. Rel. Top. **2** (1999), 131–153.
10. D. M. Chung, U. C. Ji and K. Saitô: *Notes on a C_0-group generated by the Lévy Laplacian*, Proc. Amer. Math. Soc. **130** (2001), 1197–1206.
11. I.M. Gel'fand and N. Ya. Vilenkin: "Generalized Functions Vol. 4," Academic Press, 1964.
12. K. Harada: Infinite Dimensional Laplacians and associated Partial Differential Equations, Dr. Thesis, Nagoya University (2011).
13. T. Hida: "Analysis of Brownian Functionals," Carleton Math. Lect. Notes no. **13**, Carleton University, Ottawa, 1975.
14. T. Hida, H.-H. Kuo, J. Potthoff and L. Streit: "White Noise: An Infinite Dimensional Calculus," Kluwer Academic Publishers, 1993.
15. E. Hille and R. S. Phillips: "Functional Analysis and Semi-Groups", Amer. Math. Soc. Colloquium Publications **31**, 1957.
16. U. C. Ji: *Integral kernel operators on regular generalized white noise functions*, Bull. Korean Math. Soc. **37** (2000), 601–618.

17. U. C. Ji and K. Saitô: *A similarity between the Gross Laplacian and the Lévy Laplacian*, Infin. Dimen. Anal. Quantum Probab. Rel. Top. **10** (2007), 261–276.

18. I. Kubo and S. Takenaka: *Calculus on Gaussian white noise I–IV*, Proc. Japan Acad. **56A** (1980), 376–380; **56A** (1980), 411–416; **57A** (1981), 433–436; **58A** (1982), 186–189.

19. H.-H. Kuo: "White Noise Distribution Theory," CRC Press, 1996.

20. H.-H. Kuo, N. Obata and K. Saitô: *Lévy Laplacian of generalized functions on a nuclear space*, J. Funct. Anal. **94** (1990), 74–92.

21. H.-H. Kuo, N. Obata and K. Saitô: *Diagonalization of the Lévy Laplacian and related stable processes*, Infin. Dimen. Anal. Quantum Probab. Rel. Top. **5** (2002), 317–331.

22. P. Lévy: "Lecons d'analyse fonctionnelle," Gauthier-Villars, Paris, 1922.

23. P. Lévy: "Problèms concrets d'analyse fonctionnelle," Gauthier-Villars, Paris, 1951.

24. N. Obata: *A characterization of the Lévy Laplacian in terms of infinite dimensional rotation groups*, Nagoya Math. J. **118** (1990), 111–132.

25. N. Obata: "White Noise Calculus and Fock Space," Lect. Notes in Math. Vol. **1577**, Springer–Verlag, 1994.

26. J. Potthoff and L. Streit: *A characterization of Hida distributions*, J. Funct. Anal. **101** (1991), 212–229.

27. K. Saitô: *Itô's formula and Lévy Laplacian I and II*, Nagoya Math. J. **108** (1987), 67–76; **123** (1991), 153–169.

28. K. Saitô: *A (C₀)-group generated by the Lévy Laplacian II*, Infin. Dimen. Anal. Quantum Probab. Rel. Top. **1** (1998), 425–437.

29. K. Saitô: *A stochastic process generated by the Lévy Laplacian*, Acta Appl. Math. **63** (2000), 363–373.

A NEW NOISE DEPENDING ON A SPACE PARAMETER AND ITS TRANSFORMATIONS

SI SI

Faculty of Information Science and Technology, Aichi Prefectural University
Nagakute, Aichi-ken, 480-1198, Japan
E-mail: sisi@ist.aichi-pu.ac.jp

We introduce a new noise $P'(u, \lambda)$, with a space parameter u and the intensity λ, noting that the characteristic of Poisson type process depends on the intensity. A new noise $P'(u, \lambda)$ is different with the time derivative $\dot{P}(t)$ of Poisson process $P(t)$. In this paper, the invariance of the intensity λ under the transformations : dilation and reflection is discussed. In addition, it is interesting to see how the intensity changes when the exponent of a stable process changes to its reciprocal.

Keywords: White noise theory

1. Introduction

A noise is a system of idealized elemental random variables which is a system of idealized (or generalized), independent, identically distributed (i.i.d) random variables parametrized by an ordered set.

If the parameter set is discrete, say $\mathbf{Z}$, then we simply have i.i.d. random variables $\{X_n, n \in \mathbf{Z}\}$, which can be easily dealt with. While the parameter set is continuous, say R^1, then we have to be careful with the definition and analysis of functionals of the noise.

In this paper, we shall discuss possible noises and give a classification of them.

We consider only noises depending on the continuous parameter. They are classified by the type. It is noted that two noises are said to be the same type if and only if the probability distributions are the same type.

1.1. *Two classes of noise*

We classify the well-known noises according to the choice of parameter. A noise usually consists of continuously many *idealized* random variables, and they are parametrized by an ordered set. There are time dependent noises and space dependent noises.

a) Time dependent noises

They are (Gaussian) White Noise $\dot{B}(t)$ and Poisson Noise $\dot{P}(t)$, which are well known.

b) Space dependent noise

It is $P'(u), u > 0$, which is the system what we are going to discuss in this paper.

2. The birth of noises

We wish to find possible noises by approximation under reasonable assumptions. We take unit interval I as a representative of continuous parameter set. Approximation should be uniform in $u \in I$, and the idealized variables should be i.i.d, so is for the approximating variables. There are two cases that one is that approximating variables are having a finite variance, and another one is that they are atomic.

Consider a linear parameter. To fix the idea we take the unit interval $I = [0,1]$. Let $\mathbf{\Delta}^n = \{\Delta_j^n, 1 \le j \le 2^n\}$ be the partition of I. To fix the idea, we assume that $|\Delta_j^n| = 2^{-n}$. Associate a random variable X_j^n to each subinterval Δ_j^n. Assume that $\{X_j^n\}$ are independent identically distributed (i.i.d.).

Case I. Assume that $\{X_j^n\}$ has mean 0 and finite variance v. Let n be large. Then, we can appeal to the central limit theorem to have a standard Gaussian distribution $N(0,1)$ as the limit of the distribution of

$$S_n = \frac{\sum_1^{2^n} X_j^n}{\sqrt{2^n v}}.$$

If we take subinterval $[a,b]$ of I, then the same trick gives us a Gaussian distribution $N(0, b-a)$, which is, for the later purpose, denoted by $N(a,b)$. The family

$$\mathcal{N} = \{N(a,b); 0 \le a, b \le 1\}$$

forms a consistent system of distributions, so that the family $\mathcal{N}$ defines a stochastic process. In reality it is a Brownian motion $B(t), t \in [0,1]$.

Case II. Let X_j^n be subject to a simple probability distribution such that

$$P(X_j^n = 1) = p_n, \ P(X_j^n = 0) = 1 - p_n.$$

Let p_n be smaller as n is getting larger keeping the relation

$$2^n p_n = \lambda,$$

for some positive constant $\lambda > 0$. Then, by the *law of small probabilities of Poisson* (this term comes from Lévy[9]) we are given the Poisson distribution $P(\lambda)$ with intensity λ.

Take a sub-interval $[a, b]$ of I. As in the case I, we can define a Poisson distribution with intensity $(b - a)\lambda$, denoted by $P([a, b], \lambda)$. We, therefore, have a consistent family of Poisson distributions $\mathcal{P} = \{P([a, b], \lambda)\}$. Hence $\mathcal{P}$ determines a Poisson process with intensity λ.

Here are important notes.

1) We have a freedom to choose the constant λ whatever we wish, so far as it is positive.

2) The positive constant λ is the expectation of a random variable $P(\lambda) = P([0, 1], \lambda)$. It can be viewed as *scale* or *space* variable.

We now understand that a noise, which is taken to be a realization of the randomness due to reduction, can be eventually created as a space variable.

Case III. Next step of our study is concerned with a general new noise depending on a space parameter. Keep random variables X_i^n as above and divide the sum S_n into partial sums :

$$S_n = \sum_j S_n^j,$$

where

$$S_n^j = \sum_{k_j+1}^{k_{j+1}} X_j^n,$$

with

$$1 = k_0 < k_1 < k_2 < \cdots < k_m = 2^n.$$

We assume that $k_{j+1} - k_j \to \infty$ as $n \to \infty$ for every j and that each ratio $\frac{k_{j+1} - k_j}{2^n}$ converges to $\frac{\lambda_j}{\lambda}$, respectively.

Theorem 2.1. *Let S_n, S_n^j and λ be as above. Let $P(\lambda_j), 1 \le j \le m$ be mutually independent Poisson random variables with intensity λ_j, respectively. Then,*

i) S_n and S_n^j converges to $P(\lambda)$ and $P(\lambda_j)$ in law, respectively.

ii) $P(\lambda_j)$ is a realization, in terms of distribution, of the limit (in law) of the S_n^j. Let the u_j's be linearly independent real numbers over Z, and let $P = \sum_j u_j P(\lambda_j)$. Then, if we know the values of P, we can determine the values of each $P(\lambda_j)$ which can be approximated in law by S_n^j.

Proof. It is easy by observing the characteristic functions $\varphi_n^j(z)$ and $\varphi(\mathbf{z}_n)$ of S_n^j and S_n, respectively. The evaluation gives

$$\varphi_n^j(z_j) = \left(1 + \frac{\lambda_j}{(k_{j+1} - k_j)}(e^{iz_j u_j} - 1)\right)^{n_j}, \tag{1}$$

and

$$\varphi_n(z) = \prod_{j=1}^{m}\left(1 + \frac{\lambda_j}{(k_{j+1} - k_j)}(e^{iz_j u_j} - 1)\right)^{n_j}. \tag{2}$$

Thus, $\varphi_n^j(z)$ tends to $e^{\lambda_j(e^{iz_j u_j} - 1)}$ which is the characteristic function of $P(\lambda_j)$ and $\varphi(z_n)$ tends to

$$\varphi(\mathbf{z}) = \prod_{j=1}^{m} e^{\lambda_j(e^{iz} - 1)}, \tag{3}$$

which is the characteristic function of $P(\lambda)$.

This proves the assertion.

Corollary 2.1. *The characteristic function $\varphi(z)$ of P given by the above theorem is expressed in the form*

$$\varphi(\mathbf{z}) = e^{\sum_k \lambda_k (e^{iu_k z_k} - 1)}.$$

Before we come to the next topic, we pause to see the following facts. (The type of probability distributions)

Definition 2.1. If X and $cX, c > 0$ are having the same probability distributions then their distributions are called the same type of distribution.

All the Gaussian distributions are the same type (the exceptional Gaussian variable is excluded. It is just excluded in Case I).

On the other hand, Poisson type distributions with different intensities are not the same type. This can be proved by the formula of characteristic function. Poisson type distribution means a distribution of $uP(\lambda) + c$, c may be ignored. Namely, we can compare two characteristicn functions :

$$\varphi_1(z) = e^{\lambda_1(e^{iz}-1)},$$
$$\varphi_2(z) = e^{\lambda_2(e^{iz}-1)}.$$

These two functions of z can not be exchanged by any affine transformation of z if $\lambda_1 \neq \lambda_2$.

We can say that, concerning the Poisson type,

1) We have a freedom to choose intensity arbitrary. Hence we can form, by the sum of i.i.d.random variables, continuously many Poisson type random variables with different type.

2) The intensity is a parameter, different from the time parameter, to be remarkable. The intensity is viewed as a space parameter. The above construction shows that it is additive in λ.

3) Multiplication by a constant to Poisson type variable keeps the type. The constant, however, can play the role of a *label*. So, take a constant $u = u(\lambda)$ as a *label* of the intensity λ. The function $u(\lambda)$ is therefore univalent. In view of this fact, we can form an inverse function $\lambda = \lambda(u)$ which is to be monotone.

With the remark made above, we change our eyes towards multidimensional view. We consider a vector consisting independent componenets

$$P(\lambda) = (P(\lambda_1), P(\lambda_2), \cdots, P(\lambda_n))$$

and its characteristic function

$$\varphi(\mathbf{z}) = \prod_{k=1}^{n} e^{\lambda_k(e^{iz_k u_k}-1)},$$

where $\mathbf{z} = (z_k)$.

We now wish to identify every component $P(\lambda_k)$, so that we give a label to each $P(\lambda_k)$, say different real number u_k. Now let us have passage from digital $\{k\}$ to real $u > 0$. The characteristic function $\varphi(\mathbf{z})$ turns into a a functional of ξ in some function space E, expressed in the form

$$C(\xi) = \exp[\int \lambda(u)(e^{iu\xi(u)} - 1)du].$$

What we have done is that starting from a higher dimensional characteristic function of Poisson type distribution, we have its limit $C(\xi)$.

We now claim

Theorem 2.2. *Let E be a nuclear space which is a dense subspace of $L^2([0, \infty))$. Then, $C(\xi)$ obtained above is a characteristic functional of a generalized stochastic process with parameter set $[0, \infty)$.*

The proof comes from the monograph 1 chap. III, Š4.

3. A noise of new type

We have found a new noise which we shall call the new noise. It is, of course, in line with the study of random functions using i.e.r.v.'s.

The idea is to choose another kind of parameter, say space variable, instead of time. Motivation comes from the decomposition of a compound Poisson process

By discussions in the previous section lead us to consider a functional $C^P(\xi)$, where the variable ξ runs through a certain nuclear space $E \subset C^\infty((0, \infty))$, say isomorphic to the Schwartz space $\mathcal{S}$;

$$C^P(\xi) = \exp\left[\int_{R^+} (e^{iu\xi(u)} - 1)\lambda(u)du\right], \tag{4}$$

where $R^+ = (0, \infty)$ and where $dn(u)$ is a measure on $(0, \infty)$ to be specified.

We assume that $dn(u) = \lambda(u)du$ with $\lambda(u)$ positive almost everywhere and $\frac{u^2}{1+u^2}\lambda(u)$ is integrable.

Theorem 3.1. *Under these assumptions, the functional $C^P(\xi)$ is a characteristic functional. That is*

i) $C^P(\xi)$ is continuous in ξ, with a suitable choice of a nuclear space E.
ii) $C^P(0) = 1$.
iii) $C^P(\xi)$ is positive definite.

Applying the Bochner-Minlos theorem, we can see that there exists a probability measure ν^P on E^* such that

$$C^P(\xi) = \int_{E^*} e^{i\langle x, \xi \rangle} d\nu^P(x). \tag{5}$$

We introduce a notation $P'(u, \lambda(u))$ or write it simply $P'(u)$. We understand that ν^P-almost all $x \in E^*, x = x(u), u \in R^1$ is a sample function of a generalized stochastic process $P'(u), u \in R^+$.

Theorem 3.2. *The $\{P'(u), u \in R^+\}$ defines a system of i.e.r.v., that is a noise.*

Proof. Suppose ξ_1 and ξ_2 have disjoint support; let them be denoted by A_1 and A_2. Then, the integral in the expression (4) can be expressed as a sum

$$\int_{A_1} (e^{iu\xi(u)} - 1)dn(u) + \int_{A_2} (e^{iu\xi(u)} - 1)dn(u).$$

This proves the equality

$$C^P(\xi_1 + \xi_2) = C^P(\xi_1)C^P(\xi_2).$$

This shows that $P'(u)$ has independent value at every point u. Also follow the atomic property and others.

The bilinear form $\langle P', \xi \rangle, \xi \in E$ is a random variable with mean

$$\int u\xi(u)\lambda(u)du.$$

It is, of course, not permitted to put $\xi = \delta(u_0)$ to say $E(P'(u_0)) = u_0\lambda(u_0)$.

The variance of $\langle P', \xi \rangle$ is

$$\int u^2\xi(u)^2\lambda(u)du.$$

Hence $\langle P', \xi \rangle$ extends to $\langle P', f \rangle$ under the condition $uf(u) \in L^2((0, \infty), \lambda(u)du)$.

If $uf(u)$ and $ug(u)$ are orthogonal in $L^2(\lambda(u)du)$, then $\langle P', f \rangle$ and $\langle P', g \rangle$ are uncorrelated. Thus, we can form a random measure and hence, we can define the space $\mathcal{H}_1(P)$ like $\mathcal{H}_1$ in the case of Gaussian white noise. The space $\mathcal{H}_1(P)$ can also be extended to a space $\mathcal{H}_1^{(-1)}(P)$ of generalized linear functionals of $P'(u)$'s. Note that there we can give an *identity* to $P'(u)$ for any u.

Our conclusion is that single noise $P'(u)$ with the parameter u can be found in $\mathcal{H}_1^{(-1)}(P)$.

Remark 3.1. The case where the parameter u runs through the negative interval $(-\infty, 0)$ can be discussed in the similar manner. It is, however, noted that the single point mass at $u = 0$ is omitted.

We now focus our attention to the intensity $\lambda(u)$. There is an obvious assertion.

Remark 3.2. Let $X(\lambda)$ be subject to a Poisson distribution with intensity λ. Then the probability distribution of $X(\lambda)$ and $uX(\lambda), u > 0$ are the same type, while $X(\lambda)$ and $X(\lambda')$ have distribution of different type.

Proof comes from the formulas of characteristic functions.

We now come to next topic. Since u is supposed to be the space variable running through R_+, it is natural to consider the dilation of u

$$u \to au,$$

in place of the shift in the time parameter case, $t \in R$.

The $P'(u)$ is said to be self-similar if for any $\alpha > 0$ there exists $d(a)$ such that

$$P'(au) \sim d(a)P'(u),$$

where $\sim$ means the same distribution.

Theorem 3.3. *If $P'(u)$ is self similar, then*

$$\lambda(u) = \frac{c}{u^{1+\alpha}},$$

where c is a positive constant and $0 < \alpha < 1$.

The above case corresponds to the stable noiseand α is the exponent of the noise.

Since we exclude $u = 0$, and since we may restrict our attention to the positive parameter set, i.e. $(0, \infty)$. We now consider the one-parameter group

$$\mathcal{G} = \{g_t, t \in R^1\}$$

of dilations, where

$$g_t : u \rightarrow g_t u = e^{at} u, \ a > 0.$$

There, we have

$$g_t g_s = g_{t+s}.$$

Remark 3.3.

1) We can claim the same result for the case $1 \leq \alpha < 2$ by a slight modification of the characteristic functional.

2) The group $\mathcal{G}$ characterizes the class of the stable noises. More details shall be discussed in the forthcoming paper.

4. Invariance

Like white noise, we can see invariance with respect to dilation and reflection here too.

1) Isotropic dilation

Since the parameter u runs through R_+, the dilation is the basic transformation.

Let

$$\mathcal{G} = \{g_a, g_a u = au\}.$$

There, we have

$$g_a g_b = g_{a+b},$$

so $\mathcal{G}$ is the group of dilations.

Remark 4.1. The group $\mathcal{G}$ characterizes the class of the stable noises.

2) Reflection with respect to the unit sphere S^1

Consider the reflection

$$u \longmapsto \frac{1}{u},$$

then

$$\lambda(u) = \frac{1}{u^{\alpha+1}} \longmapsto \lambda(\frac{1}{u}) = u^{\alpha+1}.$$

The following (modified) reflection changes the intensity for a stable process as follows :

$$\lambda(u) \longmapsto \lambda(\frac{1}{u}) \cdot \frac{1}{u^3} = \frac{1}{u^{2-\alpha}} = \frac{1}{u^{(1-\alpha)+1}}.$$

That is, when α changes to $1 - \alpha$, the range of α changes as

$$0 < \alpha < 1 \quad \longleftrightarrow \quad 1 > 1 - \alpha > 0.$$

Theorem 4.1. *The family $\{P'_\alpha(\lambda, u), 0 < \alpha < 1\}$ is invariant under the reflection*

$$\lambda(u) \longmapsto \lambda(\frac{1}{u}) \cdot \frac{1}{u^3},$$

where $\alpha \to 1 - \alpha$.

3) Reciprocal transformation of exponent

We first discuss the motivation of discussing this transformation. That is the inverse transform of Brownian motion. Let

$B(t)$: Brownian motion

$M(t)$: maximum of Brownian motion

$T(y)$: inverse of $M(t)$

$$B(t) \longmapsto M(t) \longmapsto T(y)$$

Then, the exponent α of $B(t)$ is 2 and the exponent α of $T(y)$ is $\frac{1}{2}$. That is, α changes to $\frac{1}{\alpha}$.

Proposition .1. *By changing the exponent α of a new noise to its reciprocal $\frac{1}{\alpha}$, then the corresponding intensity is*

$$\lambda_{\frac{1}{\alpha}}(u) = (\lambda_\alpha(u))^{\frac{1}{\alpha}}.$$

Proof Change α to $\frac{1}{\alpha}$ then the intensity changes λ_α to $\lambda_{\frac{1}{\alpha}}$.

Then we have

$$\lambda_\alpha(u) = \frac{1}{u^{\alpha+1}} \longmapsto \lambda_{\frac{1}{\alpha}}(u) = \frac{1}{u^{\frac{1}{\alpha}+1}} = (\frac{1}{u^{\alpha+1}})^{\frac{1}{\alpha}}$$

Thus, we have

$$\lambda_{\frac{1}{\alpha}}(u) = (\lambda_\alpha(u))^{\frac{1}{\alpha}}.$$

We now generalize this mapping $\alpha \longmapsto \frac{1}{\alpha}$ where α is the exponent of stable distribution.

Let $X_\alpha(t)$ be a stable process with exponent α and $T(y)$ be the inverse function of $M_\alpha(t) = \max_{s \leq t} X_\alpha(s)$.

Theorem 4.2. *The probability distribution of $y^{-\alpha} T(y)$ is independent of y.*

Proof is given by Lévy.[9]

The above theorem gives $y^{-\alpha} T(y)$ and $T(1)$ has the same distribution. The characteristic function of $T(1)$ is

$$\varphi(z) = E(e^{izT(1)}) = e^{\psi(z)}.$$

On the other hand, the characteristic function of $y^{-\alpha} T(y)$ is

$$E(e^{izy^{-\alpha}T(y)}) = e^{\psi(zy^{-\alpha})y},$$

since $T(y)$ is a Lévy process. Thus, we have

$$\psi(z) = \psi(zy^{-\alpha})y.$$

Letting $z = y^\alpha$,

$$\psi(y^\alpha) = \psi(1)y.$$

That is,

$$\psi(x) = \psi(1)x^{\frac{1}{\alpha}}.$$

In other words

$$\varphi(z) = e^{cz^{\frac{1}{\alpha}}}.$$

Hence we have a stable process $T_\alpha(y)$ of exponent $\frac{1}{\alpha}$.

Theorem 4.3. *Let X_α be a stable process with exponent α and $T(y)$ be the inverse function of $M_\alpha(t) = \max_{s \leq t} X_\alpha(s)$. Then, $T(y)$ is the stable process with exponent $\frac{1}{\alpha}$.*

We say that α stable distribution and $\frac{1}{\alpha}$ stable distribution are α conjugate.

Remark 4.2. Cauchy distribution is self conjugate. Intensity does not change for every label.

References

1. I. M. Gelgand and N. Ya. Vilenkin, Generalized functions Vol.4. Academic Press 1964.
2. T. Hida, Stationary stochastic processes. Princeton Univ. Press, 1970.
3. T. Hida, White noise approach to some statistical problems. (to appear in Professor Louis Chen's Volume.)
4. T. Hida and Si Si, Lectures on white noise functionals. World Sci. Pub. Co. 2008.
5. T. Hida, Si Si and W.W. Htay, A noise of new type and its application (in press)
6. P. Lévy, Notice sur les Travaux Scientifiques.
7. P. Lévy, Sur les exponentialles de polynômes et sur l'arithmétique des products de lois de Poisson. Ann. Ecole Normale Sup. 54 (1937), 231-292. (in particular Chap. 2).
8. P. Lévy, Théorie de l'addition des variables aléatoires. Gauthier-Villars, 1938. 2éme éd. 1954.
9. P. Lévy, Sur certains processus stochastiques homogénes, Compositio mathematica, Noordhoff-Groningen.-7,fasc.2,1939, P283à339.
10. P. Lévy, Problèmes concrets d'analyse fonctionnelle. Gauthier-Villars, 1951.
11. S.I. Resnick, Heavy-tail phenomena. Probabilistic and Statistical modeling. Springer, 2007.
12. Si Si, Introduction to Hida distributions. World Sci. Pub. Co. 2001.
13. Si Si, A.H. Tsoi and W.W. Htay, Jump finding of a stable process. Quantum Information. vol. V, World Sci. Pub. Co. (2000), 193-201.

NOTE ON COMPLEXITIES FOR GAUSSIAN COMMUNICATION PROCESSES

NOBORU WATANABE

Department of Information Sciences,
Tokyo University of Sciences,
Noda City, Chiba 278-8510, Japan
E-mail: watanabe@is.noda.tus.ac.jp

In order to study several complex systems synthetically, Prof. Ohya introduced a new concept so-called Information Dynamics synthesizing the dynamics of state change and complexity of state. In Information Dynamics, two kind of complexities, that is, a complexity of state of system itself and a transmitted complexity between two systems, are used. In Shannon's information theory, the efficiency of the information communication processes is studied by using Shannon's entropy and the mutual entropy (information) based on the classical probability theory. These entropies correspond to two complexities in Information Dynamics, respectively. In quantum information theory, one can discuss the efficiency of the quantum communication processes using by von Neumann entropy (or S-mixing entropy[8]) and Ohya mutual entropy. Those entropies are also examples of two complexities of Information Dynamics. An important axiom of complexities in Information Dynamics is given by inequalities of two complexities as follows: "Transmitted complexity with respect to an input state and a communication channel is less than a complexity of the input state." In Shannon's information theory, it is called a Shannon's type inequalities, which shows that the mutual entropy represents the quantity of information exactly sending from the input system to the output system through the communication channel. By the way, for a simple model, the Shannon's type inequalities do not satisfied in Gaussian communication processes. In Ref. 12, we introduced two complexities under three assumptions to solve this difficulty. In this paper, we propose two complexities under two assumptions by using a equivalent class in order to treat the Gaussian communication processes according to the concept of Information Dynamics.

Keywords: Information Dynamics, Complexities, Gaussian Communication processes, Quantum Communicationn Theory

1. Two Complexities of Information Dynamics

In Ref. 10, Prof. Ohya introduced a new concept so-called Information Dynamics synthesizing dynamics of state change and complexity of state in order to treat several physical and nonphysical complex systems. In Information Dynamics, two complexities $C^{\mathcal{S}}(\varphi)$, $T^{\mathcal{S}}(\varphi; \Lambda^*)$ are used to study the complex systems.

1.1. Complexity $C^{\mathcal{S}}(\varphi)$ of state φ

Let $(\mathcal{A}_1, \mathfrak{S}(\mathcal{A}_1))$ and $(\mathcal{A}_2, \mathfrak{S}(\mathcal{A}_2))$ be input and output systems, respectively. $\mathcal{A}_k$ is given by the mathematical framework of systems. $\mathfrak{S}(\mathcal{A}_k)$ is the set of all (normal) states on $\mathcal{A}_k$. For example, one can put $\mathcal{A}_k$ and $\mathfrak{S}(\mathcal{A}_k)$ by $\mathbf{B}(\mathcal{H}_k)$ (i.e., the set of all bounded linear operators on a separable Hilbert space $\mathcal{H}_k$) and $\mathfrak{S}(\mathcal{H}_k)$ (i.e., the set of all density operators on $\mathcal{H}_k$), respectively. Let $\mathcal{S}_k$ and $\widetilde{\mathcal{S}}$ be subsets of $\mathfrak{S}(\mathcal{A}_k)$ and $\mathfrak{S}(\mathcal{A}_1 \otimes \mathcal{A}_2)$. In Information Dynamics, a complexity of state φ measured from $\mathcal{S}$ is denoted by $C^{\mathcal{S}}(\varphi)$. It satisfies the following axiom:

  (1) For any $\varphi_k \in \mathcal{S}_k \subset \mathfrak{S}(\mathcal{A}_k)$, $C^{\mathcal{S}_k}(\varphi_k) \geq 0$ is hold.

  (2) If there exists a bijection j_k from $ex\mathfrak{S}(\mathcal{A}_k)$ to $ex\mathfrak{S}(\mathcal{A}_k)$, then $C^{\mathcal{S}_k}(\varphi_k) = C^{\mathcal{S}_k}(j_k(\varphi_k))$ is satisfied.

  (3) If $\widetilde{\varphi} = \varphi_1 \otimes \varphi_2 \in \mathfrak{S}(\mathcal{A}_1 \otimes \mathcal{A}_2)$, $\varphi_1 \in \mathfrak{S}(\mathcal{A}_1)$, $\varphi_2 \in \mathfrak{S}(\mathcal{A}_2)$, then $C^{\widetilde{\mathcal{S}}}(\widetilde{\varphi}) = C^{\widetilde{\mathcal{S}}}(\varphi_1 \otimes \varphi_2) = C^{\mathcal{S}_1}(\varphi_1) + C^{\mathcal{S}_2}(\varphi_2)$ is hold.

1.2. Transmitted complexity $T^{\mathcal{S}}(\varphi; \Lambda^*)$ with respect to state φ and channel Λ^*

Let Λ^* be a channel (i.e., unital c.p. map) from an input state space $\mathfrak{S}(\mathcal{A}_1)$ to an output state space $\mathfrak{S}(\mathcal{A}_2)$. A transmitted complexity associated with the input state φ_1 measured from $\mathcal{S}_1$ and the channel Λ^* is described by $T^{\mathcal{S}_1}(\varphi_1; \Lambda^*)$, which holds the following axiom.

  (4) A fundamental inequalities $0 \leq T^{\mathcal{S}_1}(\varphi_1; \Lambda^*) \leq C^{\mathcal{S}_1}(\varphi_1)$ is hold for any input state $\varphi_1 \in \mathcal{S}_1 \subset \mathfrak{S}(\mathcal{A}_1)$ and the channel Λ^*.

  (5) If Λ^* is an identity channel $\Lambda^* = id$, then $T^{\mathcal{S}_1}(\varphi_1; id) = C^{\mathcal{S}_1}(\varphi_1)$ is satisfied.

  For classical discrete probability spaces, Shannon's entropy and the mutual entropy satisfy the above axiom. Especially, the inequalities in the above (5) is called a Shannon's fundamental inequalities. Let us review these complexities for quantum communication processes.

1.3. Example of $C^{\mathcal{S}}(\varphi)$ and $T^{\mathcal{S}}(\varphi; \Lambda^*)$

1.3.1. von Neumann Entropy and Ohya Mutual Entropy

In quantum information theory, a certain complexity with respect to any density operators $\rho \in \mathfrak{S}(\mathcal{H}_1)(= \mathcal{S})$ is given by the von Neumann entropy

$S(\rho)$ defined by

$$C^{\mathcal{S}}(\rho) = S(\rho) = -tr\rho \log \rho.$$

The complexity $C^{\mathcal{S}}(\rho)$ given by the von Neumann entropy $S(\rho)$ holds the above axiom (1),(2),(3). Quantum channel Λ^* is a mapping from $\mathfrak{S}(\mathcal{H}_1)$ to $\mathfrak{S}(\mathcal{H}_2)$. Λ^* is called a linear channel if $\Lambda^*(\lambda\rho_1 + (1-\lambda)\rho_2) = \lambda\Lambda^*(\rho_1) + (1-\lambda)\Lambda^*(\rho_2)$ holds for any $\rho_1, \rho_2 \in \mathfrak{S}(\mathcal{H}_1)$ and any $\lambda \in [0,1]$. Λ^* is called a completely positive channel if Λ^* is linear and its dual $\Lambda : \mathbf{B}(\mathcal{H}_2) \to \mathbf{B}(\mathcal{H}_1)$ satisfies $\sum_{i,j=1}^{n} A_i^*\Lambda(B_i^*B_j)A_j \geq 0$ for any $n \in \mathbf{N}$, any $\{B_i\} \subset \mathbf{B}(\mathcal{H}_2)$ and any $\{A_i\} \subset \mathbf{B}(\mathcal{H}_1)$, where the dual map Λ of Λ^* is defined by $tr\Lambda^*(\rho)B = tr\rho\Lambda(B)$, for any $\rho \in \mathfrak{S}(\mathcal{H}_1)$ and any $B \in \mathbf{B}(\mathcal{H}_2)$. Almost all physical transformation can be denoted by the CP channel.[4,6,7,11,12] One of the example of the quantum communication channel is noisy optical channel introduced in Refs. 14,15. Lifting $\mathcal{E}_0^*$ from $\mathfrak{S}(\mathcal{H})$ to $\mathfrak{S}(\mathcal{H}\otimes\mathcal{K})$ in the sense of Accardi and Ohya[1] is denoted by $\mathcal{E}_0^*(|\theta\rangle\langle\theta|) = |\alpha\theta\rangle\langle\alpha\theta| \otimes |\beta\theta\rangle\langle\beta\theta|$, $\left(|\alpha|^2 + |\beta|^2 = 1\right)$. $\mathcal{E}_0^*$ is called a beam splitting. Based on liftings, the beam splitting was investigated by Accardi - Ohya and Fichtner - Freudenberg - Libsher.[3] The transmitted complexity $T^{\mathcal{S}}(\rho; \Lambda^*)$ associated with the input state ρ and a quantum channel Λ^* is given by the Ohya mutual entropy $I(\rho; \Lambda^*)$ (see Refs. 7,8) with respect to ρ and Λ^* defined by

$$T^{\mathcal{S}}(\rho; \Lambda^*) = I(\rho; \Lambda^*) \equiv \sup\left\{\sum_n S(\sigma_E, \rho \otimes \Lambda^*\rho),\ \rho = \sum_n \lambda_n E_n\right\}, \quad (1)$$

where $\rho = \sum_n \lambda_n E_n$ denotes a Schatten-von Neumann (one dimensional spectral) decomposition[16] of the input state ρ and σ_E is Ohya compound state described by $\sigma_E = \sum_n \lambda_n E_n \otimes \Lambda^* E_n$. $S(\cdot, \cdot)$ represents the Umegaki's relative entropy such as

$$S(\rho, \sigma) \equiv \begin{cases} tr\rho(\log \rho - \log \sigma) & (\overline{ran\rho} \subset \overline{ran\sigma}) \\ \infty & (\text{else}), \end{cases} \quad (2)$$

which shows a certain difference between ρ and σ. The relative entropy was extended to more general systems by Araki and Uhlmann.[2,10–12,18] The transmitted complexity $T^{\mathcal{S}}(\rho; \Lambda^*)$ given by the Ohya mutual entropy $I(\rho, \Lambda^*)$ satisfies the above axiom (5) and the Shannon's type fundamental inequality (4) such as

$$0 \leq T^{\mathcal{S}}(\rho; \Lambda^*) \leq C^{\mathcal{S}}(\rho).$$

2. Gaussian Communication Channels

Let $\mathcal{H}_k$ be real separable Hilbert spaces of an input ($k = 1$) and output ($k = 2$) systems, respectively. $\mathbf{T}(\mathcal{H}_k)_+$ is defined by $\mathbf{T}(\mathcal{H}_k)_+ \equiv \{\rho \in \mathbf{B}(\mathcal{H}_k) ; \ \rho \geq 0, \ \rho = \rho^*, \ tr\rho < \infty\}$. If μ is a Borel measure such that for $x \in \mathcal{H}_k$, there exist real numbers m_x and σ_x (> 0) satisfying

$$\mu\left(\{y \in \mathcal{H}_k; \ < y, x > \leq a\}\right) = \int_{-\infty}^{a} \frac{1}{\sqrt{2\pi}\sigma_x} \exp\left\{\frac{-(t - m_x)^2}{2\sigma_x}\right\} dt,$$

then μ is called a Gaussian measure in $\mathcal{H}_k$. For $R_\mu \in \mathbf{T}(\mathcal{H}_k)_+$, the characteristic function $\hat{\mu}$ of μ is denoted by

$$\hat{\mu}(x) = \exp\left\{i\langle x, m_x\rangle - \frac{1}{2}\langle x, R_\mu x\rangle\right\}.$$

Let a notation $\mu = [m, R]$ be a Gaussian measure μ with a mean vector m and a covariant operator R. We denote the Borel σ-field of $\mathcal{H}_k$ by $\mathcal{B}_k$ ($k = 1, 2$). μ is a Borel probability measure on $\mathcal{B}_k$ holding $\int_{\mathcal{H}_k} \|x\|^2 d\mu(x) < \infty$. We represents the set of all Borel probability measures on $\mathcal{H}_k$ by $\mathbf{P}(\mathcal{H}_k)$. For a given μ, the mean vector $m_\mu \in \mathcal{H}_k$ is given by $\langle x_1, m_\mu\rangle = \int_{\mathcal{H}_k} \langle x_1, y\rangle \mu(dy)$ for any $x_1, y \in \mathcal{H}_k$ and the covariance operator R_μ of μ is defined by $\langle x_1, R_\mu x_2\rangle = \int_{\mathcal{H}_k} \langle x_1, y - m_\mu\rangle \langle y - m_\mu, x_2\rangle \mu(dy)$ for any $x_1, x_2, y \in \mathcal{H}_k$ ($k = 1, 2$).

Based on Ref. 13, we briefly review a mathematical treatment of Gaussian communication process.

Let $(\mathcal{H}_k, \mathcal{B}_k)$ be an input ($k = 1$) and output ($k = 2$) spaces, respectively. Let $\mathbf{P}_G^{(1)}$ be the set of all Gaussian probability measures on $(\mathcal{H}_k, \mathcal{B}_k)$ ($k = 1, 2$). $\mu_1 \in \mathbf{P}_G^{(1)}$ is a Gaussian probability measure of the input space and $\mu_0 \in \mathbf{P}_G^{(2)}$ is a Gaussian probability measure of the output space associated to an additive noise of the channel. Γ^* is a mapping from $\mathbf{P}_G^{(1)}$ to $\mathbf{P}_G^{(2)}$ given by the Gaussian channel $\lambda : \mathcal{H}_1 \times \mathcal{B}_2 \to [0, 1]$ such as

$$\Gamma^*(\mu_1)(Q) \equiv \int_{\mathcal{H}_1} \lambda(x, Q) d\mu_1(x)$$

$$\lambda(x, Q) \equiv \mu_0(\{y \in \mathcal{H}_2; \ Ax + y \in Q\}), \ x \in \mathcal{H}_1, \ Q \in \mathcal{B}_2,$$

where A is a linear mapping from $\mathcal{H}_1$ to $\mathcal{H}_2$, λ satisfies (1) $\lambda(x, \bullet)$ is a Gaussian probability measure of the output space for each fixed $x \in \mathcal{H}_1$ and (2) $\lambda(\bullet, Q)$ is a measurable function on $(\mathcal{H}_1, \mathcal{B}_1)$ for each fixed $Q \in \mathcal{B}_2$. μ_{12} is the compound measure derived from the input measure μ_1 and the output measure μ_2 getting by $\mu_{12}(Q_1 \times Q_2) = \int_{Q_1} \lambda(x, Q_2) d\mu_1(x)$ for any $Q_1 \in \mathcal{B}_1$ and $Q_2 \in \mathcal{B}_2$. Let $\frac{d\mu_{12}}{d\mu_1 \otimes \mu_2}$ be the Radon - Nikodym derivative of

μ_{12} with respect to $\mu_1 \otimes \mu_2$. Then the transmitted complexity $T^{\mathcal{S}}(\mu_1;\Gamma^*)$ $(\mathcal{S} = \mathbf{P}_G^{(1)})$ with respect to the input Gaussian measure μ_1 and the mapping Γ^* is given by the Kullback - Leibler information $I(\mu_1;\lambda)$ such as

$$
\begin{aligned}
T^{\mathcal{S}}(\mu_1;\Gamma^*) &= I(\mu_1;\lambda) \\
&= S(\mu_{12}|\mu_1 \otimes \mu_2) \\
&= \begin{cases} \int_{\mathcal{H}_1 \times \mathcal{H}_2} \frac{d\mu_{12}}{d\mu_1 \otimes \mu_2} \log \frac{d\mu_{12}}{d\mu_1 \otimes \mu_2} d\mu_1 \otimes \mu_2 & (\mu_{12} \ll \mu_1 \otimes \mu_2) \\ \infty & \text{else} \end{cases}
\end{aligned}
$$

We demonstrated in Ref. 13 that (a) if the complexity $C^{\mathcal{S}}(\mu_1)$ $(\mathcal{S} = \mathbf{P}_G^{(1)})$ of the input Gaussian probability measure μ_1 is given by the differential entropy $S(\mu_1)$

$$
C^{\mathcal{S}}(\mu_1) = S(\mu_1) = -\int_{\mathbf{R}^2} \frac{d\mu_1}{dm} \log \frac{d\mu_1}{dm} dm,
$$

then the above axiom (4) does not hold, namely, $C^{\mathcal{S}}(\mu_1) < T^{\mathcal{S}}(\mu_1;\Gamma^*)$ is satisfied for a simple model of the Gaussian communication process, and (b) if the complexity $C^{\mathcal{S}}(\mu_1)$ of μ_1 is given by the entropy $S(\mu_1)$ of the discrete probability distribution by means of the set of all finite partitions of the input Gaussian space, then $C^{\mathcal{S}}(\mu_1) = +\infty$ is always hold. It is difficult to distinguish the input Gaussian measure μ_1 from other Gaussian measures.

3. A Treatment of Gaussian Communication Process Based on Complexities of Information Dynamics

According to Ref. 13, we will simply explain a mathematical treatment of Gaussian communication process.

Let $\mathbf{P}_{G,1}^{(k)}$ be the set of Gaussian measure $\mu = [0,R]$ of $\mathbf{P}_G^{(k)}$ with $trR = 1$, i.e., $\mathbf{P}_{G,1}^{(k)} = \left\{ \mu = [0,R] \in \mathbf{P}_G^{(k)}; trR = 1 \right\}$ $(k = 1,2)$. Suppose that $A^*A = (1 - trR_0)I_1$ holds for the covariant operator R_0 of μ_0. A mapping $\Gamma^* : \mathbf{P}_{G,1}^{(1)} \to \mathbf{P}_{G,1}^{(2)}$ with respect to the Gaussian channel λ is given by

$$
(\Gamma^*\mu_1)(Q) = \int_{\mathcal{H}_1} \lambda(x,Q)\mu_1(dx)
$$

for any $\mu_1 \in \mathbf{P}_{G,1}^{(1)}$ and any $Q \in \mathcal{B}_2$. $\Gamma^*(\mu_1)$ is denoted by $\Gamma^*(\mu_1) = [0, A\rho_1 A^* + R_0]$ for any $\mu_1 = [0, \rho_1] \in \mathbf{P}_{G,1}^{(1)}$. Then, there exists a bijection Ξ_k^* from $\mathbf{P}_{G,1}^{(k)}$ to $\mathfrak{S}(\mathcal{H}_k)$ given by

$$
tr\Xi_k^*(\mu_k) A_k = \int_{H_k} \langle \xi, A_k\xi \rangle \mu_k(d\xi)
$$

for any $A_k \in \mathbf{B}(\mathcal{H}_k)$ and any $\mu_k \in \mathbf{P}_{G,1}^{(k)}$ $(k = 1, 2)$. $\Lambda^* :\mathfrak{S}(\mathcal{H}_1) \to \mathfrak{S}(\mathcal{H}_2)$ is composite map of Ξ_1^*, Ξ_2^*, Γ^* by $\Lambda^* \rho_1 = \Xi_2^* \circ \Gamma^* \circ (\Xi_1^*)^{-1} \rho_1 = A\rho_1 A^* + R_0$ for any $\rho_1 \in \mathfrak{S}(\mathcal{H}_1)$.

In Ref. 13, we obtained the following results:

Theorem 3.1.[13] Λ^* is a completely positive channel from $\mathfrak{S}(\mathcal{H}_1)$ to $\mathfrak{S}(\mathcal{H}_2)$

Theorem 3.2.[13] The Gaussian measure $\mu = [0, \sigma_E]$ is a compound state (measure) derived from the input measure $\mu_1 = [0, \Xi_1^*(\mu_1)]$ on $\mathcal{H}_1$ and the output measure $\mu_2 = [0, \Lambda^* \circ \Xi_1^*(\mu_1)]$ on $\mathcal{H}_2$ in the sense that

$$\bar{\mu}_1(A) = \bar{\mu}(A \otimes \mathcal{H}_2) \quad \text{and} \quad \bar{\mu}_2(B) = \bar{\mu}(\mathcal{H}_1 \otimes B)$$

$$\text{for any subspace } A \text{ in } \mathcal{B}_1 \text{ and } B \text{ in } \mathcal{B}_2$$

where $\bar{\mu}_k(A) = \int_A \|\xi\|^2 d\mu_k(\xi)$ for any $A \in \mathcal{B}_k$ and any $\mu_k \in \mathbf{P}_{G,1}^{(k)}$ $(k = 1, 2)$.

Based on Information Dynamics, one can define the complexity $C^{\mathcal{S}}(\mu_1)$ given by the entropy type functional $\tilde{S}(\mu_1)$ of the input Gaussian measure $\mu_1 = [0, \Xi_1^*(\mu_1)]$. Let σ_E be the Ohya compound state associated with $\Xi_1^*(\mu_1)$ and Λ^*. (2) the transmitted complexity $T^{\mathcal{S}}(\mu_1; \Gamma^*)$ given by the mutual entropy type functional $\tilde{I}(\mu_1; \lambda)$ with respect to the input Gaussian measure μ_1 and the Gaussian channel λ such as

$$C^{\mathcal{S}}(\mu_1) = \tilde{S}(\mu_1) = -tr\Xi_1^*(\mu_1) \log \Xi_1^*(\mu_1),$$

$$T^{\mathcal{S}}(\mu_1; \Gamma^*) = \tilde{I}(\mu_1; \lambda) = \sup_E S(\sigma_E, \Xi_1^*(\mu_1) \otimes \Lambda^* \circ \Xi_1^*(\mu_1)),$$

Thus we had the following theorem.[13]

Theorem 3.3.[13]   For any $\mu_1 \in \mathbf{P}_{G,1}^{(k)}$ and for some Gaussian channel λ, one obtain the Shannon's type fundamental inequalities:

$$0 \le T^{\mathcal{S}}(\mu_1; \Gamma^*) \le C^{\mathcal{S}}(\mu_1).$$

Based on Ref. 13, we suppose three conditions (1) Linearity condition (linear approximation), (2) Trace preserving condition, (3) Normality condition (trace of covariance operator is equal to one.).

In this paper, we assume two conditions (1) Linearity condition (linear approximation), (2) Trace preserving condition.

Let us consider a mapping $\Pi^* : \mathbf{T}(\mathcal{H}_1)_+ \to \mathbf{T}(\mathcal{H}_2)_+$ is composite map of Ξ_1^*, Ξ_2^*, Γ^* defined by

$$\Pi^* R_1 = \Xi_2^* \circ \Gamma^* \circ (\Xi_1^*)^{-1} R_1 = AR_1 A^* + R_0, \quad (\forall R_1 \in \mathbf{T}(\mathcal{H}_1)_+).$$

The structure equivalent class in the Gaussian communication processes is defined as follows:

Definition 3.1. Structure equivalent of $\mathbf{T}(\mathcal{H}_k)_+$ and $\mathbf{P}_G^{(k)}$

(1) R_k and R'_k are structure equivalent (i.e., $R_k \overset{s}{\sim} R'_k$) if there exists a positive number $\lambda > 0$ such that $R_k = \lambda R'_k$ holds,

(2) $\mu_k = [0, R_k]$ and $\mu'_k = \left[0, R'_k\right]$ are structure equivalent (i.e., $\mu_k \overset{s}{\sim} \mu'_k$) if $R_k \overset{s}{\sim} R'_k$ is satisfied.

Definition 3.2. Structure equivalent classes of $\mathbf{T}(\mathcal{H}_k)_+$ and $\mathbf{P}_G^{(k)}$

(1) $\widetilde{R_k} \equiv \left\{ R \in \mathbf{T}(\mathcal{H}_k)_+; \quad R_k \overset{s}{\sim} R \right\}$,

(2) $\widetilde{\mu_k} \equiv \left\{ \mu \in \mathbf{P}_G^{(k)}; \quad \mu_k \overset{s}{\sim} \mu \right\}$.

Definition 3.3. Quotient sets of $\mathbf{T}(\mathcal{H}_1)_+$ and $\mathbf{P}_G^{(1)}$

(1) $\mathbf{T}(\mathcal{H}_k)_+ \diagup \overset{s}{\sim}, \quad \mathbf{T}(\mathcal{H}_k)_+ = \bigcup_{\widetilde{R_k} \in \mathbf{T}(\mathcal{H}_k)_+ \diagup \overset{s}{\sim}} \widetilde{R_k}$

(2) $\mathbf{P}_G^{(k)} \diagup \overset{s}{\sim}, \quad \mathbf{P}_G^{(k)} = \bigcup_{\widetilde{\mu_k} \in \mathbf{P}_G^{(k)} \diagup \overset{s}{\sim}} \widetilde{\mu_k}$

Definition 3.4. Mappings $\widetilde{\Pi}^*$ and $\widetilde{\Gamma}^*$

(1) $\widetilde{\Pi}^*$ is a mapping from $\mathbf{T}(\mathcal{H}_1)_+ \diagup \overset{s}{\sim}$ to $\mathbf{T}(\mathcal{H}_2)_+ \diagup \overset{s}{\sim}$.

(2) $\widetilde{\Gamma}^*$ is a mapping from $\mathbf{P}_G^{(1)} \diagup \overset{s}{\sim}$ to $\mathbf{P}_G^{(2)} \diagup \overset{s}{\sim}$.

We obtain the following theorems.

Theorem 3.4.

(1) Π^* is a completely positive map from $\mathbf{T}(\mathcal{H}_1)_+$ to $\mathbf{T}(\mathcal{H}_2)_+$.

(2) $\widetilde{\Pi}^*$ is a completely positive map from $\mathbf{T}(\mathcal{H}_1)_+ \diagup \overset{s}{\sim}$ to $\mathbf{T}(\mathcal{H}_2)_+ \diagup \overset{s}{\sim}$.

Theorem 3.5. *The Gaussian measure $\bar{\mu} = [0, \sigma_E]$ is a compound state (measure) given by the input measure $\mu_1 = [0, \Xi_1^*(\mu_1)]$ on $\mathcal{H}_1$ and the output measure $\mu_2 = [0, \Pi^* \circ \Xi_1^*(\mu_1)]$ on $\mathcal{H}_2$ in the sense that*

$$\bar{\mu}_1(A) = \bar{\mu}(A \otimes \mathcal{H}_2) \quad \text{for any subspace } A \text{ in } \mathcal{B}_1,$$

$$\bar{\mu}_2(B) = \bar{\mu}(\mathcal{H}_1 \otimes B) \quad \text{for any subspace } B \text{ in } \mathcal{B}_2,$$

where $\bar{\mu}_k(A) = \int_A \|\xi\|^2 d\mu_k(\xi)$ for any $A \in \mathcal{B}_k$ and any $\mu_k \in \mathbf{P}_G^{(k)}$ $(k = 1, 2)$.

Based on Information Dynamics, we define the complexity $C^{\mathcal{S}}(\mu_1)$ by the entropy type functional $\tilde{S}_{SE}(\widetilde{\mu_1})$

$$C^{\mathcal{S}}(\mu_1) = \tilde{S}_{SE}(\widetilde{\mu_1}) \equiv -tr \frac{\Xi_1^*(\mu_1)}{tr\,[\Xi_1^*(\mu_1)]} \log \frac{\Xi_1^*(\mu_1)}{tr\,[\Xi_1^*(\mu_1)]},$$

of structure equivalent for the input Gaussian measure $\widetilde{\mu_1} = \left[0, \widetilde{\Xi_1^*(\mu_1)}\right]$ and (2) the transmitted complexity $T^{\mathcal{S}}(\mu_1;\Gamma^*)$ given by the mutual entropy type functional $\tilde{I}_{SE}\left(\widetilde{\mu_1;\Gamma^*}\right)$

$$T^{\mathcal{S}}(\mu_1;\Gamma^*) = \tilde{I}_{SE}\left(\widetilde{\mu_1;\Gamma^*}\right) \equiv \sup_E S\left(\sigma_E, \frac{\Xi_1^*(\mu_1) \otimes \Pi^* \circ \Xi_1^*(\mu_1)}{tr\,[\Xi_1^*(\mu_1) \otimes \Pi^* \circ \Xi_1^*(\mu_1)]}\right)$$

of structure equivalent by means of the input Gaussian measure $\widetilde{\mu_1} = \left[0, \widetilde{\Xi_1^*(\mu_1)}\right]$ and the Gaussian channel $\widetilde{\Gamma^*}$, where σ_E is the Ohya compound state with respect to $\widetilde{\Xi_1^*(\mu_1)}$ and $\widetilde{\Pi^*}$. The structure equivalent for the compound states and the input states are given by

- **Structure Equivalent of Compound states**

  (1) $\Xi_1^*(\mu_1) \otimes \Pi^* \circ \Xi_1^*(\mu_1) \quad \overset{s}{\sim} \quad \sigma_0 = \frac{\Xi_1^*(\mu_1) \otimes \Pi^* \circ \Xi_1^*(\mu_1)}{tr[\Xi_1^*(\mu_1) \otimes \Pi^* \circ \Xi_1^*(\mu_1)]}$

  (2) $\sum_n \tau_n E_n \otimes \Pi^*(E_n) \quad \overset{s}{\sim} \quad \sigma_E = \frac{\sum_n \tau_n E_n}{tr[\Xi_1^*(\mu_1)]} \otimes \frac{\Pi^*(E_n)}{tr[\Pi^*(E_n)]}$

- **Structure Equivalent of the input state**

  (3) $\Xi_1^*(\mu_1) \quad \overset{s}{\sim} \quad \frac{\Xi_1^*(\mu_1)}{tr[\Xi_1^*(\mu_1)]}$

By using the complexities of Information Dynamics, one has the following theorem.

Theorem 3.6. *The axiom (4) (i.e, Shannon's type fundamental inequalities)*

$$0 \le T^{\mathcal{S}}(\mu_1;\Gamma^*) \le C^{\mathcal{S}}(\mu_1)$$

is satisfied for any $\mu_1 \in \mathbf{P}_G^{(k)}$ and for some Gaussian channel Γ^*.

References

1. Accardi, L., and Ohya, M., Compound channels, transition expectation and liftings, Appl. Math. Optim., 39, 33-59 (1999).
2. Araki, H., Relative entropy for states of von Neumann algebras, Publ. RIMS Kyoto Univ., 11, 809–833 (1976).
3. Fichtner, K.H., Freudenberg, W., and Liebscher, V., Beam splittings and time evolutions of Boson systems, Fakultat fur Mathematik und Informatik, Math/ Inf/96/ 39, Jena, 105 (1996).

4. Ingarden, R.S., Kossakowski, A., and Ohya, M., Information Dynamics and Open Systems, Kluwer, (1997).

5. von Neumann, J., *Die Mathematischen Grundlagen der Quantenmechanik*, Springer-Berlin, (1932).

6. Ohya, M., Quantum ergodic channels in operator algebras , J. Math. Anal. Appl., **84**, 318-328, (1981).

7. Ohya, M., On compound state and mutual information in quantum information theory, IEEE Trans. Information Theory, 29, 770-774 (1983).

8. Ohya, M., Note on quantum probability, L. Nuovo Cimento, **38**, 402-404, (1983).

9. Ohya, M., Some aspects of quantum information theory and their applications to irreversible processes, Rep. Math. Phys. **27**, 19–47, (1989).

10. Ohya, M., Information dynamics and its application to optical communication processes, Springer Lecture Note in Physics, 378, 81-92, (1991).

11. Ohya, M., and Petz, D., *Quantum Entropy and its Use*, Springer, Berlin, (1993).

12. Ohya, M., and Volovich, I., *Mathematical Foundations of Quantum Information and Computation and Its Applications to Nano- and Bio-systems*, Theoretical and Mathematical Physics, Springer, (2011).

13. Ohya, M., and Watanabe, N., A new treatment of communication processes with Gaussian channels, Japan Journal on Applied Mathematics, **3**, 197-206 (1986).

14. Ohya, M., and Watanabe, N., *Foundation of Quantum Communication Theory (in Japanese)*, Makino Pub. Co., (1998).

15. Ohya, M., and Watanabe, N., Construction and analysis of a mathematical model in quantum communication processes, Electronics and Communications in Japan, Part 1, 68, No.2, 29-34 (1985).

16. Schatten, R., Norm Ideals of Completely Continuous Operators, Springer-Verlag, 1970.

17. Umegaki, H., Conditional expectations in an operator algebra IV (entropy and information), Kodai Math. Sem. Rep., 14, 59-85 (1962).

18. Uhlmann, A., Relative entropy and the Wigner-Yanase-Dyson-Lieb concavity in interpolation theory, Commun. Math. Phys., **54**, 21–32, (1977).

AUTHOR INDEX